THE PRINTED VOICE OF
VICTORIAN POETRY

The Printed Voice of Victorian Poetry

ERIC GRIFFITHS

CLARENDON PRESS · OXFORD

1989

Oxford University Press, Walton Street, Oxford OX2 6DP
Oxford New York Toronto
Delhi Bombay Calcutta Madras Karachi
Petaling Jaya Singapore Hong Kong Tokyo
Nairobi Dar es Salaam Cape Town
Melbourne Auckland
and associated companies in
Berlin Ibadan

Oxford is a trade mark of Oxford University Press

Published in the United States
by Oxford University Press, New York

British Library Cataloguing in Publication Data
Griffiths, Eric
The printed voice of Victorian poetry.
1. Poetry in English, 1837–1900 – Critical
studies
I. Title
821'.809
ISBN 0–19–812989–0

Library of Congress Cataloging in Publication Data
Griffiths, Eric.
The printed voice of Victorian poetry / Eric Griffiths.
p. cm.
1. English poetry—19th century—History and criticism. 2. Speech
in literature. 3. Oral interpretation of poetry. I. Title.
PR595.S65G75 1988 821'.8'09—dc19 88–15193
ISBN 0–19–812989–0

Set by Hope Services, Abingdon
Printed in Great Britain
at the University Printing House, Oxford
by David Stanford
Printer to the University

For Eryl

PYRAMUS. I see a voice . . .

A Midsummer Night's Dream

ACKNOWLEDGEMENTS

I owe a great debt to the three colleges of the University of Cambridge in which most of the work for this book was done— Pembroke, Christ's, and Trinity.

I am grateful to many people for help, advice, criticism, or information. Particularly to: Rosemary Bechler, David Bradley, Ron Brotherhood, Fiona Cadwallader, Emma Crichton-Miller, Donald Davie, Aidan Day, Rachel Finnis, Mark Godowski, Alice Goodman, Geoffrey Hill, Kevin Jackson, Christopher Jenkins, Sheila Lawlor, Ruth Mackenzie, John Marenbon, Jeremy Maule, Adrian Poole, Tony Rick, Peter Robinson, Rod Strange, Kim Scott Walwyn, and Mary Worthington.

My greatest debt is to Christopher Ricks who started me writing this book and helped me finish it.

CONTENTS

NOTE ON THE TEXT

IN the footnotes I have given, when I could, the date of first publication of each work important to my argument. This date appears in parentheses after the first mention of the work. Where the date of first publication is much later than the known or presumed date of composition I have given instead, without parentheses, after the first mention of the work, the supposed date of composition, according to the best, recent authorities. (I do not believe that my arguments often hang, or even hinge, on dates, but I have tried to enable a reader to check up on me.)

Bibliographical details are given in full the first time a work is cited; they are, for major sources, repeated in later chapters when those sources are specially relevant to the matter of that chapter—for example, I give the details of publication of Hallam Tennyson's *Memoir* of his father in the first chapter, and I repeat them in the chapter on Tennyson but not in later chapters where the concentration of argument is only incidentally on Tennyson. There are, I hope, no ambiguities of reference within the footnotes; if there are, reference to the Bibliography should settle them.

There is much work to be done on the text of some of the writers I discuss. In the case of Tennyson, the Ricks edition gives us something to rely on. There is as yet no comparably reliable edition of all Browning's poems. Catherine Phillips has recently produced a fine edition of Hopkins's poems—which I quote throughout this book—but Phillips herself anticipates that a projected, further edition will refine again our sense of the text of Hopkins's poems. With a poet such as Coventry Patmore, there is no sensible, let alone authoritative, edition. And a writer of the stature of Newman has, apart from the work of Ian Ker on the *Idea of a University* and the *Essay in Aid of a Grammar of Assent*, and of Martin Svaglic on the *Apologia* . . ., scarcely been 'edited' at all. (I except, of course, the magisterial edition of Newman's Letters and Diaries, begun by C. S. Dessain and still in progress.) I have quoted from the texts generally accepted as standard, except where otherwise noted, but the reader should recall that many of these writers have not yet been granted a standard edition. I have made clear in the footnotes the occasions when I have preferred to cite a text in a version which is not generally accepted.

Place of publication is London unless otherwise stated.
All translations from foreign languages are my own unless otherwise
stated.

ABBREVIATIONS

Memoir   Hallam, Lord Tennyson, *Alfred Lord Tennyson: A Memoir by his Son*, 2 vols. (1897).

P   *The Poems of Robert Browning*, ed. John Pettigrew and T. J. Collins, 2 vols. (Harmondsworth, 1981).

R   *The Poems of Tennyson*, ed. Christopher Ricks (1969, 2nd edn. 3 vols., 1987).

RB/BB   *The Letters of Robert Browning and Elizabeth Barrett Browning, 1845–1846*, ed. E. Kintner, 2 vols. (Cambridge, Mass, 1969).

UE   J. H. Newman, Uniform Edition of the Works, 36 vols. (1868–81).

I

THE PRINTED VOICE

Listening to *Hamlet*

ALONE over his books, Edward FitzGerald heard voices. He wrote about what he heard to his friend, Fanny Kemble, a member of a great theatrical family, who had herself made a reputation for her readings-aloud from Shakespeare: 'I have been looking over four of Shakespeare's Plays, edited by Clark and Wright: editors of the "Cambridge Shakespeare". . . . Hamlet, Macbeth, Tempest, and Shylock—I heard them talking in my room—all alive about me.'[1] What FitzGerald heard is hard to tell. He cannot have heard Shylock talking in quite the same way as he heard the words of *The Tempest*, if only because a character even as central as Shylock does not come alive in the same way as a play lives; and it remains unclear whether he refers to the play or to the protagonist in 'Hamlet, Macbeth'.

To consider only one of these cases. What he would have seen in the text of *Hamlet* he was reading is this:

> To be, or not to be: that is the question:
> Whether 'tis nobler in the mind to suffer
> The slings and arrows of outrageous fortune,
> Or to take arms against a sea of troubles,
> And by opposing end them? To die: to sleep;
> No more; and by a sleep to say we end
> The heart-ache and the thousand natural shocks
> That flesh is heir to, 'tis a consummation
> Devoutly to be wish'd. To die, to sleep;
> To sleep: perchance to dream: ay, there's the rub . . .[2]

---

[1] Letter of 9 Apr. 1875, in *The Letters of Edward FitzGerald*, ed. A. M. and A. B. Terhune, 4 vols. (Princeton, 1980) [hereafter referred to as Terhune], iii. 575.

[2] *Hamlet*, III. i. 56 ff., in the text edited by W. G. Clark and W. A. Wright (1873). All other quotations from Shakespeare are, except where otherwise noted, from the First Folio (1623), repr. prepared by C. Hinman (1968). I have silently modernized the long 's' and the interchanged 'u'/'v', I give the standard, modern act, scene, and line reference for each citation, as well as the location in Folio.

They are the most celebrated lines in English drama, but they may not
be what Hamlet said. The First Folio puts something different in his
mouth:

> To be, or not to be, that is the Question:
> Whether 'tis Nobler in the minde to suffer
> The Slings and Arrowes of outragious Fortune,
> Or to take Armes against a Sea of troubles,
> And by opposing end them: to dye, to sleepe
> No more; and by a sleepe, to say we end
> The Heart-ake, and the thousand Naturall shockes
> That Flesh is heyre too? 'Tis a consummation
> Devoutly to be wish'd. To dye to sleepe,
> To sleepe, perchance to Dreame; I, there's the rub . . .[3]

Some differences between the texts of 1873 and 1623 can be put down
to two hundred and fifty years of changes in English spelling which
offer only the slightest aid, or the slightest impediment, to hearing
Hamlet speak (sleep/sleepe; ay/I). Even if the spelling changes were
reliably and regularly to indicate changes in pronunciation, no scholar
has managed to hark back with grounded confidence to Shakespeare's
own pronunciation; we have no strong evidence that the spelling of his
texts represents his manner of speaking, or the manner of speaking
which he approved for performance of his works. As the most recent
authority observes, 'Shakespeare's own pronunciation . . . is now lost
beyond recovery'.[4] FitzGerald, anyway, probably did not worry about
that; it was only around the time he heard his Shakespearian voices
that scholars began to speculate, and fret, and quarrel, over how
Shakespeare's English might have sounded.[5]

Other small things, though, add up to greater distances across which
Hamlet makes himself heard. The system, if it may be so called, of
punctuation in 1623 differs from the practice adopted by Clark and
Wright, and differs in large part by being less consistently governed by
syntactical considerations. Clark and Wright may have believed that
they were just translating seventeenth-century pointings into their
nineteenth-century equivalents. But consider the fact that on most
occasions when the punctuation of 1873 varies from 1623, it does so

---

[3] Ibid.; Folio, ll. 1710 ff.

[4] Fausto Cercignani, *Shakespeare's Works and Elizabethan Pronunciation* (Oxford, 1981), 1.

[5] See ibid. 2–4, where such scholarly work on these matters as White (1861), Noyes and Peirce (1864), and Ellis (1871) is described.

by being heavier, 'heavier' in the sense of implying more pauses or longer pauses (commas in 1623 become semicolons or colons in 1873, for example). The Victorian Hamlet speaks more slowly than his earlier self; he does not speak, as the Folio Hamlet asked to have his lines spoken, 'trippingly on the Tongue'.[6] As in music, so in language, the tempo affects the sense. A more ponderous delivery might imply deeper pondering—Hamlet, and this speech in particular, have acquired a reputation for profundity which they did not have at first. Hamlet has grown older along with *Hamlet*. The play and its protagonist are less readily intelligible than they were to their original audiences; more time is needed to 'put them across'. Perhaps people have come to think (or have really discovered) that there was more in them than could be guessed on first sight and hearing. Then there are the different exigencies of a text designed to be read, as FitzGerald read Clark and Wright, and a text designed to be performed; the reading text uses extra means to make itself clear, means which could be replaced in performance by tone and turn of voice, by facial expression, and by gesture. Many guesses, of which this is only a selection, can be made to account for these altered pointings.

None of the guesses is trivial in its implications. The various possible timings of the speech hint at the time in which our reception of the speech has taken place, at the history of *Hamlet*, in fact. Still, 'the history of *Hamlet*' is not the same thing as 'what Hamlet says' in 'To be, or not to be', though understanding of the one bears on understanding of the other. There is a further variant of punctuation which raises a more substantial and integral question about the speech: the placing of the question mark in the two texts. Clark and Wright give:

> To be, or not to be: that is the question:
> Whether 'tis nobler in the mind to suffer
> The slings and arrows of outrageous fortune,
> Or to take arms against a sea of troubles,
> And by opposing end them?

Folio has no question mark there but continues for three lines more before ending its sentence in interrogation:

---

[6] *Hamlet*, III. ii, 2; Folio, l. 1850. Andrew Gurr, in his *The Shakespearean Stage 1574–1642* (Cambridge, 1980), gives reasons to believe that the average performance time of Shakespearian plays was short in comparison with modern practice, probably about 'only two hours' (p. 161).

And by opposing, end them, to dye, to sleepe
No more; and by a sleepe, to say we end
The Heart-ake, and the thousand Naturall shockes
That Flesh is heyre too?

The First Quarto, on the other hand, can't hang around so long before getting to the point in question:

To be, or not to be, I there's the point,
To Die, to sleepe, is that all?

whereas the Second Quarto has no question mark at all in the opening lines quoted from the speech.[7]

These roving or vanished question marks matter for at least two reasons. First, *Hamlet* is a play of questions and has a temperamentally questioning hero;[8] so, question marks are apt places to take the pulse of his temperament. Secondly, much debate has arisen about the precise nature of '*the* question' (my emphasis) which occupies him here. Harold Jenkins in his edition of the play gives seven pages to detailing such debate and pronouncing his own views on it.[9] According to his account, the 'two fundamental disagreements' rest on whether or not Hamlet is contemplating suicide, and on whether the speech concerns Hamlet's individual plight or a more general human predicament. I should say, though, that the real question is 'how many questions does Hamlet ask?' In Wright and Clark, the question mark makes the doubt centre on how to act, whether to do this or that; in Folio, the questioning peaks on an uncertainty about whether death really is the end. The syntax of the opening lines suggests a series of parallel considerations ('or . . . whether . . . or'), but the direction of thought is not wholly clear. Hamlet begins with a disjunction between existing and not existing ('To be, or not to be'). This may or may not be a way of asking whether he should kill himself, or of wondering whether he should kill himself, or of meditating about the value of human life in some sort of abstract. Folio then gives one of its strongest stops, the colon after 'Question'. What happens during that stop is questionable, in the sense that it can be performed in several ways. The following disjunction, from 'Whether' through 'Or' to 'end them',

---

[7] First Quarto (1603, repr. Menston, 1969); Second Quarto (1604–5, repr. 1940).

[8] Adrian Poole discusses this aspect of the play in his *Tragedy: Shakespeare and the Greek Example* (Oxford, 1987). I am particularly indebted to his noticing (p. 115) that the word 'questionable' appears only in *Hamlet* of all Shakespeare's works.

[9] Harold Jenkins, *Hamlet* (1982), 484–91.

might be paraphrased, 'Is it more worthy of human dignity to show patience under affliction or actively to strive to right wrongs and remove the causes of oppression?' That is, the disjunction poses a question about stoical resignation and active virtue, or about resigning to the will of God and helping to see that God's justice is done. Evidently, there is no '*the* question' here, because there are several questions. Someone might well believe, for instance, that it was better to suffer whatever occurred and therefore not, under any circumstances, choose not to be—and he could give various, indeed mutually contradictory, reasons, Stoic or Christian, for this belief. Or someone might claim that complete acquiescence in suffering is not really being alive at all, and that 'to be' one must encounter those risks which fighting against troubles involves, risks which include the risk of one's own death. It is possible to believe that 'not to be born is the best for man' and also to forbid suicide while enjoining absolute resignation; it is equally possible to think life in itself a good but a good only in circumstances which essentially include being prepared to die for something. And so on.

Paraphrase may help us some way towards what the lines mean, but we need something more than paraphrase to hear what Hamlet means by them, to hear what he says, as FitzGerald heard him speaking. One thing that might aid such hearing is to follow the travels of the question mark. Clark and Wright mark out an orderly move of thought through the lines. In their text, Hamlet asks his opening question(s); then he is carried from 'end' into 'die' and from 'die' into 'sleep' ('and by a sleep to say we end'). The ethical debate slackens off into a longing for death, for an end to questions. Folio, however, insists across seven and a half lines on a single process of thought, in which debate and desire are syntactically interwoven; there is no opposition in Folio's text between what Hamlet thinks he ought to do and what he wants to do. No opposition, not because he so sharply distinguishes the alternatives and knows how to choose between them, but because his state of mind is so fluid and blurred. Fluid and blurred, in exactly the way internal rhymes within blank verse both assist flow and also blur contour, as the rhyme of that keen, disjunctive, 'or . . . or' with the lulling, would-be done-with 'No more' makes resolution and the longing to be dissolved one tissue.

The postponed question mark of Folio hints, as does that text's lighter punctuation, at a rapid delivery. It does so because it asks that an intonation contour to mark a question should be held over a longer

span than is required in the 1873 text. Try speaking the two versions at
the same pace; the sense of artificiality and strain in Folio presses on
the voice, and can be alleviated only by taking the passage more
rapidly, or by taking strong pauses which the pointing does not
sanction and which interrupt the contour of questioning. I do not know
how FitzGerald heard Hamlet, but I hear Hamlet according to the
description I have given of the 1623 text. It could be said that I am
'hearing things' as FitzGerald was 'hearing things'; they are merely
different things. I certainly don't hear 'To die, to sleep' with 'die'
rhyming as it does elsewhere in Shakespeare with 'boy' or 'joy' or
'annoy',[10] nor, I suppose, did FitzGerald, however strong his accent
had become after all his years in deepest Suffolk. My Shakespeare, as
his, could then be just a trick of the sound.

This objection to hearing across great distances of time suggests
that people widely separated by chronology are the most likely to
perpetrate misunderstanding and mishearing. Actually, contemporaries
are quite as great victims of history. Take the hapless scribes and
printers of the First Quarto of *Hamlet*:

> To be, or not to be, I there's the point,
> To Die, to sleepe, is that all? I all:
> No, to sleepe, to dreame, I mary there it goes,
> For in that dreame of death, when wee awake,
> And borne before an everlasting Judge,
> From whence no passanger ever retur'nd,
> The undiscovered country, at whose sight
> The happy smile, and the accursed damn'd.

The courses of English literature would have been different, had 'To
be, or not to be, I there's the point' become, in the absence of any other
surviving text of the speech, Shakespeare's greatest hit. Textual
scholars can provide arguments to show that this is a 'corrupt' text as
contrasted with the Second Quarto or with Folio, but they cannot, as
textual scholars, tell us why 'To be, or not to be, I there's the point'
would never have inscribed itself on the English ear as has 'To be, or
not to be, that is the Question'. Yet, if you listen to these lines, a reason
becomes audible. In Folio, and in the Second Quarto, the first four
lines of the soliloquy have feminine endings (Question/suffer/Fortune/
troubles), a slight, off-key tiredness on the voice which attests both to
Hamlet's efforts to think things through and to his desire to be through

[10] See Cercignani, op. cit. 246.

with thinking about things. The first Quarto, on the contrary, assaults the ear with blunt, masculine terminations throughout the extract I quote, and so lacks the lovely contrast which Folio makes between Question/suffer/Fortune/troubles and the blessed word 'sleepe'. That, I think, is one of the things most needful to hear when listening to Hamlet's voice in the text: a cadence comes through, it is the sound of an imagined person, the sound of Hamlet, in FitzGerald's casually acute phrase, 'talking in my room—all alive about me'.

Whatever reasons there might be to doubt FitzGerald's hearing, something like his provision of voices for lines of print has to be done with every text. Indeed, we cannot have a text of Shakespeare at all without such an exercise of imagination. For the meaning of the words on the page does not declare itself, nor is it separable from features of voicing. I shall give, in what follows, arguments to prove this contention. Let us clear away one possible objection to such a claim now. Somebody might say: 'You can look up each word in the dictionary, and there will be only a finite set of plausible combinations of these words: that finite set, though it will contain variant possibilities which may make the meaning ambiguous, still marks out the range of possible meanings.' The *OED* is certainly a unique resource for readers, but you have to know how to read the *OED* too: it does not declare its own meaning though it defines many meanings. (We may set aside the occasions on which a text is so corrupt that we do not know which words actually are on the page.)

Take, for example, what can be heard in the word 'Heart-ake' in Hamlet's 'To be, or not to be' soliloquy. According to the *OED*, this is the first time the word was used with the sense of 'Pain or anguish of mind, esp[ecially] that arising from disappointed hope or affection'. Certainly, there is no other occurrence of the word in Shakespeare's works. How can we know that 'Heart-ake' does not mean, at this point in *Hamlet*, what the *OED* tells us it means before Hamlet's use, that is, something equivalent to 'heart-burn', indigestion? After all, the line reads, 'The heart-ache and the thousand natural shocks' in 1873; such a line might be paraphrased, 'Indigestion, along with all other physical discomforts'. But this is a point at which Folio and the Second Quarto agree in their punctuation, and disagree with most later editions:

> The Heart-ake, and the thousand Naturall shocks   (F)

> The heart-ake, and the thousand naturall shocks   (Q2)

Why is there a comma after heart-ache? It is there to distinguish a heart-ache from natural shocks such as flesh is heir to; it makes the distinction by hinting at a pause and a subsequent contrastive intonation: 'The Heart-ake, *and* (on the other hand) the thousand Naturall shockes'.

The *OED* rightly notes this occurrence as meaning what we would now call something like 'emotional misery' but the great dictionary wrongly defines 'heart-ache' as 'pain or anguish of mind'. *Hamlet* uses the word 'heart' with especial caution and extreme poignancy; both the caution and the poignancy are, mostly, Hamlet's own. The word 'heart' stands in contrast to 'mind' in the play, in such contrast that 'heart-ache' could not mean 'pain . . . of mind'. When Hamlet wonders about which conduct is 'Nobler in the minde', he recalls his own 'how Noble in Reason' and harks forward to Ophelia's two usages of 'Noble minde' and her expansion of his 'Noble in Reason' into 'Noble, and most Soveraigne Reason'.[11] The complex structure of relation between 'noble' and 'mind' well defines one aspect of the play and its protagonist: that confidence, at once of social and intellectual status, which is so peculiarly Hamlet's. He is a Prince, and he is intelligent. His intelligence often prompts him to see through civic grandeurs, as in his mockery of empty courtliness or his reflections in the graveyard, but he stands firm on his own dignity—his speech constantly revolves around possession and inheritance. He is not the deracinated intellectual of some imaginings. Both Horatio and Ophelia, to whom, apart from his relatives, he feels closest in the play, so constantly address him with honorifics that when they do not, the effect startles. Hamlet is almost always 'your Lordship', 'my Lord', 'good my Lord', 'my honour'd Lord'. In the 'nunnery' scene with Ophelia, for instance, Ophelia has nine honorifics in her first seven speeches, and drops Hamlet's title only at the moment of his most cold brutality and her most severe distress:

> HAMLET. . . . I loved you not.
> OPHELIA. I was the more deceived.[12]

A world of disappointment speaks in her not saying 'my lord'.

'Noble' in Shakespeare's works usually means 'of noble birth'; most of the usages in the plays are of the form 'my noble lord of Lancaster', 'the noble duke', 'noble peers', 'the noble thanes', and so on. *Hamlet* has the word seventeen times (which is not a high frequency: *Antony*

---

[11] *Hamlet*, III. i. 100, 152, 159; Folio, ll. 1755, 1806, 1813.
[12] Ibid. III. i. 119–20; Folio, ll. 1774–5.

and Cleopatra has thirty-three, and *Coriolanus*, fifty-eight). Yet the play is unusual in the high proportion of significant occasions on which 'noble' has only an unclear relation to lineage; when it does refer to lineage, it particularly stresses the relation between Hamlet and his father—its first two occurrences being 'seeke for thy Noble Father in the dust' and 'my noble Fathers person'. *Hamlet* debates, Hamlet debates in 'To be, or not to be', the relations between noble fathers and noble minds. These debates on insight and inheritance are crossed with an other consideration, with thoughts of 'heart'. Though *King Lear* has that word more often than *Hamlet*, the very frequency with which Lear speaks of his own heart makes what he says of it less intimate and touched than Hamlet's way with the word. Lear brandishes his heart about; for much of the play, he speaks of his heart as if it were a seal of office. But Hamlet's first reference to his own heart comes at the end of a soliloquy, and is bitten-back, an abrupt check to eloquence which is very like him—'But breake my heart, for I must hold my tongue'. This privacy in his sadness is one of the things which makes Hamlet both so lonely a character and so close to an audience. The word comes to him often when he is alone with the audience: 'Must (like a Whore) unpacke my heart with words' or 'Oh Heart, loose not thy Nature'.[13] He reproaches Rosencrantz and Guildenstern for wanting to give his heart publicity: 'you would pluck out the heart of my Mysterie'.[14]

And yet it is the mystery of his own heart which he speaks most piercingly, in three exchanges of tenderness with Horatio, exchanges which mark the relation between Hamlet and Horatio as the most loving and trusting relation between two men in Shakespeare's work. First, when he tells Horatio—and tells himself at the same time— what he admires in a man:

> Dost thou heare,
> Since my deere Soule was Mistris of my choyse,
> And could of men distinguish, her election
> Hath seal'd thee for her selfe. For thou hast bene
> As one in suffering all, that suffers nothing.
> A man that Fortunes buffets, and Rewards
> Hath 'tane with equall Thankes. And blest are those,
> Whose Blood and Judgement are so well co-mingled,
> That they are not a Pipe for Fortunes finger,

[13] Ibid. II. ii. 581 and III. ii. 384; Folio, ll. 1626, 2264.
[14] Ibid. III. ii. 356–7; Folio, ll. 2236–7.

> To sound what stop she please. Give me that man,
> That is not Passions Slave, and I will weare him
> In my hearts Core: I, in my Heart of heart,
> As I do thee. Something too much of this.[15]

Here again, as at 'But breake my heart . . .', the word is barely out of
Hamlet's mouth before he waves it away in the princely dismissal of
'Something too much of this'. The lines look back to his quandaries
about suffering the 'Slings and Arrowes of outragious Fortune'
('suffering all, that suffers nothing. | . . . Fortunes buffets') and
forward to his indignant refusal to wear his heart on his sleeve for
Rosencrantz and Guildernstern—'you would play upon mee; you
would seeme to know my stops'[16] ('to sound what stop she please').
They come then between the solitariness of his 'Heart-ake' and the
disgust he feels at the offers of *confidants*, the solicitations to candour.
Horatio's tact shows in this passage as elsewhere when he abstains
from picking Hamlet up on such intimate remarks; he behaves as if
Hamlet had not said anything of the sort, though we do not doubt (nor
does Hamlet) that he has heard every word. Where Rosencrantz and
Guildernstern press and squeeze everything Hamlet says for symptom
or clue, Horatio lets things pass. He is as one that hearing all has heard
nothing, and that is why Hamlet can talk to him.

A later exchange of 'heart' between them begins and ends with a
similar, apparent inconsequence, discussing the wager on the duel
with Laertes:

HORATIO. You will lose this wager, my Lord.
HAMLET. I doe not thinke so, since he went into France, I have been in
    continuall practice; I shall winne at the oddes: but thou wouldest not
    thinke how ill all's heere about my heart: but it is no matter.[17]

'But breake my heart', 'Something too much of this', 'but it is no
matter'—it is the same motion again, something compounded of a
reticent dignity, despair of ever being able, even to Horatio, quite to
say what is the matter, and a flinching from the fact of his own
vulnerability. The passage extraordinarily combines the most savagely
farcical of Hamlet's jests with Hamlet in his greatest, plainest need for
kindliness. For what does 'since he went into France, I have been in
continuall practice' mean? Harold Jenkins annotates 'This, though

---

[15] Ibid. III. ii. 62–74; Folio, ll. 1913–25.
[16] Ibid. III. ii. 355–6; Folio, l. 2235.
[17] Ibid. v. ii. 205–9; Folio, ll. 3658–62.

appropriate here, contradicts II. ii. 296–7',[18] contradicts, that is,
Hamlet's claim to Rosencrantz and Guildenstern that he has 'forgone
all custome of exercise'. Indeed it does, but this misses the point, for
the 'practice' that Hamlet refers to, practice in swordsmanship, is the
murder of Laertes' father, Polonius. The grim boast implicit in the
remark is that having done for the father, he'll be a match for the son.
The joke is insensately cruel, but what follows it is a return of that
nature Hamlet prayed his heart would not lose, an admission of 'ill',
wrong-doing, as well as sickness and 'Heart-ake'—'but thou wouldest
not thinke how ill all's heere about my heart'. Christopher Ricks has
heard these lines clearly and written well about what he hears:

We feel intensely with Hamlet, we share his pain and premonition, and the
very uttering of the words is astonishingly at one not only with such pain but
with the *hearing* of the uttered words. There can be no easy saying or hearing
of them, partly because of the perilous thickening within 'how ill all's heere',
partly because there is no natural fluency of linkage between any of the words
in the sequence 'how ill all's heere about my heart', partly because of the
precarious breathy hesitation of the alliteration on 'h'—'how', 'heere',
'heart'—groping against the vowelled cautious openings for 'ill', 'all's', 'about'.
The effect is intensely dramatic, in its combination of the most intimate
sharing of experience with the most awed sense of our being so extraneous to
Hamlet, his being—despite Horatio's love for him—so terribly isolated.[19]

The last hearts in *Hamlet* are not so terribly isolated:

HAMLET. If thou did'st ever hold me in thy heart,
  Absent thee from felicitie awhile,
  And in this harsh world draw thy breath in paine,
  To tell my Storie.
                        ...   .

                    O I dye *Horatio*:
  The potent poyson quite ore-crowes my spirit,
  I cannot live to heare the Newes from England,
  But I do prophesie th'election lights
  On *Fortinbras*, he ha's my dying voyce,
  So tell him with the occurrents more and lesse,
  Which have solicited. The rest is silence, O, o, o, o.   [*Dyes*
HORATIO. Now cracke a Noble heart:
  Goodnight sweet Prince,
  And flights of Angels sing thee to thy rest . . .[20]

---

[18] Jenkins, op. cit. 406 n.
[19] 'John Milton: Sound and Sense in *Paradise Lost*', repr. in *The Force of Poetry*
(Oxford, 1984), 62.   [20] *Hamlet*, V, ii. 351 ff.; Folio, ll. 3832 ff.

It is entirely apt that Horatio at last, at the last, has the word 'heart', that over Hamlet's corpse, he eventually replies to Hamlet with words of sincere, decided, intelligible tenderness. Shakespeare has written the completeness of Horatio's reply into the sound of his lines, as they wish for Hamlet something more than he could speak for himself, though Horatio speaks words so like Hamlet's own. 'The rest is silence': it is again that side-stepping motion, Hamlet's shrug at his own miseries, as in 'but it is no matter'. Hamlet means probably both that he has nothing more to say (and this is one reason for preferring the Second Quarto's omission of 'O, o, o, o') and also that when you're dead, you're dead, that nothing follows death. But Horatio will not leave Hamlet in a silent inexistence—'And flights of Angels sing thee to thy rest'—*not* silence but song for this 'sweet Prince'. The phrase transfigures Horatio's innumerable 'my Lord's into a single 'sweet Prince', an honorific that is also an endearment. Nor will Horatio leave Hamlet's 'the rest' alone. 'The rest' in Hamlet's mouth means 'what's left over', as in 'all the rest of it', a messy remainder one is glad to be shot of. Horatio takes the long 'i' of 'silence' and turns 'the rest' to 'thy rest', a repose, a rest in peace. The close weave of the words between them sounds a moment of intimacy even as they are separated. Someone else is with them too: Ophelia. For Horatio's 'Noble heart' recalls Ophelia's 'Noble minde', and so Horatio speaks for both the people who have at least wholly loved Hamlet in this play, though this was all they could do for him.

This is some of what may be heard in Hamlet's 'Heart-ake', and it is such that no printed page, even helped along with a dictionary, could ever make clear. No page displays a voice's pace, its dips and rises, how some words come readily to it and others only with reluctance, the ever-varying timbres of allegiance, longing, shyness, or disdain which colour utterance and give character to a voice, give voice to a character. To hear what is 'all alive' in 'talking' is to hear at least some of these things. And it is not because Shakespeare is remote from us that we have difficulty sometimes catching his words, or doubt whether we have understood them. To read Shakespeare is an experience like that of listening to Hamlet, an experience, in Ricks's words, which combines 'the most intimate sharing with the most awed sense of our being so extraneous'. Nor is it because Shakespeare's imagination is so palpably rich and the records which remain to us of that imagination so comparatively frail and unreliable, for these doubts and difficulties attach to all print just because of the constitutive relation of print to voicing. Print does not give conclusive evidence of a voice; this raises doubts about what we hear in writing but it also gives an essential pleasure of reading, for as we meet the demand a text makes on us for our voices, we are engaged in an activity of imagination which is delicately and thoroughly reciprocal. Nor, again, is the co-operativeness Shakespeare asks of us, and himself exemplifies, a result of the fact that *Hamlet*, for instance, is a play, and so requires performance, if only in imagination, in a parlour. All writing is dramatic, though not all writing is theatrical. 'Dramatic' in the sense that writing is an act of supplication to an imagined voice. Sometimes, the voice writing asks is baffled, choked, actually unvoiceable by any physical voice, so many-angled and disparate are the things which written words may be saying all at once. But even to know that one has lost one's voice, or that one's voice can do no good at some point, is an effect of voice, something that only those who have a voice, and listen to others' voices, can feel.

Recorded Sounds: Linguistics and the Voice

Hamlet is perhaps too complicated a speaker to take as an example for making basic points and raising simple questions; his voice moves too individually to provide a plain instance for comment. An instance less rich in character may bring out more sharply what is at issue here. Such as: 'PLEASE talk QUIETLY'. A sign to this effect stands in the Cambridge University Library at the head of the staircase leading from the tea-room to the areas reserved for study. Standing where it does, libraries being what they are, it means, 'If you must talk, talk quietly', and implies that silence would be preferred. If the sign were put somewhere else, it would mean something else. In a confessional, for example, or in the ante-room to the office of some great dignitary. In those places, it would read differently. In a confessional, it might encourage a penitent to speak and admonish him to do so quietly. Coming from a dignitary, it would accord a permission, allowing the person who might speak to 'feel free' to talk but asking that person to remember the circumstances in which liberty was exercised. The library's sign might less ambiguously be written, 'Please   talk quietly'; the other two signs, 'Please talk   quietly'.

Some attempt has been made in the library to write clearly what is meant: the smaller characters of 'talk', the emphatic capitals of

'PLEASE' and 'QUIETLY', employ familiar graphic conventions to indicate that the important point is volume control. Yet all this polite ingenuity cannot of itself prevent a determinedly fertile interpreter from taking the sign as not discouraging talk but asking for it. Such an interpreter might suggest that the sign should be read as a message from a shy sub-librarian, avid of conversation but wary of superiors: 'PLEASE' would then be an intimate supplication, 'talk', an urgent whisper, and 'QUIETLY' a note of warning, a written glance over the shoulder in fear of being caught. Nothing in the text of the sign precludes this interpretation, but our being conversant with what library authorities generally permit, and what sub-librarians generally dare, puts it out of the question.

The sign is essentially ambiguous until we find an appropriate setting for it in a context of intentionality. In this case, such a context will not involve the intentions of any particular individual. Indeed, the story about the sub-librarian is mistaken just in imagining that the sort of intention relevant to understanding here is personal. Perhaps it would be better to say: the sign can be shown to be ambiguous by providing it with several possible authors whose intentions in and for it may be divergent or mutually incompatible. Unless we have a reason for making such provision, we may content ourselves with the apparent context—an official notice in a library—and within that context the sign is not ambiguous at all. This is a familiar principle of economy in interpreting. Someone might object that calling it a 'principle' is a grand way of describing a habitual tendency people have to follow the interpretative line of least resistance; a little more rigour, or curious energy, might enable us to read more deeply and to recognize the principle as merely a sanction for cosy acquiescence in a world already known. Certainly, the principle operates usually with the unreflective force of a habit, but it can still be articulately justified.

Doubts about the meaning of a sign can always be created because no sign can be self-interpreting; it has its meaning 'in its place', and *what* its place is no sign can of itself determine. Even in the notation of formal logic or mathematics, individual signs have a determinate meaning only within the system as a whole which needs to be grasped and used consistently if the signs are to preserve their unequivocal character. As no sign can show fully and unequivocally how it itself is to be interpreted, we must exercise interpretative judgement. The principle of interpretative economy is one maxim of such exercise. As Wittgenstein wrote:

How does it come about that this arrow ——> *points?* . . . The arrow points only in the application that a living being makes of it.[21]

What you can make of the arrow is not entirely up to you, as someone who tried displaying the arrow in a library with the hope that people would understand it to mean 'Please talk quietly' would discover. It is, however, possible to use a conventional sign, such as the arrow, for purposes which are not purely conventional, to give a new application to a sign, as Wittgenstein does in the remark I have quoted from the *Investigations*. The arrow there appears not only as a conventional sign but also as an example of a conventional sign. A reader who took the remark as concerned only with the particular arrow represented, or with only arrows generally, would not have understood what Wittgenstein wrote.

The many people who are puzzled in the library at Cambridge don't usually puzzle about 'PLEASE talk QUIETLY', so this example, while abstractly helpful for the purposes of exposition, has little serious content. Nothing actually troubles understanding here. The footling has, though, the advantage of being transparent. If we turn to a case in real doubt, the issues involved are not so neat. When Coleridge first published in the *Monthly Magazine* the poem which eventually became 'Reflections on Having Left a Place of Retirement', he called it 'Reflections on entering into active life' and gave it the proudly humble subtitle 'A Poem which affects not to be Poetry'.[22] A reader might have taken this as 'A Poem which affects not-to-be-Poetry' ['affects' in the sense of 'pretends'] rather than what Coleridge meant, 'A Poem which affects-not to be Poetry'. After all, the inversion which expresses Coleridge's meaning is itself less colloquial or plain than the syntax which he did not intend where the negative governs the infinitival phrase. So easy is it for a declaration of unaffected sincerity to turn out looking like itself an affectation. To determine what Coleridge meant, we have to bring contextual information to bear on what he wrote. Nothing in what he wrote there tells us what his writing means there. Deciding which information is relevant, and assessing its weight, is an exercise of judgement—amongst other things, judgement about a particular value attached by Coleridge at this stage in his career

---

[21] *Philosophical Investigations* (1953; 2nd edn. 1958), trans. G. E. M. Anscombe, No. 454. Some implications of Wittgenstein's remarks about the arrow are discussed in John Casey, *The Language of Criticism* (1966), 4–9.

[22] 'Reflections on entering into active life', *Monthly Magazine* (Oct. 1796), repr. in *Coleridge: Poetical Works*, ed. E. H. Coleridge (1912), 106.

to the simple, the unadorned and the natural, judgement also of the difficulty with which such simplicity was achieved. The various tussles between the 'poetical' and the 'plain' which occupy Wordsworth and Coleridge in this period lurk within the ambiguity of the poem's subtitle. A critic who inclined to say that their ideal of purified ordinariness for poetic speech was quite as artifical as the vicious phraseologies they decried might find the subtitle subtly aware of that fact, or (if he had a lower estimate of Coleridge's manymindedness) an inadvertent give-away of the posturing within the avowed simplicity. An other reader, readier to consider naturalness of style as something more than a refinement of the *faux-naif*, could take the slight haziness of meaning as conscious testimony to the trouble Coleridge went through in extricating himself from the conventions of his day.

The subtitle is ambiguous now, and was so then, because the writing lacks a sign for the desired collocation of the words, and particularly lacks signs for those 'junctural features'[23] which indicate collocation in speech. Coleridge could have explained himself more thoroughly by writing his subtitle in one of the two ways I have suggested or, more straightforwardly, by rephrasing it as 'which does not affect to be poetry'. He may have failed to do so not because he was set on ambiguities, or because he wanted to show how declarations of sincerity can go wrong; he may just not have seen that these written words can be spoken in distinct ways. Underlining his probity by writing out the desired collocation—'A Poem which affects-not to be Poetry'—might have seemed an idiosyncratic self-righteousness, protesting too much. If the gramophone had been invented in 1795, and Coleridge had released this poem as a single rather than in a magazine, he would have been able to speak the poem's subtitle in such a way as to settle which of the two meanings he intended (he could also, with a little more effort, have left the issue still unsettled). But, given the means available for publishing words in 1795, he was doing nothing special by leaving his written words without a clear indication of how they should be heard, even though the result is a possibly unfortunate equivoque, for English orthography normally makes little attempt to indicate in detail the manner in which a written

---

[23] The phrase is David Crystal's, and refers to such vocal features as slight pauses, variations in length, aspiration, and so on, which 'suffice to indicate unambiguously where a tone-unit boundary should go in connected speech'. See his *Prosodic Systems and Intonation in English* (Cambridge, 1969), esp. pp. 195 ff.; an extract from this is published in D. Bolinger (ed.), *Intonation* (Harmondsworth, 1972) [hereafter referred to as Bolinger], 110–37.

text is to be voiced. It is in fact standard practice not to indicate many features such as pace or intonation-contour, though in some of Coleridge's later writings the very profusion with which he resorts to such devices as italics show a writer striving to make himself heard, to make himself plain.

Even if writers, including Coleridge, tried harder in this matter, problems of interpretation would remain, because it is impossible to notate speech unambiguously in writing, in *any* system of writing. By 'unambiguously notate', here and elsewhere, I mean 'notate in such a way that, *when reasons are given for doubting* what a written sign means, the writing can always settle all such doubts'. The clause '*when reasons are given*' is important; the claim is not that writing cannot unambiguously notate in the sense of 'settle any possible ambiguity', for it is always possible (and often pointless) to think up ambiguities, even for sentences in axiomatic, formal languages or for heavily conventionalized signs. Thus, time may be passed by providing narratives which render 'PLEASE talk QUIETLY' ambiguous. To observe that no system of notation can inhibit such aimless imaginings is true enough but nugatory. Coleridge's subtitle, though, is genuinely in doubt; it is a doubt that could have struck him too; reasons which lie at the core of his enterprise raise that doubt, and ask us to settle, entertain, or learn to live with it. We may on reflection come to feel confident that all that happens in his subtitle is a slip of attention rather than a deliberate sign of the writer's predicament, but the reflection engages us with genuine questions of understanding Coleridge rather than only with abstract scepticism about what his meaning was or what the meaning of any particular sign might be. It is an essential accident in the history of literature that machines for recording speech were not invented till the second half of the nineteenth century—essential, for the possibilities of misunderstanding which arise in the absence of such machines give substantial life to those disputes of value which provoke misunderstanding, and make what might be mere technicalities of notation the very air of cultural exchange. A notation is like an explanation: 'an explanation serves to remove or avert a misunderstanding—one, that is, that would occur but for the explanation; not every one that I can imagine.'[24] I will defend the claim that writing is an inherently ambiguous notation in the third section of this chapter when I describe the importance of such ambiguity for philosophers of language. The

---

[24] Wittgenstein, op. cit., No. 87.

claim may be set aside for the moment, while we consider in more detail the fact that writing does not fully notate speaking, and some of the implications of that fact for linguistics.

In some systems of writing, this hardly matters because the spoken versions of written sentences in these systems are secondary. 'Secondary' because even if the sentences of the system were never spoken this would not lead to widespread confusion about the meaning of those sentences. Symbolic logic is such a system. The natural languages are mostly not like this, and certainly the European languages are not. Such languages exist in both written and spoken forms; it is possible to speak one of them without knowing how to write it, and vice versa, though the majority of native speakers of a European language can nowadays both speak and write them. The sentences of the written language frequently depend on the spoken language in so far as they remain indeterminate between several semantically distinct vocal utterances, and can become unambiguous only when spoken. In the case of all of these languages, it is true, as I. J. Gelb remarks, that 'normally, writing fails to indicate adequately the prosodic features'.[25] By 'prosodic features', Gelb refers to the significant features of the voice in the spoken form of the language: pitch, pace, volume, pronunciation, stress, juncture, and intonation.[26] Linguists have long been familiar with this 'failure'. The great phonetician, Henry Sweet, recorded a consensus: 'It is, I think, generally admitted that our notation of quantity, stress, and intonation is unsatisfactory.'[27] and, fervent though his enthusiasm was for a reformed spelling which would bring writing closer to speech, he saw limits to the project of graphic phoneticism: 'If it were possible to give an exhaustively, minutely accurate representation of the pronunciation of any one speaker by means of alphabetic signs, such a transcription would not be legible in the practical sense of the word: it would only be, at the best, decipherable.'[28]

Literary critics, or theorists of literature, have not always been so sensible of this. Barbara Herrnstein Smith, for example, writes that '. . . given any orthographic system, anything speakable is also

[25] I. J. Gelb, *A Study of Writing* (Chicago, 1952; rev. edn. 1963), 14.

[26] The list is not intended to be exhaustive, though linguists have, understandably, paid less attention to other prosodic features, such as vibrato and timbre, which are more expressively important for the singing than the speaking voice.

[27] *The Indispensable Foundation: A Selection from the Writings of Henry Sweet*, ed. E. J. A. Henderson (1971), 203.

[28] Ibid. 200.

inscribable'.[29] On the contrary, it is only with a highly specialized orthographic system that some readily speakable things can be inscribed, and, in Sweet's more precise terms—'exhaustively, minutely accurate representation'—much that is spoken every day cannot ever be written down. Gelb exaggerates when he claims that 'it is almost impossible to understand a writing without knowledge of the speech for which it is used',[30] but Herrnstein Smith's fantastical belief that speech can be completely inscribed in 'any orthographic system' would entail that acquaintance with the spoken form of a language was merely an optional extra for someone who wished to understand writing in that language. Imagine someone who had never heard English spoken trying to establish what was meant by the written sign 'Oh'; there is much the voice can do with the sound we write like that. This example may be thought too pat to my purpose, as 'Oh' is barely a word at all and so has no semantic content apart from its sounding. But the same is true of 'Yes' and 'No', of 'true' and 'false'.

Those who in fact have never heard English, or any other language, spoken—I mean, the deaf—can learn and understand language, but their production of the languages they learn is marked by a general deficiency in the 'prosodic features' other speakers produce, with consequences for the intelligibility of what they say:

Hudgins and Numbers . . . analyzed the speech of 192 deaf pupils from two schools for the deaf. Each subject read ten unrelated simple sentences, and intelligibility was rated by individuals familiar with the speech of deaf individuals. Two major types of error may be identified: articulatory and rhythmic. . . . However, articulation, per se, was of only secondary importance in the intelligibility of the speech of the deaf subjects as compared with other aspects of speech. For example, sentences spoken with correct rhythm were understood correctly four times more frequently than those with incorrect rhythm. In general, the speech of subjects was characterized by arhythmic patterns, poor phrasing, monotonic expression, and lack of pitch.[31]

This instance of a practical connection between the prosodic features of a language and intelligibility demonstrates a link between what might be thought of as the 'form' and the 'content' of an utterance, a link existing in the material medium of the language. Rhythm and

[29] *On the Margins of Discourse: The Relation of Literature to Language* (Chicago, 1978), 4–5.
[30] Gelb, op. cit. 223.
[31] D. F. Moores, *Educating the Deaf: Psychology, Principles, and Practices* (Boston, 1978), 225. I am indebted to Ruth Mackenzie for this information.

cadence belong to the intelligibility of utterances, and not only in the
literary uses of language which have been traditionally distinguished
from non-literary uses just by an allegedly special coherence of sounds
and senses in literature.

Linguists who concentrate on the meaning of sentences rather than
the intelligibility of utterances sometimes rely on this customary
relegation of prosodic features from the central realm of 'ordinary
language' or 'communication' to the outlying provinces of the
ornamental and the creative, and so begin their studies of these
features from a disputable demarcation between them and the primary
conveyors of meaning. Kenneth L. Pike:

all . . . lexical meanings have this in common, that they are indicated only by
the requisite consonants, vowels and stress, and a context where such a
meaning is possible; in that sense, the lexical meaning is intrinsically a part of
the word itself and not dependent upon extraneous phenomena such as pitch
produced by emotion.

The intonation meaning is quite the opposite. Rather than being a stable
inherent part of words, it is a temporary addition to their basic form and
meaning. Rather than being carried by permanent consonants and vowels, it is
carried by a transitory extrinsic pitch contour. Rather than contributing to the
intrinsic meaning of a word, it is merely a shade of meaning added to or
superimposed upon that intrinsic lexical meaning, according to the attitude of
the speaker.[32]

This is odd. What is '*the* lexical meaning' of a word (my emphasis)?—
many words have many lexical meanings. Consider the four pages
devoted to 'green' by the *OED*, or the eight given over to 'light'. Which
meaning a word has in a particular utterance depends on its place in
that utterance, and prosodic features essentially locate that place
because intonation, for example, marks out the units of connected
speech, separating one from another and sounding out the internal
links within the units. Pike wants to have it both ways: 'lexical meaning'
both derives from 'context' and 'is intrinsically a part of the word
itself', but a word's 'context' can hardly be 'intrinsically a part of the
word itself' any more than where my latch-key now is can be
intrinsically a property of my latch-key (for if that were so, I should
never lose it). This distinction between stable inherent meaning and
transitory extrinsic superadded shades of meaning falsifies the
actualities of intelligibility and meaning in linguistic utterance.

[32] Extract from *The Intonation of American English* (Ann Arbor, 1945), repr. in
Bolinger, p. 55.

A wish was father to these confusions of thought, and neither the wish nor the confusion is Pike's alone. The prosodic features of utterance present a crucial difficulty to the science of linguistics in so far as it works with Saussure's distinction of *parole* from *langue*, of 'the actual manifestations of language in speech or writing' from 'language as a system',[33] for prosodic features are, in Pike's words, 'semi-standardized'[34] and so fall across the Saussurian dividing-line between system and instance. Intonation patterns are neither purely individual nor wholly systematic, and the same is true of most other prosodic features of language. Dwight Bolinger observes that though there is 'wide agreement among linguists on the units of sound that make distinctions in word meanings' there is 'no such agreement on the units of intonation', and he suggests that the lack of agreement reflects the 'difficulty of treating intonation independently of all the other events that tend to colour it'.[35] These 'other events' include elements not strictly linguistic, such as gesture, facial expression, and the various pressures of the circumstances in which a prosodic feature plays an articulating role. The field of relevant considerations hereabouts grows so wide as to defeat the attempt to systematize it, and, by the same token, eludes strict notation. When this happens, Saussure's linguist may be in some difficulties, for Saussure insisted that 'The linguist needs above all else a means of transcribing articulated sounds that will rule out all ambiguity'.[36] In two important respects, the prosodic features of language make trouble for the project of Saussurian linguistics: they do not fit the *langue/parole* distinction and they cannot be unambiguously transcribed, even in the weaker sense of 'unambiguously' previously defined. And these difficulties cast doubt back on Saussure's notion of *parole*, 'the actual manifestations of language in speech or writing'. That 'or', 'in speech or writing', hints at an indifference between the manifestations of *langue* in speech and writing, as if *langue* turned up in both of them in the same way though, as Saussure insisted, more richly in speech, the linguist's prime quarry for extraction of materials for the study of *langue*. He was wrong to assume such an indifference, and the possibly different manifestations of *langue* in speech and in writing may account for the unattainability of

---

[33] Jonathan Culler, 'Introduction' to Ferdinand de Saussure, *Course in General Linguistics*, trans. Wade Baskin (1959; rev. edn., 1974), p. xvii. Saussure makes the distinction repeatedly in the text, see esp. pp. 14–15.
[34] Pike, in Bolinger, 53.   [35] Bolinger, 14.
[36] Saussure, op. cit. 33.

Saussure's chief requirement: 'a means of transcribing articulated sounds that will rule out all ambiguity'.

Some of the difficulties linguists face in describing and explaining the role of prosodic features in language come from the technical awkwardness of transcribing the sounds people make when talking (these difficulties survive even in the days of tape recording and oscilloscopes), but the wide diversity of notations and taxonomies employed by linguists for just one prosodic feature—intonation[37]—suggests that the state of cross-purpose in which the discipline currently toils may derive from permanent conceptual intractability in the material rather than from a mere cussedness of the equipment which we could expect will eventually be engineered away.

Take Richard Gunter's thoughtful essay on 'Intonation and Relevance'. He records some obstacles to a formalist account of the workings of intonation:

One may take an intonation in the abstract, divorced from words, and may try to assign a meaning to it, but that task is baffling. It is equally baffling to attempt to find connections between an intonation and the internal semantic or grammatical facts of the sentence with which that intonation occurs. Such connections are at best elusive, and it may be that they do not exist at all. There are sentences that can take many different intonations; there are intonations that can occur with all sorts of sentences; and—most telling of all—there is no string of words that has one necessary intonation.[38]

These acute points conduce to despair of ever rendering explicit a system which connects intonation and sentences; there seems to be no such system. A linguist might then attempt to find some regularities of rapport between intonation and utterances by moving on from 'the *internal* semantic and grammatical facts of the sentence' [my emphasis] to the, as it were, 'external' facts about sentences—where they occur, when they happen, and so on. A situated sentence is an utterance. Gunter's essay tries to show how some intonation contours make explicable contact with the situations in which they are produced: one intonation contour conveys recapitulation of what has just been said, another bears contradiction of the prompting remark. Gunter does not claim that any particular contour has to be used for any particular purpose; indeed, he finds, and finds it 'astonishing',[39] that there is no strictly predictable correlation of contour and purpose. He reaches this

---

[37] A variety discussed by Bolinger in his introduction to *Intonation*, loc. cit.
[38] Bolinger, 197–8.   [39] Ibid. 214.

not very promising conclusion through inspecting a number of schematized examples of linguistic exchange. The relation of intonation to context, of utterance to situation, appears always in terms of the way an intonation contour may respond to a single sentence prompt. His examples are all of the form

> Context: *Who is in the house?*
> Response: JOHN [giving information]

or

> Context: *John drank tea.*
> Response: WINE [contradicting the previous remark][40]

The varying intonation contours of the two responses may be drawn according to any of the many systems of notation in use by linguists. These are difficult examples to deal with; a lot needs doing to make what meets the ear here meet the eye with graphic adequacy and acceptable comment. Yet the schematization is great, though not gross. It would be, in some respects, absurd to complain that the instances of 'context' are so impoverished as to be unrecognizable, but the more the context of these minuscule dialogues is described, the more the significance as a 'response' of the intonation contour may vary. People, when discussing tone of voice, make such remarks as 'Coming from *him*, it wasn't really a very cynical asperity' or 'When you know him better, you'll realize he's joking half the time'. Intonation contour forms part of the everyday concept, 'tone of voice', but Gunter's schematism depends on a sense of 'context' which is divorced from considerations about the person speaking, and, therefore, provides us only with attenuated utterances. Not much can be expected from a transactional study of linguistic exchange which removes the traces of agency from its examples. Yet the main difficulty lies rather in the inherent ambiguity of the terms 'context' and 'response' than in the starkness of instancing; linguists have worked on much longer stretches of connected discourse which are not vulnerable to these objections. The ambiguity of 'context' and 'response' stems from the assumption—it is so widespread in linguistic studies of intonation that it has acquired the dignity of an axiom—that written and spoken utterances relate to their contexts in the same way. This is the assumption which is also questionable in Saussure's 'actual manifestations of *langue* in speech or writing'.

[40] I omit Gunter's notation of intonation curves from these examples; the notation does not affect my point. The examples occur in Bolinger, 204, 205.

If the context-dependency of meaning were different in speech and in writing, then the role of prosodic features would equally vary between the two, and such variance would result in a more than technical difficulty in the transcription of speech into writing. In the third section of this chapter, I will give some reasons for thinking that this is the case. Gunter's exiguous dialogues derive their character from being at once written and also intended purely as records of speech; they have the plausibility they have because such snippets are familiar in one kind of writing, but they are offered as means to the imagining of speech, though the speech they record is unimaginable unless we supply more than the examples give. That is, they are intelligible as written instances because they have a written context but they remain cryptic as imagined speech because they have no spoken context, even though they are supposed to be instances of such a context and the response it summons.

An achieved version of the transcriber's plight, of this phantom understanding in the hollow of an ambiguity between the contexts of writing and speech dominates much of the greatest English poetry of the nineteenth century, especially that nineteenth-century invention, the dramatic monologue, and later chapters of this book are taken up with some literary critical studies of that achievement. For the present consider, in contrast to Gunter's examples, the opening line of Browning's 'Andrea del Sarto': 'But do not let us quarrel any more'. What 'context' does that respond to, if it is a 'response'? The very incompleteness of the context in which Browning sets Andrea's words raises the question of whether Andrea responds to his wife or anything she has said. He may incline to complain, 'You never listen to a word I say' but she might retort, 'You never really talk to me, you're so wrapped up in yourself.' There are also contradictory contexts in which we must consider these words and any intonation contour they may have. As the first line of a poem by Browning, the words have a literary context in which they might be described as abrupt and surprising or as an instance of the convention of *in medias res*. Within the situation the poem asks a reader to imagine, they may not be abrupt at all, coming perhaps after a lengthy raking-over of grievances or pleading, and, within the fiction that Andrea del Sarto just happens to be saying these things to his wife one evening and is speaking not the English verse we read but Italian prose, they are not deploying a literary convention. Here we find that to call something a 'response' is already to assign to it a place in linguistic exchange with a corresponding manner of utterance; such an assignment is not a merely neutral classification in descriptive linguistics.

David Crystal has exposed the overworking to which 'context' has been subjected as an explanatory term:

It has been used to include some or all of the following: the co-occurring formal structure of an utterance (whose intonation is being analysed), the formal structure of utterances preceding or following the focal utterance, the intonation patterns preceding the focal utterance, the observable situation in which the utterance takes place, factors in the observable situation preceding the focal utterance, the presuppositions in the mind of the speaker, or hearer, and other things besides. Given such a broad interpretation, it would not be surprising for the meaning of an intonation pattern to be wholly dependent upon context; but of course this says nothing, until the specific conditioning claimed to be operating within each kind of context is explained, and some kind of criteria set up.[41]

It is easy to sympathize with Crystal's impatience in front of this untidy state of terms, but the remedy he proposes, though cleansing, is too sweeping. His remedy is to jettison (at least for the time being) all those senses of 'context', those kinds of context, which do not fall within the descriptive scope of linguistics, in particular, the context of 'co-occurring and preceding observable alterations in situation' which has formed 'the traditional core of the notion of context in intonation studies':

The choice of intonation is referred to such situationally-conditioned attitudes as politeness, surprise, anger and so on; and it is this aspect of context which has been overrated. The range of 'attitudes' to be considered under this heading is very wide, and it is the multiplicity of semantic labels and the problems of quantifying affect which have led to pessimism in the semantics of intonation and the extreme reliance on the notion of context. Doubtless it will be possible to make much progress in these areas, once we have available more adequate analyses of affective states and their linguistic categorization, so that the idea of situational context might become more explicit. But in the meantime it is important to realize that this aspect of context is by no means the cornerstone of intonational semantics . . .[42]

'Doubtless' and 'much progress' are most encouraging, however remote or tricky 'adequate analyses of affective states' seem at present. Thus heartened, Crystal sets out on the description of an exemplary dialogue:

[41] 'Prosodic Features and Linguistic Theory', in his *The English Tone of Voice: Essays in Intonation, Prosody, and Paralanguage* (1975), 30. [42] Ibid. 31.

A / you've got something in your H̄AĪR / (said in a jocular tone, thinking that it
is no more than a fallen leaf)

B / HAVE I /

The low falling tone here, where one might have expected a livelier reply (such
as a high rise) to suit A's jocular mood, provides a contrast that indicates B's
displeasure—let us call it 'offended'. . . . this example shows the importance
of . . . other contextual factors in accounting for a particular effect.

These other contextual factors, describable in the strict terms of
linguistics, now line up to assist analysis. First, the 'co-occurring
syntactic and lexical pattern of utterance', for 'with a less abrupt syntax
and lexis, there would be less likelihood of an offensive interpretation'.
Crystal does not make clear whether he thinks A or B or both abrupt,
nor whether 'offensive interpretation' means only that B is offended or
that it is also offensive to A that B takes amiable A's 'jocular tone'
amiss. Even if Crystal gave a more adequate description of the affective
states involved, his comment would remain unconvincing, for it relies
on the belief that abruptness is a feature of syntax and lexis which can
be established without reference to that messy 'aspect of context which
has been overrated'. Clearly, if B is the Archbishop of Canterbury and
A a passing vagabond not of his acquaintance, A's remark is abrupt,
but if A and B are good friends on a ramble in the woods, it may not
be. Therefore, an ampler notion of the setting in which they speak is
needed, and Crystal offers this with his second factor, 'preceding and
subsequent syntactic and lexical patterns': 'if B had been replying in
this manner in previous discourse, A would have grounds for
discounting the offensive interpretation'. This depends on what A has
inferred from B's preceding lexical patterns, for repeated brief
remarks such as 'Have I' could have revealed to A that B was in a filthy
mood, and bent on taking umbrage; the same patterns could then
confirm the 'offensive interpretation' rather than otherwise.

Of course, if A knows that this is only B's 'way', then he can
discount happily along with Crystal, but A would need more than even
very many lexical patterns to know that about B. A third factor is called
upon to assist: 'preceding and subsequent intonational patterns'.
These are to help A to observe and get a line on 'the contrast between
A's and B's intonation'. Unfortunately, they are subject to the same
ambiguities as B's preceding lexical patterns, and, what is worse, they
are the sorts of thing A is trying to understand anyway. True though it
is to say that the workings of intonation are understood by
understanding the workings of intonation, saying so does not seem to

mark a big step forward in analysis. Finally, A may have recourse to a fourth factor: 'relevant co-occurring and preceding semiotic behaviour, especially facial expression, but including all forms of kinesic and proxemic behaviour', for 'the offensive interpretation could disappear entirely if an appropriate facial expression or gesture were involved.' A fair-sized question is begged by the term 'semiotic behaviour', though it will be possible to make much progress in these areas when we have available more adequate analyses of smiles, winks, arches of eyebrows, the adoption of an odd, shambling gait, and their semiotic categorization. While we wait for these adequate analyses, we have time to reflect that this way of putting it implies that gestures and suchlike are transparent whereas things people say are not, though, in fact, we often have to interpret gesture in the light of remarks as well as vice versa. A's problems with B's inflections might equally strike him when he contemplates all the co-occurring forms of kinesic and proxemic behaviour. And 'all forms of kinesic and proxemic behaviour' [my emphasis] could cover quite as wide a range as the originally messy 'context' from which analysis sought to extricate itself. At any rate, consideration of such behaviour would have to return us to 'pre-suppositions in the mind of the speaker, or hearer' and 'situationally-conditioned attitudes' because the physical motions in themselves are ambiguous until fitted into a description of the agent and what he can be thought to have been intending in them.

Crystal recognizes this annoyance in the way of method when, in a later paper, he desires an 'ethnography of speaking' which (humorously, I suppose) he 'informally define[s] as a specification of what kinds of things one may appropriately say in what message forms to what kinds of people in what kinds of situation, and, given a set of alternatives, what consequences stem from selecting one rather than another. . .'.[43] That would come in handy, and not only for a linguist, but it brings us back to where we started, to the sprawl of all imaginable contexts. Crystal's proposal treads a vacuous circle. It was bound to round on itself in this way, for it begins by settling what it pretends to sort out, when it simply *assigns* to A 'a jocular tone'. Nothing in Crystal's four factors could ever compel B to accept that description as a purely linguistic fact, and that is one reason why Crystal cannot *notate* A's tone but only stipulate it.

Some modern linguists believe that notation is peripheral to the

---

[43] 'Prosodic and paralinguistic correlates of social categories', in *The English Tone of Voice* . . ., 84.

study of the prosodic features of language. R. P. Stockwell is sure that the question 'how can one unambiguously and efficiently write sentences down so that they can be read back as intended' is 'not important'.[44] They have come a long way from Saussure's assurance that 'the linguist needs above all else a means of transcribing articulated sounds that will rule out all ambiguity'. The earliest students of English prosodic features were, on the contrary, convinced that the opportunity said to be afforded by their systems of notation for a perfect reiteration of the utterance which they had transcribed proved the excellence of those systems, and guaranteed the under-standing of prosodic features which that notation was supposed to enshrine.

Thus, Joshua Steele claims, in the first attempt at a thorough treatment of what he calls 'the melody and measure of speech', that 'When the *cadences* of our language, either poetry or prose, are properly marked in our way, every person initiated in the practical knowledge of music, will be able to comprehend our meaning, and to read the words according to the *melody* and *rhythmus* we shall mark to them'.[45] Abraham Tucker, who tries only to transcribe pronunciation without also taking down pace, quantity, accent, and so on, as Steele does, speaks up for what his system can offer:

... I flatter myself that any person who would take the pains to be acquainted with my alphabet, would be enabled thereby to read any speech or composition in the same manner, that is, the same articulate, I do not say the same tonical, musical ... sounds, as the speaker had delivered, or the author would read it himself, and even to follow them through whatever peculiarities of utterance they may have adopted.[46]

Tucker's distinction between the 'articulate' and the 'tonical, musical ... sounds' of an utterance is an ancestor of Pike's 'lexical meaning' and 'intonation meaning'. They both regard pronunciation (Pike's 'requisite consonants, vowels and stress') as relatively more stable than other prosodic features—as, roughly speaking, it is; they both incline to regard this 'articulate' as identifiable with the semantic core of a word, an inclination which should be checked, even in rough speaking. Philosophers have alerted us to the need to suspect such long-standing

[44] 'The Role of Intonation: Reconsiderations and other Considerations', Bolinger, 90.
[45] *An Essay towards establishing the Melody and Measure of Speech* (1775, repr., Menston, 1969), 18.
[46] *Vocal Sounds* (1773, repr., Menston, 1969), 45.

inclinations, stemming as they do from contestable assumptions about meaning in language.[47] Indeed, it is now usual to find a philosopher maintaining, as J. R. Searle does, that 'The unit of linguistic communication is not, as has generally been supposed, the symbol, word or sentence, or even the token of the symbol or word or sentence, but rather the production or issuance of the symbol or word or sentence in the performance of the speech act.'[48]

Prosodic features are amongst the major elements which individuate various utterances of the same sentence, or various productions of the same 'symbol or word or sentence'. Intonation can easily reverse the apparent drift of a remark (sample the variety of vocal issuance possible with 'This is a fascinating argument') and so prosodic features, far from being 'superadded' as Pike would have it, are essential to the meaning of utterances. They constitute the sound of agency in the production of a symbol and bear the impress on an utterance of its context. Steele and Tucker were right in thinking that a test of their notations was the ability correctly to guide reiteration of an utterance; such guidance would demonstrate the degree to which an act of speech could be re-created from the transcription.

Curious though Steele's attempted transcription of Garrick's performance of 'To be or not to be' now looks, his aim and his sense of its importance were not odd at all:

... as the practice of ... the art of reducing the *melody* and *measure* of speech to practicable and legible notes (if it was ever compleat) ... [has] lain, as it were, in a *terra incognita*, for at least a thousand years past, I think, these small specimens produced, may be our vouchers to prove, that we have discovered the land, and marked out the route which may be followed by others. . .[49]

His dependence on musical notation and terminology for his treatise made his task difficult, for as he himself observes '. . . the *melody of speech* moves rapidly up or down by *slides*, wherein no graduated distinction of tones or semitones can be measured by the ear',[50] and so his use of the musical stave is misleading (as are rests to indicate pauses because the timing of speech is not strictly *tempo*). More fundamentally, musical notes and phrases are not themselves un-

---

[47] See for instance J. L. Austin's classic essay, 'The Meaning of a Word' in his *Philosophical Papers*, ed. J. O. Urmson and G. J. Warnock (Oxford, 1961, 2nd edn., 1970). The topic is dealt with at length and in depth throughout the opening remarks of the *Philosophical Investigations*.
[48] *Speech Acts* (Cambridge, 1969), 16.
[49] Steele, op. cit. 46.   [50] Ibid. 4.

ambiguous, and require a tradition of understanding performance to sound again. There is, then, a sad truth in Steele's claim that 'If the method, here essayed, can be brought into familiar use, the types of modern elocution may be transmitted to posterity as accurately as we have received the musical compositions of Corelli.'[51] 'As accurately . . .', but only that accurately. Corelli (d. 1713) was more in earshot for Steele than he is for us, and poor Steele, whose method was not adopted, who never acquired the needed 'familiar use', for whom there were to be no manuals of performance technique or other such aids to musical resurrection, is now quite out of our hearing, and Garrick along with him.

Garrick is not the only lost type of modern elocution. Boswell made a request to his readers which can be met only in and with imagination:

I cannot too frequently request of my readers, while they peruse my account of Johnson's conversation, to endeavour to keep in mind his deliberate and strong utterance. His mode of speaking was indeed very impressive; and I wish it could be preserved as musick is written, according to the very ingenious method of Mr. Steele . . .[52]

Things, it might seem, could have been otherwise. Traditions of performance, and the interpreting communities sustained in those traditions, may be actually un-notated, but that would not prove that no system could have been invented to encompass them. The task of such invention is daunting—it might have the dimensions of Crystal's 'ethnography of speech'—but not evidently impossible until we consider some further features of meaningful speech (which have similarities with elements of musical composition and performance), features such as 'implicature'. The term is H. P. Grice's[53] and refers to those informal implications which a speaker can produce by breaking a normal, communicative convention. Such conventions are pretty loose but still powerfully operative, as, for example, the convention that in reply to a question I give neither more nor less information than is required to answer it. Thus, if you ask me what time it is and I reply,

[51] Ibid. 14.
[52] Boswell's Life of Johnson, ed. G. B. Hill, rev. and enlarged by L. F. Powell, 6 vols. (1934; 1964–71), ii. 326–7.
[53] See particularly his 'Utterer's Meaning and Intention', Philosophical Review (1969); 'Logic and Conversation', in P. Cole and J. Morgan (eds.), Syntax and Semantics, iii: Speech Acts (New York, 1965); 'Presupposition and Conversational Implicature', in P. Cole (ed.), Radical Pragmatics (New York, 1981). D. Sperber and D. Wilson have a helpful summary of Grice's account of implicature in their Relevance (Oxford, 1986), 31–8.

'Midnight, and I must be up early tomorrow', the surplus 'and I must be up early tomorrow' may effect an implicature with the force 'It's time you went home'.

Combining an elastic and inexplicit logic with great variability of suggestion according to context, implicatures elude systematic notation, as Marilyn M. Cooper writes:

In communicating their meanings, speakers and writers always depend on context much more than they do on the linguistic shape of their utterances or their texts. Implicatures are simply the most radical cases, in which speakers and writers draw on and exploit mutual knowledge of beliefs and conventions in order to communicate meanings that are in no way encoded in linguistic forms or that contradict the conventional meanings of the linguistic forms used.[54]

Implicatures are specially pertinent to the problem of notating speech in writing, for a principal use of the prosodic features of a language is to effect implicatures without the flagrant informational redundancy of the example given above. The most consequential implicature in English literature comes in *Othello*:

IAGO. Did *Michael Cassio*
  When he woo'd my Lady, know of your love?
OTHELLO. He did, from first to last:
  Why dost thou aske?
IAGO. But for a satisfaction of my Thought,
  No further harme.
OTHELLO.   Why of thy thought, *Iago*?
IAGO. I did not thinke he had bin acquainted with hir.
OTHELLO. O yes, and went betweene us very oft.
IAGO. Indeed?
OTHELLO. Indeed? I indeed. Discern'st thou ought in that?
  Is he not honest?
IAGO.   Honest, my Lord?
OTHELLO. Honest? I, Honest.
IAGO. My Lord, for ought I know.
OTHELLO. What do'st thou thinke?
IAGO.   Thinke, my Lord?
OTHELLO. Thinke, my Lord? Alas, thou ecchos't me;
  As if there were some Monster in thy thought

---

[54] 'Context as Vehicle: Implicatures in Writing', in Martin Nystrand (ed.), *What Writers Know: The Language, Process, and Structure of Written Discourse* (New York, 1982), 119.

Too hideous to be shewne. Thou dost mean somthing:
I heard thee say even now, thou lik'st not that,
When *Cassio* left my wife. What didd'st not like?
And when I told thee, he was of my Counsaile,
Of my whole course of wooing; thou cried'st, Indeede
And didd'st contract, and purse thy brow together,
As if thou then hadd'st shut up in thy Braine
Some horrible Conceite. If thou do'st love me,
Shew me thy thought.

IAGO.   My Lord, you know I love you.

OTHELLO. I think thou do'st:
And for I know thou'rt full of Love, and Honestie,
And weigh'st thy words before thou giv'st them breath,
Therefore these stops of thine, fright me the more:
For such things in a false disloyall Knave
Are trickes of Custome: but in a man that's just,
They're close dilations, working from the heart,
That Passion cannot rule.

IAGO.   For *Michael Cassio*,
I dare be sworne, I thinke that he is honest.

OTHELLO. I thinke so too.

IAGO. Men should be what they seeme,
Or those that be not, would they might seeme none.[55]

The implicature of Iago's 'Indeed?' steals in on the voice, interloping amidst their exchanges in which it is assumed by Othello that conventions are honestly employed; his plot begins with a rapid yet ineffaceable trick of the voice, as Othello half-realizes when he puzzles why Iago 'cried'st, "Indeede?" '. 'Cried'st': speaking a fraction too loud, at an unusual pitch in what Othello takes to be the circumstances, and with an exaggerated intonation contour, Iago begins to voice an allegation he dare not speak. The whole passage demonstrates Iago's mastery of conversational implicature with the conscious social dexterity which that mastery implies in him. He does not need to make anything explicit because he can rely on a repertoire of expectations about correct procedure in conversation, expectations which, when flouted, produce of themselves the effect he desires. For example, it is an informal rule of talk that one speaker does not repeatedly echo previous remarks, though he may do so once in a while to express astonishment, but here Iago sticks on 'indeed' and 'honest' and 'think' and 'love' to the extent that Othello himself recognizes the trick of the

[55] III. iii. 100 ff.; Folio, ll. 1696 ff.

voice—'Alas, thou ecchos't me'—and draws the inference Iago wishes, though Othello does not recognize that it is a trick. The horrible echoing in the passage, like the looming of an obsession, spreads eventually throughout the play and already begins to infect Othello himself in these lines, as he reverts to Iago's 'my thought' with 'thy thought', and as Iago's phrase 'your love'—which refers to Othello's love for Desdemona, slowly mutates into Othello's 'if thou do'st love me' and 'full of Love, and Honestie'. As the words 'thought' and 'love' pass from Iago's lips to Othello's, so too Iago begins to take hold of Othello's mind. This he does the more easily because Othello has a desire to trust in the machineries of social behaviour. He thinks that such things as furrowed brows, flusters of vocal control, and convulsive gestures must be 'close dilations, working from the heart' in Iago, though he knows that such apparent spontaneities can sometimes be no more than 'trickes of Custome'.

It is of the nature of Othello's crucial relation to Venetian society that he has adopted its *mores* only in mid-life. 'Venetian', in the broad sense of the language and social behaviour of the maritime republic, is not his mother-tongue. He is, then, all the more ready to make the mistake (to which even native speakers in the play are subject) of not realizing that Iago actually *performs* rather than behaves; indeed, what Iago performs *is* 'behaviour', the accepted conduct of an honest Venetian. It is a terrible irony in this passage, as well as a mark of Iago's exceptional cunning, that Iago himself pronounces the maxim which governs proper conversation and which he himself so cruelly exploits: 'Men should be what they seeme'. This maxim is, of course, an ethical imperative in a wide sense but it also operates as a major element in interpreting intonation. That is one reason why there can be no systematic correlation between intonation contour and speaker-meaning. As I argued before, a conception of the speaker, of the kind of person he is, must influence a description of his utterance; what is brusque or fulsome or despondent from one person would not seem so from another. Coming from Iago, Othello thinks tricks a close dilation.

'Men should be what they seeme' is a prime rule of hman agency and speaking is a form of action. Iago traps Othello partly by using the fact that any set of conventional forms—and all societies possess such sets—can reverse function if the prime maxim is disobeyed; any rule can become malign in effect if one only plays at playing by it. And this is a nemesis which attaches to social groups in the proportion to which they are self-conscious, proud, and sophisticated, as is the Venice of

Othello. It is the Venice of *Othello*, though, and not Othello's Venice, for he has the regard for that city-state of a foster-child, deeply attached but finally not quite inward. That inwardness appears to be something which really belongs to Iago, knowing as he is about all things Venetian, but Iago's devilry is to *act* inwardness, to render inwardness itself the most extrinsic of shows. He does this with his voice. In a way, Othello is right when, later, in his fit, he declares, 'It is not words that shakes me thus. . . .'[56] Not words, indeed, but still something spoken.

The inherent difficulties for structural linguists in their dealings with prosodic features encourage socio-linguists to take the context-dependence of, say, implicatures as evidence of the need for an 'interactional' approach to meaning in natural languages. As far as concerns the relations between written and spoken languages, however, the distinction socio-linguists like to draw between 'oral' and 'literate' cultures, useful as it proves in anthropology or the study of ancient literatures,[57] muddies reflection on the status of oral discourse in literate societies by leading to an exaggerated emphasis on those features which oral discourse may display in pre-literate societies. Deborah Tannen, for example, writes that '. . . strategies that have been associated with orality grow out of emphasis on interpersonal involvement between speaker/writer and audience, and . . . strategies that have been associated with literacy grow out of focus on content.'[58] The distinction between 'interpersonal involvement' and 'content' is no more tenable than analogous distinctions in the terms of structural linguists, especially not if it implies either that in oral discourse speaker-meaning can be perceived without grasp of content, or that in literate discourse content can be understood apart from speaker-meaning. (It is safe to presume that 'interpersonal involvement' must mean, at least, 'recognition of what a speaker intends by his words' and 'content' must include 'what a speaker's words mean irrespective of individual intention'.) The study of examples does not make the distinction more convincing, for it may well be that 'in an oral tradition . . . it does not matter whether one says "I could care less" or "I

---

[56] Ibid., IV. i. 41: Folio l. 2418.
[57] See, for instance, E. A. Havelock, *Preface to Plato* (Cambridge, Mass., 1963); A. B. Lord, *The Singer of Tales* (Cambridge, Mass., 1960); W. J. Ong, *The Presence of the Word* (New Haven, 1967).
[58] Introduction to D. Tannen (ed.), *Spoken and Written Language: Exploring Orality and Literacy* (Norwood, NJ, 1982), p. xv.

couldn't care less" '[59] (as in spoken American) but this fails to show that oral discourse works by interactive perception of speaker-meaning unmediated by the sort of semantic considerations that operate in written discourse. It equally does not matter in certain kinds of written text whether one writes 'I could care less' or 'I couldn't care less', whereas it matters in some spoken exchanges which phrase one uses (as in spoken English). So too, the interpretation of double negatives goes on in quite the same way in oral and written discourse.

An extreme version of this slant amongst theorists of orality results in a championing of the spoken as against the written. Oral discourse shines forth as 'interpersonal', 'somatic', 'situational', and writing is cast into the shade as 'autonomous', 'minimally dependent on the contribution of background information' on the part of the reader.[60] Walter J. Ong becomes eloquent in contemplation of speech: 'The oral word . . . never exists in a simply verbal context, as a written word does. Spoken words are always modifications of a total, existential situation, which always engages the body.' Or again:

Spoken words are always modifications of a total situation which is more than verbal. They never occur alone, in a context simply of words.

Yet words are alone in a text. Moreover, in composing a text, in 'writing' something, the one producing the written utterance is also alone. Writing is a solipsistic operation.[61]

These lyrical antitheses don't bear examination: written words exist in more than a 'simply verbal context', as the sign in the library and Coleridge's subtitle show; a note from a bank manager can modify a 'total, existential situation' as thoroughly as an interview with him; Chinese calligraphy engages the body of the writer and a love-letter may involve the body of its reader. The loneliness of a writer is a cultural phenomenon of some importance, but it is not a constitutive feature of writing itself—we may exchange notes during a committee meeting, I know you are watching me write, waiting to see what I have written, and your anticipation suffuses my writing. Nor is loneliness solipsism.

None the less, socio-linguists of oral discourse give much substance to Freud's concise insight: 'Writing was in its origin the

[59] D. Tannen, 'The Oral/Literate Continuum in Discourse', in Tannen, op. cit. 2.
[60] See e.g. Paul Kay, 'Language Evolution and Speech Style', in B. G. Blount and M. Sanches (eds.), *Sociocultural Dimensions of Language Change* (New York, 1977), esp. 21–2, 30.
[61] *Orality and Literacy: The Technologizing of the Word* (1982), 67, 101–2.

voice of an absent person.'[62] And this is an origin which has not been left behind; writing is still this 'voice of an absent person', re-originated as it is in the numerous writings of every day. Many socio-linguistic accounts of the vivid interplay of conversation which is withdrawn from most writing—opportunities for gesture, faint implicatures in vocal timbre, uptake and letting-on in swift, reciprocal interanimation—provide further witness to that lack of shared and comprehensive air in their writings which imaginative writers often lament. But literature aspires to re-create in a sublimed atmosphere the conditions of speech; most of the literature we know, though coming to us in print, envisages and reaches towards a rediscovery of some characters of speech. To do so, it searches out from print something more precise but as elusive and taxing as 'interpersonal involvement'—prosodic features, tones of absent voice.

Keats wrote to his brother and sister-in-law in the autumn of 1819:

Writing has this disadvan[ta]ge of speaking. one cannot write a wink, or a nod, or a grin, or a purse of the Lips, or a *smile—O law*! One can-[not] put ones finger to one's nose, or yerk ye in the ribs, or lay hold of your button in writing—but in all the most lively and titterly parts of my Letter you must not fail to imagine me as the epic poets say—now here, now there, now with one foot pointed at the ceiling, now with another—now with my pen on my ear, now with my elbow in my mouth—O my friends you loose the action—and attitude is every thing . . . And yet does not the word mum! go for ones finger beside the nose—I hope it does.[63]

Zestful and judicious, this writing, as it regrets its discarnation, fashions a newly articulate body in the imagination of its readers, partly by such unbuttoned irregularities as the inconsistency about whether to use the apostrophe or not ('put ones finger to one's nose'), partly by the rapid exuberance of its specifications ('a wink, or a nod, or a grin, or a purse of the Lips'), or by the ranginess of the syntax ('now here, now there, now . . ., now . . .,—now . . ., now . . .') hetting itself up with swerves of reference, going in imagination all over the place, fit almost to burst with the comical desire to make its writer present, to *join* the reader, till Keats brings himself to a sober self-recall, the attention to loss in 'O my friends you loose the action', a loss it seems even the energy of this writing cannot repair but only, on the rebound, render

---

[62] *Civilization and its Discontents* (Vienna, 1930), trans. Joan Riviere, rev. and ed. James Strachey (1963), 28.

[63] Letters of about 20 Sept. 1819, in *The Letters of John Keats 1814–1821*, ed. Hyder E. Rollins, 2 vols. (Cambridge, Mass., 1958), ii. 205.

the more evident, but he will not stop at despondency, for the curve of thought is capped and balanced by that hope glimpsed at the last— 'And yet does not the word mum! go for ones finger beside the nose— I hope it does.' It is literature's hope as well as Keats's.

Speech Acts and Acts of Writing: The Philosophy of Language

The description I have just given to Keats's letter does not seem inhibited by the worries I previously mentioned about the incomplete transcription of speech into writing. I think this freedom from worry is a tribute to the lively exactness of what he wrote, but someone might pay me the double-edged compliment of attributing it rather to my excitable imagination. Certainly, such a reading participates in the creativity of what it reads, but it participates in the discovery of how to respond to a voice implicit in the text and not in an act of ventriloquism, to lend one's voice being something different from throwing it. Keats was not asking George and Georgina to impersonate but to imagine him. If the splendid tact and verve of his letter are to become evident, a reader now needs to imagine them reading as well as him writing to them. There is a drama to Keats's liveliness of written voice, the drama that evidence—even only stylistic evidence— of vitality would have for this family after Tom's death and after Keats had observed in himself the signs of the same fatal disease.

The lack of unambiguous notation for Keats's speech does not preclude this sort of reading but requires it; unambiguous notation would render this kind of critical description superfluous. And a function of such criticism is to furnish and extend just that interpretative community which ambiguous notations must call for if they are to have significant life. There is perhaps an analogy between the traditions of musical performance through which we require the hold we have on the scores of past composers and this form of critical description, of suggestions towards voicing. With regard to such traditions—of musical performance, of critical description—it would be misleading to say that their texts are 'inadequate' though they are ambiguous, while for other purposes, such as those of a musicologist, or a descriptive linguist attempting a theory of prosodic features, the texts would indeed fail to yield them the precisions they require. Keats's letter would not have seemed inadequate to George and

Georgina, though they had personal reasons (as a reader now does not) to wish they could have seen him speaking instead. Reading the letter now need not be an activity different in kind from reading it then; indeed, the closeness of the imaginative activity Keats asked of his brother and sister-in-law to what is asked of a reader now preserves an intimacy with his writing across the time since the letter was sent.

Abstract scepticism about the possibility of correct critical description is possible but as empty as abstract epistemological scepticism.[64] One always may be wrong in particular descriptions of what one reads, or in judgements of what one has seen, but that does not show that all one's descriptions or judgements may be wrong, nor that no notion of correctness can apply here. On the contrary, specific doubts can rationally arise only when abstract scepticism has been given up. My account of Keats's letter may be thick with errors, but its errors would be demonstrated by other, better instances of this kind of description, not by abandoning such description altogether.

Quentin Skinner has written that 'the notion of illocutionary re-description lies at the heart of literary-critical procedures',[65] and my account is such an attempted 'illocutionary re-description'. I agree with Skinner on this point, because illocutionary re-descriptions make explicit that imaginative voicing which turns readers into an audience; they are the heart of literary criticism by being ways in which literature comes to life. J. L. Austin invented the term 'illocutionary' to refer to 'the performance of an act *in* saying something as opposed to performance of an act *of* saying something'.[66] We may wonder what an utterance means, we may equally wonder what was meant on some occasion *in* that utterance. I may say to you, 'I want you to be good,' and mean by this to warn you of dire consequences if you fail (I am your father taking you to visit a wealthy aunt) or to illustrate the concept of 'illocutionary force' (I am J. L. Austin, you are a seeker after truth). The same locutionary act of saying, 'I want you to be good,' may have various illocutionary forces. Editors of literary texts try to establish which locutionary acts have been performed, literary critics are concerned, amongst other things, with the illocutionary force of

---

[64] The assertions in this paragraph derive from Wittgenstein's arguments about scepticism in *On Certainty*, ed. G. E. M. Anscombe and G. H. von Wright, trans. Denis Paul and G. E. M. Anscombe (Oxford, 1974); see esp. the remarks numbered 23, 115, 164–6, 310, 378.

[65] 'Hermeneutics and the Role of History', *New Literary History*, 7/1 (Autumn 1975), 221.

[66] *How to Do Things with Words*, ed. J. O. Urmson (Oxford, 1962), 99.

those locutionary acts, and the interdependence of locution and illocutionary force will often, therefore, require critics to tackle editorial problems and editors to exercise critical tact.

An interest in the illocutionary potential of a locutionary act is not just an optional 'extra' in the study of language. Recognition of the truth in Searle's claim that 'the unit of linguistic communication is . . . the production or issuance of the symbol or word or sentence in the performance of the speech act' brings illocution to the centre of attention, for, in situated utterances, people do not simply emit sentences, things are at issue in their issuances, and illocutionary force directs us to those things.

Michael Dummett describes the centrality of illocutionary considerations:

The notions of sense and reference do not suffice for a complete account of language. If we know of a language only what sense the expressions which occur in it have, and thereby their reference, we know nothing which can tell us the significance of uttering an expression of this language: the *point* of doing so.[67]

He offers the helpful analogy of a chess game. We might know all the rules of chess, the permitted moves and the possible ends of the game—White mates Black, Black mates White, stalemate, as well as resigning in anticipation of one of these outcomes—and yet the account we could give with only this knowledge would not tell us what it is to play chess. Dummett writes of such an account:

It will not . . . suffice by itself to provide us with a 'theory' of chess as an activity: it is not enough to tell anyone what it is to play chess. This can be seen from the fact that there could be a large number of variant games each sharing the same formal description: for instance, the game in which each player tries to force his opponent to checkmate him, or again the game in which it is White's object to produce a checkmate (of either side by the other) and Black's to achieve stalemate. The difference between these various games lies not in the initial position or what constitutes a legitimate move, but in what winning consists of. From the formal description it is impossible to tell what, in playing chess, a player is trying to do . . .[68]

Searle takes Dummett's point further. A theory of language in terms of sense and reference alone could not perform even the limited task Dummett assigns to it because 'in general the meaning of a sentence

[67] *Frege: Philosophy of Language* (1973; 2nd edn., 1981), 295.
[68] Ibid. 296.

only has application (it only, for example, determines a set of truth conditions) against a background of assumptions and practices that are not representable as a part of the meaning'.[69] Dummett's view depends on thinking that 'He plays chess' is a sentence which determines a set of conditions for its truth by virtue of its sense and reference (though what it might be for a particular referent of 'he' in an utterance of that sentence to play chess, and what might be the point of someone's saying this about him, could be decided only by a supplementary account of human purposes in playing chess and saying things about people who play chess.) Searle argues, in effect, that the sentence 'He plays chess' determines no truth-conditions until such an account is given, and he provides an elegant example in demonstration of the insufficiency of sense and reference to determine truth-conditions for a particular sentence.

Consider a set of sentences, each of which contains the word 'cut', such as:

Bill cut the grass.

The barber cut Tom's hair.

Sally cut the cake.

Searle observes that, though 'cut' is literal in all these sentences (as it is not, usually, in 'I cut him dead at the party'), on each occasion it determines a different set of truth-conditions for the sentence in which it occurs:

The sort of thing that constitutes cutting the grass is quite different from, e.g., the sort of thing that constitutes cutting a cake. One way to see this is to imagine what constitutes obeying the order to cut something. If someone tells me to cut the grass and I rush out and stab it with a knife, or if I am ordered to cut the cake and I run over it with a lawnmower, in each case I will have failed to obey the order.[70]

A pioneer seeking the gold of unambiguous utterance might wish to eliminate such vagaries by inventing a series of different verbs, one for each of the distinct truth-condition-determining senses of 'cut' (and similarly for all other words which 'moonlight' semantically). Searle doubts the prospects of such clarity:

Assuming that the contextual dependence of the applicability of semantic content is as widespread as I claim, why not simply get rid of it; why not for

[69] 'The Background of Meaning', in J. R. Searle, F. Kiefer, and M. Bierwisch (eds.), *Speech Act Theory and Pragmatics* (Dordrecht, 1980), 221.

[70] Ibid. 223.

example invent a language that would not in this way have contextually dependent semantic contents? The answer is that the features we have cited are features not just of semantic contents but of representations generally, in particular they are features of intentional states, and since meaning is always a derived form of intentionality, contextual dependency is ineliminable.[71]

This argument disposes of one objection to the view that literary description centres on illocutionary re-description. The objection runs: why concern oneself with what Keats meant by what he wrote in his letter? It is enough to study the meaning of the text. But it is not easy to see what the phrase 'the meaning of the text' means, given the ineliminable contextual dependence of semantic contents, a dependence deriving from the intentional character of meaning. Take the remark 'O my friends you loose the action'. The word 'loose' is a Keatsian misspelling for 'lose'; the meaning of 'O my friends you loose the action' appears only when we know what he meant by it. He spells the word in this way because he is recalling the text of Hamlet's words in 'To be or not to be'—'loose the name of Action'. A critic might not accept this interpretation, and insist that 'loose' correctly spells the intended word 'loose'. He would then need to explain either the sense the word was meant to make at this point or explain why Keats momentarily stops making sense. A very abstemious reader might think the issue between these two views undecidable. All three positions, however, involve a reference to what is meant by the word, even the abstemious 'don't know' must explain his abstention by arguing that we cannot know which of the two alternatives was meant by Keats. Anti-intentionalist arguments which appeal to 'the meaning of the text' as determinable apart from its illocutionary force rely on an inadequate theory of meaning. The illocutionary force of a text or utterance is, as it were, what links its meaning to its context; it is what the remark means on its particular occasion, in its particular circumstances. Illocutionary force individuates sentences as utterances. J. L. Austin wrote some years ago: 'for some years we have been realizing more and more clearly that the occasion of an utterance matters seriously, and that the words used are to some extent to be 'explained' by the 'context' in which they are 'designed' to be or have actually been spoken in a linguistic exchange.'[72]

Austin's phrase 'spoken in a linguistic exchange' reveals an unexpected kinship with Saussure and with the linguist's assumption

[71] Ibid. 231.   [72] *How to do Things with Words*, 100.

that *langue* manifests itself in the same way in speech and writing, for Austin too takes it as read that speech is the type of all linguistic exchange—as do the philosophers who employ 'speech act' to cover all utterance. A philosophical study of meaning can perhaps afford not to occupy itself with the differences between written and spoken language in so far as it seeks a theory of meaning which will hold good for both forms of language. That search is not evidently futile, for the possibility of translating without complete loss of meaning a written into a spoken utterance, and vice versa, proves that there is a semantic dimension common to both. Yet a theory which concentrates on utterances in context could be asked to give more thought to possible distinctions between the context-dependency of meaning in speech and in writing. If utterance (or 'production' or 'issuance') is the unit of concern, and illocutionary force semantically individuates sentences as utterances, while prosodic features linguistically individuate various utterances of the same sentence, then interest in how language is used by people to mean things, interest in the meeting of the semantic with the linguistic, might properly focus on the relation between illocutionary acts and prosodic features. Putting it sharply, intonation is the sound of intention and should attract the attentions of those who believe that 'meaning is always a derived form of intentionality'. The comparative absence from the written language of signs for spoken prosodic features would then set questions about the application of speech-act theory to written documents and also enable us better to formulate what is distinct, as well as what is shared, in the dependencies of writing and speech on their context.

Austin provided some methods for establishing the illocutionary force of an utterance: grammatical mood of the verb; tone of voice, cadence, emphasis; adverbs or adverbial phrases; connecting particles; accompaniments of the utterance; circumstances of the utterance.[73] Let us consider these methods, and what he claims for them, in turn.

1. Austin claims that, for example, in speech the use of the imperative mood of the verb makes an utterance a command or some other member of the family of commands (exhortations, warnings, and so on). This is not so, for the reasons previously given against sharing Crystal's reliance on 'co-occurring syntactic and lexical patterns' to establish the significance of intonation contours. If you ask me, 'How do you translate "Tais-toi"?' and I reply 'Shut up', I may or may not

[73] Ibid. 73–6.

have ordered you to shut up. You would not be able to tell what I had
done, nor what I had intended to do, in saying 'Shut up' from analysing
the grammar of my response. Any connection between the mood of a
verb and the illocutionary force of an utterance in which it appears may
be overridden by the context of the utterance; the same is true for all
grammatical forms.

2. Tone of voice, cadence, emphasis, like grammatical form, can
contribute, but only indecisively, to establish illocutionary force. Again
like grammatical forms, their assistance is not guaranteed by any strict
connection between prosodic feature and illocutionary force, though in
natural languages some semi-standardized intonation contours generally
function to indicate some kinds of illocutionary force. However, tone
of voice, cadence, and emphasis cannot assist us to identify illocutionary
force in written texts, because they are not unambiguously represented
in writing. Very often it is our judgement of what the illocutionary
force of a written sentence is which guides our vocalization of it, and
not the other way about. The inference goes: 'Given this locution from
this person in this context, what illocutionary force should be ascribed
to it? And which intonation of these words would best convey that
force?'

Austin himself instanced the quandaries which populate this area
when he gave examples of attempts to indicate illocutionary force by
written marks for intonation. Of a bull:

It's going to charge! (a warning);
It's going to charge? (a question);
It's going to charge!? (a protest).

These features of spoken language are not reproducible readily in written
language . . . we have tried to convey the tone of voice, cadence, and emphasis
of a protest by the use of an exclamation mark and a question mark (but this is
very jejune). Punctuation, italics, and word order may help, but they are rather
crude.[74]

'Jejune' and 'rather crude' look as if progress in refinement might
eventually deliver a written language in which these features of speech
would be readily reproducible, but Austin does not espouse this hope,
and is right not to do so for the reasons which I gave in the second
section of this chapter when considering notation of prosodic features
by linguists. This is not to say that, given the rich subtlety with which

[74] Ibid. 74.

writing can evoke a voice in literature, Austin's words here are not harsh, but then he was probably not thinking about literature.

3. The paragraph Austin devotes to 'connecting particles' lumps together, in an uncharacteristically indiscriminate embrace, such devices as 'therefore' to indicate a conclusion and 'hereby' to effect a performative. It is also said that 'A very similar purpose is served by the use of titles such as Manifesto, Act, Proclamation, or the sub-heading "A Novel . . .".'[75] 'Hereby' does have a special force but only in a highly delimited sort of document (those issued by authorities empowered to do what they claim 'hereby' in the document to do). Most written documents fall outside this class. The remarks about titles are a jumble. 'Act' and 'Proclamation' may be classed along with 'hereby'; 'Manifesto' does not belong with these terms because anybody can issue a manifesto, and the title 'Manifesto' does not govern the illocutionary force of the component sentences of a manifesto as 'Act' does the component sentences of an act, though it may govern the force of the entire document. I have no idea what sort of illocutionary force Austin thought was conveyed by 'A Novel . . .', apart from such unhelpful generalities as 'not all the sentences in what follows are statements of fact'. This must be one of his jokes.

4. Utterances may be accompanied by 'gestures (winks, pointings, shruggings, frowns, &c.) or by ceremonial non-verbal actions'.[76] Written texts are almost always unaccompanied in this manner. The utterance and the gesture must also always be interpreted together, there being no necessary, governing primacy to, say, a wink over the significance of the utterance it accompanies. What I say may determine the force of my gesture just as what I do may determine the force of what I say, though, in practice, the gestural accompaniment tends to be the indicator of force added to the utterance. It is still at times rational to suspect some people's smiles.

5. Austin has some interesting words about the circumstance of the utterance:

An exceedingly important aid is the circumstance of the utterance. Thus we may say 'coming from *him*, I took it as an order, not as a request'; similarly the context of the words 'I shall die some day', 'I shall leave you my watch', in particular the health of the speaker, make a difference how we shall understand them.

---

[75] Ibid. 75.
[76] Ibid.

But in a way these resources are over-rich: they lend themselves to equivocation and inadequate discrimination . . .[77]

As a philosopher, Austin was right to be embarrassed by these riches. Some of those who develop his work lack his apprehension. Manfred Bierwisch, for example, writes: 'To sum up, what is needed for a full analysis of explicit performatives is nothing but a truth theoretical account of their ordinary semantic structure, and a theory of social interaction providing interactional settings and communicative senses'.[78] 'To sum up' and 'nothing but' look the prospect that cowed Austin squarely, indeed breezily, in the face. Roughly translated, the quoted sentence appears to mean: 'what is needed for a full explanation of the working of utterances like 'I promise to pay you back' is simply an account of how their meaning determines the conditions under which they are true or false along with an account of human behaviour which will comprehend all possible types of relation between people and give rules for the ways in which those relations affect the meaning of the things they say to each other.' This amounts to a philosophical craving for Crystal's 'ethnography of speaking', and seems a lot to ask. Actually, even this would not be enough, because 'a truth theoretical account of . . . semantic structure' should not just co-exist, as Bierwisch's 'and' makes out, with 'a theory of social interaction'. For the 'full analysis' desired, a theory of the connection between the facts governed by the first two theories would also be necessary.

The circumstances of an utterance are both 'exceedingly important' in helping us to determine illocutionary force and also 'over-rich', productive of 'equivocation and inadequate discrimination'. This is unfortunate, and it says a good deal for Austin's candour as a philosopher that he admits it (it says even more for his prudence that he does not pursue it). In fact, indicators of the fifth type can override all the other kinds of indicator; it is because this is so that there is no connection between grammatical or prosodic features *per se* and the force of an utterance. Each helpful facet of a locution of course assists that practical reasoning in language by which we elicit illocutionary force but we must remember that we exercise what insight or tact we have in such judgements and not falsify our cultural practice by crediting any facet of a locution, whether singly or in combination, with the power to determine such force. The supremacy of circumstance, of

[77] Ibid. 76.
[78] 'Semantic Structure and Illocutionary Force', in Searle, Kiefer, and Bierwisch, op. cit. 15.

the 'background of assumptions and practices that are not representable as a part of the meaning'[79] of a sentence, makes trouble for the philosopher and it might seem to make life not worth living for a literary critic. Dealing with writing as the critic does, he is short of clear indicators of types 2 and 4, and indicators of types 1 and 3 are unreliable in themselves as well as subject to indicators of types 2 and 5.

Worse still, the critic may not know what the circumstances of a particular piece of writing are. Knowing the circumstances of the writer at the time of writing need not amount to knowing the circumstances of the writing. We know that Ezra Pound was confined in a cage near Pisa during the composition of the Pisan Cantos but the writing furnishes other circumstances too for the self which is to be imagined speaking in the poems—an English cottage, Paris in the twenties, an Elizabethan concert. To identify the circumstances of Pound's writing with Pound's circumstances would be to cancel the imaginative work of the writing; to think there was no relation between the two sets of circumstance would be to ignore the work that went into imagining.

If we take a short lyric such as Wordsworth's 'She dwelt among the untrodden ways', the circumstances of the writer tell us almost nothing about the circumstances of the writing. And this is not just because we do not know anything about Lucy or when exactly Wordsworth wrote the poems which are grouped under her name but because this kind of poem deliberately so abstracts the circumstances of the writing that they fail to provide cues for establishing the illocutionary force of either parts or the whole of the text. We cannot properly say more of the situation in which 'She dwelt . . .' is to be imagined as uttered than that a female acquaintance of the speaker of the poem has died (and saying that seems at once too much and much too little). Such an abstract circumstance does not assist us to settle the intonational ambiguities of the written words of the poem. An interpretation of the poem may then ask whether there are reasons for its created indeterminacy. We might say that the speaker of the poem was fond of the 'Maid' who has died, but that statement depends on our already having alighted on an intonation and force for those last two, grievously reticent lines: 'But she is in her grave, and, oh, | The difference to me!',[80] whereas the deep tact of the piece is, by keeping those last lines

---

[79] Searle, 'The Background of Meaning', loc. cit.
[80] 'She dwelt among the untrodden ways', in *Lyrical Ballads* (1800). Here as elsewhere, unless otherwise stated, I quote Wordsworth's poems from the text of

tacit, to convey Wordsworth's own blankness before his loss and convey too any reader's distance—of puzzlement and commiseration—before such not entirely articulable grief. If we feel we know the stress and intonation of 'oh' and 'The difference to me', we imagine a situation to fit the voice we think we hear in the lines; we have little evidence for these imaginings which is why they are so tempting and also why they are to be resisted. This is of the essence of such a lyric. Our guides here can be only the literary circumstances of the poem; we call on an informed sense of rhythm and diction, on preceding and subsequent poems which we judge to be similar or significantly different. We might, for instance, consider poems written in similar quatrains, or other elegiac writings which sharply contrast with Wordsworth's reticence in the matter of eulogy here. Particularly, in the case of 'She dwelt . . .', we call on other poems by Wordsworth which tradition groups as the 'Lucy' poems. Doing so, we make legitimate recourse to information which may assist us to hear what Wordsworth might have meant by the poem. Criticism cannot escape this move because meaning is always a derived form of intentionality. Those who presume that because the author is missing from the text he is therefore dead are over-hasty; his very absence may be something he creates, and so a sign of life.

It would be possible to surround this 'Lucy' poem with other texts not by Wordsworth and give many different reasons for one's selection of literary circumstance. There is, though, a crucial difference between the furnishing of circumstance according to what we believe to be the known author's intention and any other supply of circumstance to guide interpretation of illocutionary force, and hence of meaning. Intention determines relevant context. That is, we can halt our description of contextual features only by reference to what the agent in the context can be conceived to have been doing (this need not be identical with what he conceived himself to have been doing, but nor can it normally be entirely independent of the agent's self-conception). I sit at my typewriter and write these words. Where do the circumstances of this act of writing end? The weather is not bad, my health is all that could be expected, there was no mail this morning, and I look forward to tonight's dinner-party with less than usual dread. These are some circumstances of me writing (there are all too many more of them) but none of them is a circumstance of my writing

J. O. Hayden, *William Wordsworth: Poems*, 2 vols (Harmondsworth, 1977) [hereafter referred to as Hayden], i. 366.

because the intention I have *here*—in these pages—has nothing to do with such thick circumambience. The circumstances of my writing could better be given by mentioning the names of other writers, discussions I have had, recently and long ago, pleasures felt in reading and puzzles consequent on those pleasures.

In this, writing resembles other intentional acts. The actual circumstances are innumerable, the number of relevant circumstances usually finite, and, with patience, they may be stated; the criteria for distinguishing the relevant from the actual must include reference to what the agent can plausibly be described as having thought himself to be up to.[81] It is my task, writing, to create the grounds for an estimation of what I intend in writing, an estimation which will permit a reader to tell those circumstances of my writing which bear on its interpretation from those which do not. This does not amount to a demand that a writer must always speak his mind and never be in two minds about anything; he may be reticent about his intentions, or have the intention to be reticent. The indeterminacy of intonation in writing permits, as later chapters of this book will detail, great reserve, multiplicity, self-checking, and pliability of intention in writers, as the instance of 'She dwelt . . .' briefly suggests. Though the reader of that poem neither hears nor sees a person who might wink and nudge helpfully, whose history is intimately known, and whose purposes are made clear by circumstances shared by utterer and auditor, circumstances and purpose, which satisfy all but abstract scepticism that the poet employs the devices of illocutionary indication in a way we can understand, though, in short, we are not in conversation with Wordsworth, his writing speaks to us.

Jacques Derrida has been writing vividly against such a view of the relations between graphic signs and vocalization for many years, and no consideration of the topic can ignore his work, difficult though it is to state its claims without accepting its terms. Trying to write about Derrida in an idiom distinct from his own, one fears that one will misrepresent him. The fear can be partly allayed by recalling that, as Derrida appears to suspect the notion of a unitary subject of writing and to deny that writing can be a representation of consciousness,

---

[81] Compare John Casey's remark, 'Our conception of the role in which a man is acting, or our assessment of his character, will considerably affect what we can describe him as doing in a particular situation.' The discussion of agency from which this remark comes is throughout pertinent to my argument here. See 'Actions and Consequences', in John Casey (ed.), *Morality and Moral Reasoning* (1971), esp. 168–95.

there is perhaps nobody to complain of misrepresentation, and Derrida (whoever he may be), as he cannot be represented, cannot be misrepresented either.

Derrida has a sharpened version of Austin's point about the richness and equivocality of circumstances or context of utterance in governing illocutionary force. He writes that context can never be 'exhaustively determinable',[82] from which it would follow that meaning, if it is indeed governed by context, also lacks complete determinability. This is certainly true in the abstract, but it does not matter. Derrida himself provides a reason for this when he remarks: 'I am convinced that speech act theory is fundamentally and in its most fecund, most rigorous, and most interesting aspects . . . a theory of right or law, of convention, of political ethics or of politics as ethics.'[83] Arguments like those of Searle and Austin do indeed return language to the curacy of practical wisdom, and, though this may be unwelcome to the more systematically-minded of speech act theorists and structural linguists, the conclusion is both inescapable and undismaying. As the instance of Iago's corruption of Othello which I described in the second section of this chapter shows, the fact of linguistic agency in itself demands that any full account of how people talk, and what they do to each other when they talk, will become an ethical account, a story of what may be done under and with the maxim 'Men should be what they seeme'. The kind of judgement which interpretation requires to make the living being's application of the sign works only in communities; it is, in that sense, inescapably an ethical exercise.

But Derrida makes a further claim about writing in particular: 'a written sign carries with it a force that breaks with its context, that is, with the collectivity of presences organizing the moment of its inscription. This breaking force . . . is not an accidental predicate but the very structure of the written text.'[84] Other writers have put this breaking with the context in similar ways; Austin: 'written utterances are not tethered to their origin in the way spoken ones are';[85] Ricœur and Herrnstein Smith apply the observation particularly to literature in which we encounter 'a meaning that has broken its moorings to the psychology of its author' or in which 'the poem is unmoored from any

[82] 'Signature Event Context' (1971), trans. Samuel Weber and Jeffrey Mehlman, *Glyph*, 1 (Baltimore, 1977), 192. [I follow Derrida hereafter in referring to this article as SEC.]

[83] 'LIMITED INC a b c . . .', trans. Samuel Weber, *Glyph*, 2 (Baltimore, 1977), 240 [hereafter referred to as INC].

[84] SEC 182.   [85] *How to do Things with Words*, 61.

specific occasion in the world of objects and events. . . .[86] These
nautical metaphors have a tang to them but they are not very clear.
Yeats's 'Easter 1916', for example, has some connection with a specific
occasion in the world of events, even if the connection does not
constitute a 'mooring'. The connection holds firm enough for
statements which contain facts about occurrences in Dublin in 1916 to
be among the truth-conditions of lines in the poem or, at least, if, as
some maintain, poems do not have truth-conditions, among its
intelligibility-conditions in the sense of 'things you need to know to
understand it'. There is a view that would deny even this slight
attachment of poems to the world, arguing, in the case of 'Easter
1916', that the poem is 'mythopoeic' and that no facts about Padraic
Pearse and post offices have any bearing on it. This seems to be based
on a confused extension of Sidney's well-known maxim about the poet
who 'nothing affirmeth' and therefore 'never lieth'.[87] Sidney can hardly
have meant more than that the truth or falsity of apparently factual
statements in poems (about, for instance, the birth of heroes and the
causes of rain) was not itself a criterion of their poetic excellence, and
he may have meant a good deal less; from this debatable principle, it
does not follow that whether or not statements are offered as factual by
a poem is irrelevant to its meaning, to how it asks to be taken, though
its meaning cannot be settled simply by reference to the relevant facts.
Which facts are relevant must be judged from what the agent can be
conceived to have been doing in his context, and our conception of him
cannot, except in unusual circumstances, be entirely unimpressed by
what we can know of his conception of himself. Consequently, factors
which affect the agent's self-conception must bear on the meaning of
the poem because they constitute part of its ineliminable context. Facts
about the poem's occasion are amongst such factors. Even a
strenuously 'mythopoeic' Yeats can tell the difference between writing
a poem which starts off from something which happened—the Easter
Rising—and one which starts from something which didn't, or hasn't
yet—the Second Coming. Less occasional poems, such as 'She
dwelt . . .', are not wholly at sea and out of sight of 'the world of objects
and events', for the fact that there were such things as maids (as
contrasted with, say, unicorns) in the world when Wordsworth wrote
can fairly be presumed to have had some effect on Wordsworth's self-

    [86] P. Ricœur, 'The Model of the Text', *New Literary History*, 5 (Autumn 1973), 95;
Herrnstein Smith, op. cit. 10.
    [87] *The Defence of Poesie* (1595, repr. Menston, 1968). G. [i.e., p. 41].

conception in writing and so to constitute a part of the ineliminable context of the poem's meaning.

Yet for all this, writing does somehow detach itself from context, as speech does not. If I say in conversation, 'It is midnight. The rain is beating on the windows,' the truth-conditions of my utterance (provided I speak seriously with the intent to give information) include the time on the clock and the state of the weather at the moment of my speaking; my interlocutor, if he has any doubts, will guess what I am going on about by looking at the clock and the windows. When Beckett's *Molloy* ends

> Then I went back into the house and wrote, It is midnight.
> The rain is beating on the windows. It was not midnight.
> It was not raining.[88]

the writing has no spatio-temporal or metereological location for the reader to check and determine intention from. This is largely because *Molloy* is a work of fiction, and the same point would hold if it were a work of spoken fiction. But the example helps to show the power of writing to generate for itself a second context over and above, and sometimes in the entire absence of, the context of utterance. This corresponds to the distinction I drew before between 'the circumstances of me writing' and 'the circumstances of my writing'. It is from this first kind of context that writing may abstract; it does so in varying degree— sometimes hardly at all, as in the case of a written statement made to the police (in which truth-conditions would be determined in much the same way as in a spoken remark about when it was raining) and sometimes to a distance of extreme remoteness, as in *Molloy*. That power of abstraction enables it to become the voice of an *absent* person. The second context, however, retains for writing its intentionality, which enables it to remain the voice of an absent *person*.

This intentionality is sometimes interpreted as if it were a psychological state of the author at the time of writing, within the first context of utterance, a state in some way comparable to 'suddenly thinking of my mother' or 'feeling a twinge of pain'. Intention is assimilated to a sensation which accompanies utterance and action, and which is thought occurrential and 'private' in the way sensations are held to be. This appears to be Derrida's idea in his description of the context from which writing breaks as 'the collectivity of presences organizing *the moment of its inscription*' (my emphasis). So too, Ricœur

---

[88] *Molloy* (1950), trans. Samuel Beckett and Patrick Bowles (1955, repr. 1959), 176.

denies that such a kind of intention determines the meaning of a written text which has 'broken its moorings to the psychology of the author'. He is right but the reason why such a form of intention does not govern the meaning of the written text is not because of special features of writing but because the intention which governs texts is of a different form; I leave aside the question whether the kind of intention which is an occurrential, private state of a consciousness exists at all; whether it does or not, it is irrelevant here.

Consider sentences which ascribe intentions such as:

He is trying to give up smoking.
He is searching for the key to all mythologies.

These sentences may be true of the person they refer to even if he is asleep when they are uttered; they do not therefore record conscious occurrences of sensations. Or take the case of someone who says, 'I didn't mean to hurt you'—does this imply that the speaker has or has had sensations of 'meaning not to hurt' or a felt lack of sensations of 'meaning to hurt'? Nothing need have crossed the speaker's mind at all, and that may be just what his remark means. These considerations are familiar to philosophers,[89] but they do not figure as centrally as they should in those theorists of literary intention who rest their view of what it is to intend something on a philosophical tradition which runs from Descartes to Husserl and which conceives intention as an interior state of the subject's consciousness. Many arguments against intentionalism rely on the mistaken belief that this tradition offers the only possible way of construing intention and conclude that if such a construction turns out to be unsound the concept of intention, and the practices of interpretation which invoke that concept, must be rejected.

Derrida argues like this; intention is part of the myth of an absolutely self-present consciousness, it is 'absolutely actual and present intention . . . the plenitude of [the utterer's] desire to say what he means' or 'conscious intention . . . totally present and immediately transparent to itself'.[90] But I may not know exactly what I am doing and still be acting intentionally, as in the game of 'Blind Man's Buff' or as Eliot acted during the composition of *The Waste Land*. Derrida does not reject the concept of intention, but he believes himself to be

[89] I have relied on discussions of intention in G. Ryle, *The Concept of Mind* (1949; Harmondsworth, 1963), G. E. M. Anscombe, *Intention* (Oxford, 1957, 2nd edn., 1963), A. Kenny, *Action, Emotion and Will* (1963). The basic notions of my account derive from the remarks on intending and meaning in Wittgenstein's *Philosophical Investigations*.
[90] SEC 181, 192.

effecting a major relocation of its place in a theory of meaning: 'the category of intention will not disappear; it will have its place, but from that place it will no longer be able to govern the entire scene and system of utterance'.[91] We need not accept that intention has ever ruled so absolutely, nor that all philosophical accounts of meaning rely on such a self-present, 'private' intention-as-consciousness. Wittgenstein: 'Can I say "bububu" and mean "If it doesn't rain I shall go for a walk"?—It is only in a language that I can mean something by something. This shews clearly that the grammar of "to mean" is not like that of the expression "to imagine" and the like.'[92] My failure to make 'bububu' mean 'If it doesn't rain I shall go for a walk' has nothing to do with my lack of will power. Derrida urges that 'the criterion of intention . . . is a necessary recourse in order that the "serious" and the "literal" be defined',[93] which is too strong because the distinction between literal and metaphorical uses of some locutions is partly conventional, and may not require recourse to intention (though it may do—'I cut him dead at the party' can be literally true). He is certainly right to claim that the only way of telling whether a locution is seriously uttered or nor on a particular occasion is by some interpretation of the intention with which it is uttered. There is no unequivocally defined sign for 'serious utterance' in a natural language to correspond to, say, the 'assertion-sign' which Frege thought might be introduced into his symbolic notation[94] and it may be that there could not be such a sign. Yet Derrida reveals the inadequate theory of intention he trusts in when he continues: 'this intention [which distinguishes the 'serious' from the 'non-serious'] must indeed . . . be situated "behind" the phenomenal utterance (in the sense of the "visible" or "audible" signs, and of the phono-linguistic manifestation as a whole): no criterion that is simply *inherent* in the manifest utterance is capable of distinguishing an utterance when it is serious from the same utterance when it is not.'[95] He writes as if intention must be 'behind' the utterance if it is not 'in' it, and he assumes that this 'behind' can be characterized only as a Husserlian immediate self-presence. These views are taken for granted in Derrida's discussions of intention, and they account for his belief that any intentionalist theory of meaning must make certain metaphysical presuppositions which he considers illicit. Nothing is

---

[91] Ibid. 192.   [92] *Philosophical Investigations*, 18.   [93] INC 208.
[94] See Michael Dummett's discussion of the assertion-sign in Dummett, op. cit. 314–16, 328–30.
[95] INC 208.

said to explain why intention may not be a function of convention and circumstance, why it may not be, as it were, 'between' utterances rather than 'behind' them, a function of the place of the utterance in an interpretable pattern of other utterances and actions, a matter of perceptible design rather than occulted inner experience; nor is any reason given for believing that the only possible account of self-conscious intentionality is the Husserlian one.

Derrida presents a further move specifically against the thesis that linguistic meaning is a derived form of intentionality. It runs like this:

And this is the possibility on which I want to insist: the possibility of disengagement and citational graft which belongs to the structure of every mark, spoken or written, and which constitutes every mark in writing before and outside of every horizon of semio-linguistic communication; in writing, which is to say in the possibility of its functioning being cut off, at a certain point, from its 'original' desire-to-say-what-one-means . . . and from its participation in a saturable and constraining context. Every sign, linguistic or non-linguistic, spoken or written (in the current sense of this opposition), in a small or large unit, can be *cited*, put between quotations marks; in so doing it can break with every given context, engendering an infinity of new contexts in a manner which is absolutely illimitable.[96]

The point seems to be that Shakespeare may write 'I have yet | Roome for six scotches more'[97] in the text of *Antony and Cleopatra* and that I may then quote this, meaning that I am ready for an other drink, which is not the ' "original" desire-to-say-what-one-means'. From this possibility of quotation, inherent in any sign, Derrida infers that neither original context nor original intention can control the possible engendering of illimitably new contexts by citation, and that, therefore, the system of language escapes entirely the governance of intentional contexts. The point would be blunt indeed if there was not a sense of the verbs 'quote' or 'cite' in which I do something different when I quote or cite from merely repeating a linguistic formula which you have also used. As Searle has argued against Derrida, 'any linguistic element written or spoken, indeed any rule-governed element in any system of representation at all must be repeatable, otherwise the rules would have no scope of application.'[98]

We need a way to distinguish between the case in which (a) you write 'x' and I write 'x', quoting you, and (b) you write 'x' and I write 'x', not quoting you. There are instances of case (b), such as the 'Dear

---

[96] SEC 185. [97] IV. vii. 9–10; Folio, ll. 2635–6.
[98] 'Reiterating the Differences: a Reply to Derrida', *Glyph*, I 199.

Sir' at the start of letters, which are not instances of mass inter-quotation. If you write to me, 'Dear Sir, I was appalled . . .' and I reply, 'Dear Sir, I was sorry to hear you had been "appalled". . .', then I do not quote your 'Dear Sir', though I do quote your 'appalled'. (I do not *repeat* your 'Dear Sir' either, though the phrase appears in both letters, just as I do not repeat your performance of Hamlet when I play the role later, even though I say the same lines. That *Hamlet* is repeatedly performed does not show that each performance of *Hamlet* is a repetition of some previous performance.) How, then, are we to distinguish between a citation and a repetition? Just the features of context-dependency and intention within a context which Derrida believes citation destroys provide the criteria which distinguish a citation from a repetition; they provide the criteria of identity which need to hold so that what I say repeats and quotes *your utterance* (rather than merely using the same words again). Someone who did not know that 'I have yet | Roome for six scotches more' occurs in Shakespeare could not make the amusing joke of quoting it in a new sense; he could do no more than coincidentally use the same set of words. In Derrida's own powerful and playful response to Searle's criticisms, we find him constantly moving from Searle's written text to an imagined intending author, and he does so by reconstituting the tone he catches in the writing, for all the world as if he were a literary critic re-describing illocutionary force—so that Searle's text is said 'loudly' and 'nervously' to 'denounce' Derrida's errors, to result from a 'passionate and exacerbated struggle', to show signs that its author is 'growing angry', and so on.[99]

It would be misleading to neglect the fact that Derrida's interest in 'writing' covers more than the questions about written in relation to spoken language which concern me. Derrida begins his 'deconstruction' of Western metaphysics by rejecting any account of writing as merely a transcription of the spoken language, any attempt to 'confine writing to a secondary and instrumental function: translator of a full speech that was fully *present* (present to itself, to its signified, to the other, the very condition of the theme of presence in general), technics in the service of language, *spokesman*, interpreter of an originary speech itself shielded from interpretation'.[100] It is possible to believe that writing generally works as an ambiguous transcription of imaginable speech without believing that speech itself does not require interpretation, but

[99] INC 175, 176, 179.
[100] *Of Grammatology* (Paris, 1967), trans. G. C. Spivak (Baltimore, 1976), 8.

Derrida is having none of that. He takes off from Saussure's dictum
that the linguistic sign is essentially arbitrary,[101] and arrives at his key
insistence:

Now from the moment that one considers the totality of determined signs,
spoken, and a fortiori written, as unmotivated institutions, one must exclude
any relationship of natural subordination, any natural hierarchy among
signifiers or orders of signifiers. If 'writing' signifies inscription and especially
the durable institution of a sign (and that is the only irreducible kernel of the
concept of writing), writing in general covers the entire field of linguistic signs.
In that field a certain sort of instituted signifiers may then appear, 'graphic' in
the narrow and derivative sense of the word, ordered by a certain relationship
with other instituted—hence 'written', even if they are 'phonic'—signifiers.
The very idea of institution—hence of the arbitrariness of the sign—is
unthinkable before the possibility of writing and outside of its horizon.[102]

This passage is a tissue of inconsequence. The arbitrariness of the
linguistic sign, in Saussure's sense, does not preclude a hierarchy of
orders of signifiers; it precludes only a 'natural' hierarchy. Other non-
natural hierarchies remain possible—social and historical ones, for
example. Derrida tries to make 'arbitrary' tell against 'hierarchy'
whereas it tells only against 'natural'.
   He equally takes for granted Saussure's claim that the linguistic sign
is arbitrary, a notable superstition for Derrida can be sharply critical of
the conceptual equipment of modern linguistics. He treats, as
Saussure treated, 'arbitrary' and 'instituted' as synonyms, though they
need not be so treated. Wittgenstein again:

I am not saying: if such-and-such facts of nature were different people would
have different concepts (in the sense of a hypothesis). But: if anyone believes
that certain concepts are absolutely the correct ones, and that having different
ones would mean not realizing something that we realize—then let him
imagine certain very general facts of nature to be different from what we are
used to, and the formation of concepts different from the usual ones will
become intelligible to him.
   Compare a concept with a style of painting. For is even our style of painting
arbitrary? Can we choose one at pleasure? (The Egyptian, for instance.)[103]

Imagine that the phenomena now known as 'pain' and as 'rain' were to
exchange certain features of their occurrence, so that pain was
distributed geographically as rain now is, falling more heavily on some

---

[101] *Course in General Linguistics*, 'The bond between the signifier and the signified is
arbitrary', 67.
[102] *Of Grammatology*, 44.   [103] *Philosophical Investigations*, 230.

regions than others and at some times of year than others, and that
bouts of pain were forecast of an evening on the television as showers
of rain now are. The grammar of the concept of 'pain' would be
different then. Sentences which strike us now as bizarre, such as 'This
has been the most painful August for thirty years' or (scanning the grey
horizon) 'Looks like pain', would find an intelligible place in our
language but we could not just *decide* now to exchange our concepts of
pain and rain and use them in the way imagined in the example.[104]
This shows that, though instituted, the concepts are not arbitrary.
Linguistic signs are similarly conventional but not arbitrary; of a
convention we may ask what are our grounds of assent to it, or what is
involved in our adoption of it—thereby demonstrating that our
institutions need not, by the mere fact of being institutions, be
'unmotivated', though it is only within a fabric of institutions that
human beings can have recognizable motives.

Derrida's chief contention—that 'writing' needs to be extended to
cover the field of all linguistic signs—takes place in the second
sentence of the passage quoted above: 'If "writing" signifies inscription
and especially the durable institution of a sign (and that is the only
irreducible kernel of the concept of writing), writing in general covers
the entire field of linguistic signs.' This looks as if it contained an
argument of the form:

All writing is the creation of durable signs.
All linguistic signs are durable.
Therefore, all linguistic signs are part of 'writing', are 'written'.
This argument is invalid. Compare:
All four-legged creatures are mortal.
All men are mortal.
Therefore, all men have four legs.

Derrida offers no other reason for extending the concept of writing to
cover all linguistic signs, and so he offers no reason for the prime move
in his deconstruction of Western metaphysics as the repression of
writing. The further account of language as the play of 'difference'
('the thesis of *difference* as the source of linguistic value' which is
regarded by Derrida as laying the 'foundations' for 'the thesis of the
arbitrariness of the sign'[105]) coheres no better. Derrida claims that the
notion of 'meaning' should be replaced by a recognition of the play of

---

[104] I am indebted here to P. M. S. Hacker's *Insight and Illusion: Wittgenstein on
Philosophy and the Metaphysics of Experience* (Oxford, 1972).
[105] *Of Grammatology*, 52.

difference which is all language is: 'One could call *play* the absence of
the transcendental signified as limitlessness of play . . .'.[106] But this
'thesis of *difference* as the source of linguistic value' is unsustainable.
Derrida takes it from Saussure's 'in language there are only
differences'.[107] Saussure's claim does not tell us much about language
because it provides no method for distinguishing between linguistically
significant and linguistically insignificant differences. No two utterances
of a sentence, for example, will produce exactly the same pattern on an
oscilloscope, but the differences may make no difference to the
meaning of the sentence, or even to its 'linguistic value' to use more
strictly Saussurian terms. Even if we go down to the level of the
smallest semantically distinctive unit, the phoneme, we cannot account
for linguistic operations with the concept of difference alone, though
Saussure asserted otherwise—'Phonemes are characterized not, as
one might think, by their own positive quality but simply by the fact
that they are distinct.'[108] Again, two utterances of the same phoneme
may be acoustically distinct, but the distinction may have no linguistic
significance. Saussure makes a remark parallel to his description of
phonemes when he says of written letters that their value is 'purely
negative and differential'.[109] The instance he gives to demonstrate this
contradicts his claim: he shows three different ways of writing the
letter 't' and rightly states that the differences between them do not
matter, that 't' may be written in any way at all so long as it remains
distinguishable from, say, 'l' and 'd'. Exactly, and it is because
linguistic value is something more than difference *per se* that 't' may
without loss of orthographic identity be written in these various ways.

The abstract power and scope of Derrida's writings fail to
compensate for these essential weaknesses in his conception of the
relations of language and consciousness, and his unjustified extension
of the concept of 'writing' to cover all linguistic activity (indeed, all
experience, Derrida has claimed at times). His pages combine passion
of intellectual commitment with friskiness in an unusual manner but
they do not contain the radical critique of 'Western metaphysics' which
they discourse upon so largely. Nor would it follow from such a
critique, even if it existed, that the practice and language of literary
criticism had been criticized to the core, as some who have managed to

---

[106] *Of Grammatology*, 50.
[107] *Course in General Linguistics*, 120.
[108] Ibid. 119.
[109] Ibid.

accept Derrida's writings believe.[110] Oddly, Derrida's critique of
origins itself massively exploits a model of intellectual practices in
which all other discourses have one origin or foundation—metaphysics. It
is natural for a philosopher or an anti-philosopher to enjoy believing,
for example, that the literary-critical concept of 'author', and the
practice of literary criticism with that concept, as in the matter of
illocutionary re-description, rest upon a basis of philosophical
propositions, and that when this foundation is shaken the edifice of
'secondary' practices must also topple. But take the analogy of the legal
concept of the person. Can we believe that this concept depends upon
a philosophical thesis such as the Husserlian self-presence of
intentional consciousness, 'depends upon' in a sense so strict that
nobody can consistently disbelieve the philosophical thesis and
continue to employ the legal concept? This assumes an absurdly
schematic relation of human practices to their philosophical articulations.

It is quite possible to think of the relation between the legal concept
with its attendant practice and the philosophical thesis as the converse
of what the Derridan critique suggests—to think that philosophy is
itself only a secondary elaboration of the rationality of what is already
actual, a fallible attempt to bring practices to articulated self-
consciousness, whose failures may reveal the difficulty with which we
make forms of life articulate. In such attempts, we may commit
philosophical mistakes without showing up all philosophy as mistaken,
and the discovery of those mistakes is part of the history of philosophy
as one of our practices rather than a demonstration of the
groundlessness of our behaving. Some philosophers have also held
this, or a similar, view—Aristotle, Hegel, and Wittgenstein, for
example.

The Printed Voice in the Nineteenth Century

John Hollander marks the critical importance of illocutionary re-
description when he writes that 'speaking and writing are both
language . . . it is the region between them which poetry inhabits'.[111]
He is thinking of poetry which is not purely oral, the poetry which
Yeats believed to be distinctively English: '. . . English literature, alone

---

[110] See e.g. the inferences drawn in Colin MacCabe's *James Joyce and the Revolution of
the Word* (1978), 76–9.
[111] *Vision and Resonance* (New York, 1975), p. x.

of great literatures, because the newest of them all, has all but completely shaped itself in the printing-press'.[112] Yeats's testimony about the history of literatures is not entirely reliable, and his 'all but completely' (which manages to make a reservation sound like a redoubled assertion) conceals several major questions. English literature does not stand alone in having shaped itself in the printing-press. Despite its exaggerations, Yeats's insight remains considerable, and considering it, even in application only to poetry, involves a great deal.

The spoken voice, more richly than writing, conveys, in emphasis and intonation, by stress and juncture, and so on, much of the illocutionary force of any utterance. Not that the wealth of such features in vocal utterance is always only an advantage; creative opportunities may also arise from the comparative absence of vocal quality in print. The semantic role of prosodic features, as I have argued so far, renders them integral to the study, whether linguistic or philosophical, of utterances. For literary critics, I shall now argue, acknowledgement of the printed character of the poetic voice in most texts involves a recognition of the very conditions of poetic meaning. The reader must inform writing with a sense of the writer it calls up— an ideal body, a plausible voice. How this is done repays study, as Leavis thought it might when he reviewed Eliot's recording of *Four Quartets*: 'These records should call attention to the problem of reading *Four Quartets* out. The problem deserves a great deal of attention, and to tackle it would be very educational.'[113] Whatever else poetry may be, it is certainly a use of language that works with the sounds of words, and so the absence of clearly indicated sound from the silence of the written word creates a double nature in printed poetry, making it both itself and something other—a text of hints at voicing, whose centre in utterance lies outside itself, and also an achieved pattern on the page, salvaged from the evanescence of the voice in air. Browning names this double nature in a phrase from *The Ring and the Book*—'the printed voice'.

He describes the odd system of trial which his poem records, a trial in which written depositions were made to the judges but

> . . . there properly was no judgment-bar,
> No bringing of accuser and accused,

---

[112] 'Literature and the Living Voice' (1906), repr. in *Explorations*, selected by Mrs W. B. Yeats (1962), 206.
[113] 'Poet as Executant', *Scrutiny*, 15/1 (1947), 80.

And whoso judged both parties, face to face
Before some court, as we conceive of courts.[114]

In literature shaped by the printing-press, writer and reader do not 'properly' face each other. But this sense of a lost community, felt as a form of death by some writers when their voice fails to be manifest in print, is the germ of a further community and a new life; it prompts the reader to interpret and resuscitate. Browning notes that the absence of voices from the trial he reports allows 'the trial | Itself, to all intents,' to be 'then as now | Here in the book and nowise out of it'; his poem truly perpetuates in its own phantasmal character the meaning of what then went on:

> 'T was the so-styled Fisc began,
> Pleaded (and since he only spoke in print
> The printed voice of him lives now as then)[115]

We may come to find the same thing true of the voices of poets.

They were often baffled themselves. Yeats wrote, remembering early days: 'I spoke them [the lines of Yeats's poems] slowly as I wrote and only discovered when I read them to somebody else that there was no common music, no prosody.'[116] Though Browning had confidence that the voice of the Fisc spoke in his lines, G. H. Lewes felt that Browning's own voice failed to ring in the poems: 'And respecting his versification, it appears as if he consulted his own ease more than the reader's; and if by any arbitrary distribution of accents he could make the verse satisfy his own ear, it must necessarily satisfy the ear of another.'[117] Poets were not always good at hearing each other. Coleridge thought the young Tennyson had no metrical sense: 'The misfortune is, that he has begun to write verses without very well understanding what metre is . . . I can scarcely scan some of his verses'[118] (this after Tennyson had published the book of poems which contains such masterpieces of versification as 'The Lady of Shalott' and 'To J. S.').

Problems of translating the intended music of a voice into the scant

---

[114] *The Ring and the Book*, I. 167 and I. 155–8. I quote from the edition of R. D. Altick (Harmondsworth, 1971), hereafter referred to as Altick.

[115] Ibid. I. 165–7.

[116] *Reveries over Childhood and Youth* (1915), repr. in *Autobiographies* (1955), 67.

[117] *British Quarterly Review* (1847), repr. in B. Litzinger and D. Smalley (eds.), *Browning: The Critical Heritage* (1970), 122.

[118] Entry for 24 Apr. 1833 in *Specimens of the Table Talk of Samuel Taylor Coleridge*, ed. H. N. Coleridge (1837, 3rd edn. 1851), 236–7.

notation of the written word posed themselves even in such rudiments of writing as the punctuation of a text. Masters of the language felt at times like schoolboys. Yeats wrote to Bridges:

I chiefly remember you asked me about my stops and commas. Do what you will. I do not understand stops. I write my work so completely for the ear that I feel helpless when I have to measure pauses by stops and commas.[119]

Perhaps the 'Do what you will' sounds more lordly than helpless, but the problem remains, though attitudes to it may vary. Wordsworth's letter to Humphry Davy on the same subject strikes a different note:

You would greatly oblige me by looking over the enclosed poems and correcting any thing you find amiss in the punctuation, a business at which I am ashamed to say I am no adept.[120]

The different characters of the two poets show in their distinct approaches to the task of getting your point across by getting your pointings right: Yeats faces the academically correct Bridges, feels conscious of his lack of an education in the punctilio of written language (he had not been to university) and yet makes that lack a blazon of what he wishes to consider the direct and rooted nature of his own poetry, direct in its address to the ear, rooted in the community of his imagined Irish speech;[121] Wordsworth is humbly formal, exasperated, no doubt, by the mechanics of committing his voice to print—a tinge of that exasperation in the phrase 'a business' and in the slightly flowery 'adept' with its mild intimation of sarcasm about such expertise, but submitting none the less to correction. It is not surprising that much should hinge, for poets, and so for readers of poetry on these minutiae; they compose the fibre of communication.

Francis Berry in his pioneering book, *Poetry and the Physical Voice*, insists: 'Most poetry is vocal sound . . . and not the signs *for* vocal sound printed on a page. . .'.[122] Whatever most poetry is, most poems arc in fact 'signs *for* vocal sound printed on a page'. Berry recognizes the intonational ambiguity of many written lincs but thinks such

---

[119] Letter of ? July 1915, in *The Letters of W. B. Yeats*, ed. A. Wade (1954), 598.

[120] Letter of 29 Sept. 1800, in *The Letters of William and Dorothy Wordsworth: The Early Years 1787–1805*, ed. E. de Selincourt (1935), 2nd edn., rev. C. L. Shaver, 2 vols. (1967), i. 289; the poems were 'Hart-Leap Well', 'There was a Boy', 'Ellen Irwin', and part of 'The Brothers'.

[121] See, for example, Yeats's comments in 'A Literary Causerie' (1893), repr. in *W. B. Yeats: Uncollected Prose*, vol. i, ed. J. P. Frayne (1970), 288 and in *Explorations*, 206; also the conversation recorded (not without malice) in G. Moore's *Ave* (1911), ch. 1.

[122] (1962), 34.

uncertainty of voice only a hurdle and never a resource: 'Normally, the qualities of a voice, which are resistant to mensuration and notation, should declare themselves in a valid poem without attempting the inadequate aid of marks usually employed in the scoring of music, the whole context determining those qualities.'[123] Wordsworth's 'She dwelt . . .', for example, does not work 'normally', but only in the sense that it is an unusually good poem. Berry expects the 'whole context' to compel vocal quality to appear and he fails to allow that sometimes contexts are not so readily to hand or mouth—because they are in doubt (the writer's and/or the reader's doubt). 'She dwelt . . .' creates a sense of 'not being sure what to say', a reflective absence of an instantly declared tone. It may be, roughly speaking, that 'through the totality of the printed signs on the page the poet conveys his voice',[124] but he may also convey the absence of his voice which it can be quite as important for us to hear as that individual physical timbre which Berry celebrates.

His argument gets into a tangle. At times, it seems that he wants the reader to try to hear 'the actual physical voice of the poet',[125] at others, the poet's voice matters not 'as it actually was, as it actually sounded to others, that is, but as the poet himself heard it, or experienced it, or . . . as he supposed it to sound to others when he was employing it to good effect'.[126] The many coils of Berry's qualifications show how hard it is to hear a voice, whether your own or someone else's, and that it is not only notation which makes it hard but also a web of hopes and fears between the self and others. We will never hear another's voice as it sounded to him if only because we cannot hear from inside the acoustic of another's body, but then the voice too intimately sounds of the self for us to be certain that we hear it always 'as it actually was' and the speaker always only as it sounds to him. Uncertainties of voice are amongst the excursions of the self, and a voice, poetic and otherwise, can no more be strictly identified with the physical voice than a person with his body. Berry's thesis comes adrift on two connected snags: his underestimate of the inherent problems of vocal transcription, and his failure to recognize these problems as problems of individuality and the voice's role as a sign of that individuality, the linguistic impress of an agent.

Valéry gives these general considerations a particular twist in two of his remarks about voice and poetic voice:

Le Moi, c'est la Voix.
[The Self is the Voice.]

[123] Ibid. 178.   [124] Ibid. 193.   [125] Ibid. 3.   [126] Ibid. 189.

. . . cette voix ne doit faire imaginer quelque homme qui parle. Si elle le fait, ce
n'est pas elle.

[. . . this voice must not make us think of a particular man talking. If it does so,
it is not the poetic voice.][127]

The second stipulation evidently aligns Valéry with the *symboliste*
ambitions clearly sounded by Mallarmé's desire for 'la disparition
élocutoire du poëte' ['the disappearance of the poet-as-elocutionist'],[128]
and so might be thought only part of a specific literary movement.
Symboliste 'disparition élocutoire' did indeed rely strongly on an
assertion of a poem's graphic as well as its vocal existence, but the
printed voice was not discovered by the *symbolistes*. The permanent
possibility of creatively exploiting the region between writing and
speech stems rather from those relations of a self to its voice which
Valéry's glitteringly concise remarks express. My self is my voice, the
uttered agent in a significant medium, but the medium exacts, as the
cost of utterance, a loss of immediate physical particularity, of that
imaginable self-possession of 'quelque homme qui parle'. Just as my
self is not only 'how I seem to myself', nor only 'how I seem to others',
nor yet the result of adding those two appearances together, the voice
in poetry is not only the voice 'as it actually sounded to others', nor
only the voice 'as the poet himself heard it', nor yet some unspecified
compound of these sounds. For the other is in my self, as I come to
know my self in terms not of my own invention. The poetic voice
sounds essentially in the hearing of others, in a delight within
shareable language. Valéry considered these attributes as uniquely
belonging to the poetic voice, but this is not so—if it were, only poets
could speak or understand poetry (and not even poets are poets all the
time). Rather, the poetic activity takes and gives exemplary pleasure in
a common condition. The poetic voice, as any self in utterance, will
have, in Eliot's phrase, a 'mid-way reality',[129] such as Valéry delineated
when he asked, coyly philosophizing, 'Le présence serait-elle oscillatoire?'
['And is presence then an oscillation?'].[130] The scene of that oscillation,
for the poet, is the page and what it asks in the way of responsive
listening-in from the reader.

---

[127] *Paul Valéry: Cahiers*, 29 vols. (Paris, 1957–61), xiv. 390 and vi. 176. I am indebted
to Christine Crow's *Paul Valéry and the Poetry of Voice* (Cambridge, 1982).
[128] 'Crise de vers' (1886–96), in *Œuvres complètes*, ed. H. Mondor and G. Jean-Aubry
(1945), 366.
[129] *Knowledge and Experience in the Philosophy of F. H. Bradley*, written 1916 (1964), 79.
[130] *Valéry: Cahiers*, xvi. 160.

It was, then, more than a technical difficulty which prevented Yvor Winters from getting what he wanted out of poetry. In defence of his desires, he offered

rules for poet and reader alike: (1) There should be no conflict between rhetorical stress and metrical stress, but in so far as possible the metrical stress should point the meaning; (2) where the mechanical potentialities of the language indicate the possibility of a stress in either of two directions, the grammatical structure should be so definite that a certain rhetorical stress will be unmistakable and will force the metrical interpretation in the right direction. . .[131]

Winters makes the business sound easy when he writes that metrical stress should point the meaning, as if meaning were a high road and stress the occasional signpost along the way. But stress may compose meaning, in poetry as in speech 'where', as Browning wrote, 'an accent's change gives each | The other's soul'.[132] Winters's idea of the bearing of stress on meaning sounds like Pike's untenable distinction between 'transitory extrinsic pitch contour' and 'lexical meaning'. The 'grammatical stress' (whatever Winters means by that phrase, for there is, really, no such thing in English, though there is something which might be called 'grammatical stress' in French) can never be 'so definite' as to make 'rhetorical stress', which is Winters's term for the indication of illocutionary force in prosodic features, 'unmistakable'; this was shown before in discussion of Crystal's attempt to establish the significance of intonation contours by reference to syntactic and lexical patterns. Such unmistakability cannot always be achieved, and this is no disaster because unmistakability as a general feature of writing is not simply and solely desirable, for reasons which glimmer behind Winters's 'force the metrical interpretation in the right direction' (my emphasis). Force and forcing wreck the community in and for which a poet writes, turning the humane conditions of trust into a system of domination. The poet cannot dictate to his reader because he cannot always dictate his voice into print. And the failure to do so need not be a 'failure' at all, but a discovery of the conditions of essential reciprocity in the exchanges which take place between writers and readers. Winters's prescription that there should be no conflict between rhetorical stress and metrical stress expresses no more than a preference for one type of poetry over another. Poems in which such a

---

[131] 'The Audible Reading of Poetry', *Hudson Review*, 4/3 (Autumn 1951), 440.
[132] *Sordello*, v. 636–7; I quote from P. ii. 265.

conflict exists, as those of Browning in which there often occurs a 'struggle between the metrical bars and the sense-rhythm or natural emphasis of thought',[133] do not necessarily display incompetence on the part of the poet—they may exploit one more dimension of the expressive resource which is contained in the loose fit of writing on speech, of the formal institutions of language-practice on the various actuality of the things people can say.

John Hollander gives an excellent example of intonational ambiguity arising from uncertain stress in Thomas Weelkes's musical setting of 'Hark, all ye lovely saints above', and particularly in the line 'Then cease, fair ladies, why weep ye?' Hollander writes of this line:

. . . the last three syllables pose a rhetorical problem: contrastively accented, they can be '*why* weep ye?' or 'why *weep* ye?' or 'why weep *ye*?'—that is, 'what are you crying for?' or 'why are you crying about it?' or 'why *you*, of all people?' It is the last of these that is certainly suggested by the metrical position of the phrase, reinforced by rhyme. But in the context of the song, it should, if possible, be shown to mean all three. This is just what Weelkes does, by setting the line in a faintly polyphonic texture, with every voice repeating the phrase at least once, and, among the different parts, the various rhythmic patterns generating three different stressings of the text.[134]

Each of the three questions is a good question to ask of and about fair ladies; a polyphonic setting which enables us to hear the three questions asked almost at once has the advantage of encouraging us to consider the relations between all three, to wonder whether, for example, fair ladies are in fact or only as a graceful convention specially insured against causes to lament. The intonational ambiguity of a written text may create a mute polyphony through which we see rather than hear alternatively possible voicings, and are led by such vision to reflect on the inter-resonance of those voicings.

Hollander puts this very well: 'the poem is song and picture at once, and the relation of scanning and hearing lies at the heart of all textual (rather than purely oral) poetry.'[135] Geoffrey Hill makes a related point, when talking about that crucial circumstance in which a poet reads aloud the poems he has written:

The true realisation of the poet's voice comes from a blending or a marriage of the silent and the spoken forms. If we put this into the shape of a figure of

[133] The phrase is quoted in J. B. Oldham, 'On the Difficulties and Obscurities Encountered in a Study of Browning's Poems', *Browning Society Papers*, ii (1889–90), 345.

[134] Hollander, op. cit. 35.   [135] Ibid. 8.

speech, if we conceived of the voice as it reads the poem as being on the horizontal plane, and if we thought of the text on the page, as it were, going down vertically, then I think that the listener should follow the spoken poem in the way that a listener follows a string quartet with a score. I think only by being most keenly sensitive to that moment when the horizontal of the spoken voice comes into contact with the formalities, with the restraints, with the restrictions that are there printed in the text, only by recognising with immediate sensitivity those moments of contact, of harmony or of hostility, only then can the reader, the listener, truly appreciate how the poet's voice is being realised in the most minute, intimate, and yet profoundly rich, prosodic forms.[136]

The poet's voice issues from 'a blending or a marriage of the silent and the spoken forms'; it cannot be identified with the sounds an individual's larynx may produce, even when that individual is identical with the individual whose name appears on the title-page of the book of poems which he is reading aloud. The poet's voice is not the voice of the person who is the poet.[137] This is one reason why we are able to hear poets' voices after the people they were are dead. It is also a reason why it is irrelevant to complain that a search for the voice of a poetic text is conducted under the aegis of a false belief in the existence of a speech which surpasses the ambiguity of writing by not needing to be interpreted. That Derridan cry fails to take the point of what poetic voice is; there is no myth of intention immediately transparent to itself involved here, as may be seen by considering the fact that a poet may fail, and feel himself to fail, to read his own poems properly aloud. As Donald Wesling writes, 'literary voice simultaneously affirms writing and puts it into question; voice is the undecidable between person and text.'[138] 'Undecidable' comes on too strong; say rather that voice is that which is decided *in reading* a text, and that such a decision is not a matter of the 'choice' of a particular reader any more than it lies at the disposal of any particular writer.

If my general thesis about the incapacity of writing unambiguously to transcribe speech is true, these decisions in reading will figure to some degree in all written texts. There might, nevertheless, be a history of writers' and readers' consciousness of such matters. This

[136] Interviewed by the present writer for a BBC Radio Three programme, 'The Composed Voice', first broadcast 14 July 1981. I am grateful to Geoffrey Hill for his permission to quote from a transcript of this interview.

[137] Hence, many of the problems with Francis Berry's arguments.

[138] 'Difficulties of the Bardic: Literature and the Human Voice', *Critical Inquiry*, 8/1 (Autumn 1981), 80.

book gives part of that history for Victorian poetry. The relations between writing and speaking on which I concentrate appear, in various ways, in earlier poetry too, and the indicative force of those relations for a study of poetic self-consciousness is greater than I have been able to describe. The book, then, takes a conventional, though not, I hope, an arbitrary starting-point for towards the end of the eighteenth century a new philosophical articulation of self-consciousness was given by Kant and Hegel and, about the same time, 'the rise of the popular press along with efforts to secure mass literacy in British and American schools fostered a decline of oratory along with a concomitant rise of writing as the primary mode of rhetoric in these societies. A major effect of this shift was an abstraction of audience'.[139] This abstraction of the audience asked of poets a vigilant and concrete consciousness of the actual disparities in their public, a consciousness of disparity breaking in on the notion of 'human nature', a notion already beset diversely by empirical psychology, comparative linguistics, anthropology, and the other attempted sciences of human diversity.

Wordsworth is the great discoverer in English lyric poetry of these conditions of 'human nature'. He could no more do without the term than earlier writers, but he felt the shock it got, and he contributed to giving it that shock. He wrote to John Wilson, after Wilson had complained of 'The Idiot Boy' that its story was 'almost unnatural' because 'We are unable to enter into [Betty Foy's] feelings; we cannot conceive ourselves actuated by the same feelings . . .':

People in our rank of life are perpetually falling into one sad mistake, namely, that of supposing that human nature and the persons they associate with are one and the same thing. Whom do we generally associate with? Gentlemen, persons of fortune, professional men, ladies persons who can afford to buy or can easily procure, books of half a guinea price, hot-pressed, and printed upon superfine paper. These persons are, it is true, a part of human nature, but we err lamentably if we suppose them to be fair representatives of the vast mass of human existence.[140]

The prose shivers with an indignation it can hardly contain or render plain. Consider the shade of distinction between 'People in our rank of life' and 'the persons they associate with', 'persons of fortune', 'persons who can afford', and 'These persons'. What is the difference between people and persons? None, it seems, except the degree of

---

[139] Nystrand, 'Introduction' to *What Writers Know*, 4.
[140] Letter of 7 June 1802, in *Wordsworth Letters: Early Years*, i. 355.

kindliness Wordsworth can muster in contemplating them. The sentence beginning 'Gentlemen' gathers annoyance as it goes along, huffing and puffing rather ('half . . . price', 'hot-pressed') as Wordsworth thinks of the complicity between social arrogance and the institution of literature. This kind of reader, it seems, filches literature from Wordsworth, degrading it from its task of reflective apprehension to the level of a luxurious commodity ('superfine paper') with a novelty value ('hot-pressed') and a use as a symbol of status ('can afford to buy') in a well-oiled machine of supply and demand ('easily procure'). He checks the momentum of his vexation—'These persons are, it is true, a part of human nature'—because it will not do to discard the polite world as not truly human. He was, after all, a gentleman himself and not the writer to forsake his own past, the culture in which his personality had grown. Indeed, when Wordsworth in this letter turns from the irritatingly familiar 'persons' whose manners he knows so well that he can insinuate them into his own prose with great economy and dexterity of language, he faces something quite blank—'the vast mass of human existence'. The difficulty of much of his prose writing, and especially the Prefaces to *Lyrical Ballads*, faces the task of a precarious adjustment between 'human nature' and 'the vast mass of human existence', as he takes the strain put on the concept of human nature by the recognition of the diversities of this incalculable 'vast mass'. Particularly, that strain sets him the question of how poets might be 'fair representatives' of the 'vast mass'.

William Empson succinctly imagined what happens to a writer when he tries to write as a mediator between distinct, and sometimes mutually uncomprehending, groups within a society:

Large societies need to include a variety of groups with different moral codes or scales of value, and it is part of the business of a writer to act as a go-between; so their differences are liable to become a conflict within himself.[141]

He also brilliantly suggests that these distinctions, introjected into the practical consciousness of the poet, may produce a condition of 'Neurotic Guilt'; 'The Rime of the Ancient Mariner' is, Empson claims, 'the first major study of that condition, recognized as such'.[142] Such introjection sets in high relief the separation between the actual voice of the person who is a poet and the poetic voice in the sense I

---

[141] Introduction to *Coleridge's Verse: A Selection*, ed. W. Empson and D. Pirie (1972), 39.

[142] Ibid.

have previously outlined. For our voices greatly bear the marks of
social particularity—gender, regional and class origin, diversities of
cult, degree of education, and so forth. A go-between who speaks too
completely in the accents of one group may thereby move less well
between groups. Trying to speak, if only briefly, in the accents of
neither party, the poet develops a style which relies on the reticences of
the written word to postpone the identification of his own socially
fractional position and give time (a breathing space on the page) for
ampler sympathies to be felt.

Coleridge dramatizes the predicament of poetic mediation, the
practical task of speaking for other people in their variety, of being, as a
poet, a 'fair representative', at that moment in his poem when the
Mariner shies from utterance:

> And I quak'd to think my own voice
> How frightful it would be![143]

This draws on lines from Cowper's poem about the eloquence of
castaways, the 'Verses, supposed to be written by Alexander Selkirk,
during his solitary abode in the Island of Juan Fernandez':

> I am out of humanity's reach,
> I must finish my journey alone,
> Never hear the sweet music of speech,
> I start at the sound of my own.[144]

Here too, the loss of a hold on humanity troubles the person's relation
to his own voice. The printed page which retains the poetic voice
('retains' in the double sense of 'keeps back' and 'preserves') becomes
the dramatic scene of this searched and searching utterance; we can, as
it were, see the blank space around Selkirk's self-startled words.
Cowper's poem was praised by Wordsworth: '. . . the Reader has an
exquisite pleasure in seeing such natural language so naturally
connected with metre.'[145] It may be just casualness that made
Wordsworth write 'seeing' rather than 'hearing' at this point, but the
word is felicitous, for the ear might be less gratified by a recitation of
the piece, its slightly bland mellifluousness consorts uneasily with the
subject of isolation from speech. The poem's tune seems more

---

[143] 'The Rime of the Ancyent Marinere', 1798 text, ll. 337–8; I quote from
Wordsworth and Coleridge: Lyrical Ballads, ed. R. L. Brett and A. R. Jones (1963), 23.

[144] (1782). I quote from *The Poems of William Cowper*, ed. John D. Baird and Charles
Ryskamp, 2 vols. (Oxford, 1980), i. 403.

[145] 'Appendix' to *Lyrical Ballads* (1802), in Brett and Jones, op. cit. 318.

sociable than its tenor, whereas on the page the merely potential tune has a redolence of longing about it which beautifully imagines how Selkirk might like to speak, were there anyone to hear.

I'm not suggesting that Cowper's poem should be seen and not heard; it 'asks' to be read aloud. The nature of that request, though, plays a small, imaginative drama about speaking, a drama on the page and for the voice elicited from the page. Perhaps because poets like calling their works 'songs', critics tend to write as if the only important sound-quality of poems is that they should be 'grateful' to the voice, as we say that Mozart's lines are 'grateful' to a singer or 'lie' well on the voice. Coleridge was inclined to that view when he complained about some of Wordsworth's verse:

Among the possible effects of practical adherence to a theory that aims to *identify* the style of prose and verse . . . we might anticipate the following as not the least likely to occur. It will happen . . . that the metre itself, the sole acknowledged difference, will occasionally become metre to the eye only. The existence of *prosaisms*, and that they detract from the merits of a poem, *must* at length be conceded, when a number of successive lines can be rendered, even to the most delicate ear, unrecognizable as verse, or as having even been intended for verse, by simply transcribing them as prose . . .[146]

Coleridge stands in distinguished company when he makes this complaint, echoing Dr Johnson's sentiments:

The variety of pauses, so much boasted by the lovers of blank verse, changes the measures of an English poet to the periods of a declaimer; and there are only a few skilful and happy readers of Milton, who enable their audience to perceive where the lines end or begin. *Blank verse*, said an ingenious critick, *seems to be verse only to the eye*.[147]

It could fairly be asked whether Johnson and Coleridge are not right to think that metre must be audible, and that metre to the eye only must be factitiously added to words which are not really poetry at all. Well, Johnson thought Milton's verse was verse only to the eye but had the sense to admit, 'I cannot prevail on myself to wish that Milton had been a rhymer; for I cannot wish his work to be other than it is'.[148] If Milton falls victim to the argument that metre must be audible, then

[146] *Biographia Literaria* (1817). I quote from the edition of J. Engell and W. J. Bate, 2 vols. (1983), ii. 79.
[147] 'Life of Milton', in *Lives of the English Poets* (1779–81, repr. in 2 vols., 1906), i. 133.
[148] Ibid.

the argument rather than Milton may be at fault. Its fault lies in the implicit conviction it contains that unless metre is guiding vocalization it can be doing nothing creatively genuine in a poem. To check this conviction, we should refer to Wordsworth's account of the work of metre in his 1802 Preface to *Lyrical Ballads*:

The end of Poetry is to produce excitement in coexistence with an overbalance of pleasure. Now, by the supposition, excitement is an unusual and irregular state of the mind; ideas and feelings do not in that state succeed each other in accustomed order. But, if the words by which this excitement is produced are in themselves powerful, or the images and feelings have an undue proportion of pain connected with them, there is some danger that the excitement may be carried beyond its proper bounds. Now the co-presence of something regular, something to which the mind has been accustomed in various moods and in a less excited state, cannot but have great efficacy in tempering and restraining the passion by an intertexture of ordinary feeling, and of feeling not strictly and necessarily connected with the passion. This is unquestionably true, and hence, though the opinion will at first appear paradoxical, from the tendency of metre to divest language in a certain degree of its reality, and thus to throw a sort of half consciousness of unsubstantial existence over the whole composition, there can be little doubt but that more pathetic situations and sentiments, that is those which have a greater proportion of pain connected with them, may be endured in metrical composition, especially in rhyme, than in prose.[149]

Metre tends to 'divest language in a certain degree of its reality' because it provides, as it were, an alibi for the words in a poem—they are there both as their expressive and significant selves, but also elsewhere, half-absent, tokens of rhythmical units. The metrical form of a poem records the poet's compositional activity which may or may not entirely square with the drift of what is said, or the state of mind implicit in that drift. (Wordsworth neglected to observe that much the same can be said of the Augustan diction he so disliked.) An awareness of compositional activity enables us to distinguish, say, cries of pain from poetic lamentation, and thus makes 'pathetic situations and sentiments' more tolerable. This is, as Wordsworth claimed, 'unquestionably true', but the truth has often been not even questioned but simply ignored, because it runs counter to an emphasis, derived from Coleridge, on the organic unity of poems and the unity of consciousness in the poet which the wholeness of his works is called to

---

[149] Brett and Jones, op. cit. 264.

witness: 'The poet, described in *ideal* perfection, brings the whole soul of man into activity . . . .'.[150]

When Wordsworth admits into the poem 'an intertexture . . . of feeling not strictly connected with the passion', he puts in doubt the Coleridgean unity both of his work and of his self at work. The 'intertexture of feeling' will be the metrical record of composition, the 'passion', the expressive tenor of the poem. Wordsworth, then, envisages the possibility of a created double consciousness within the poem, and that doubleness may respond to divisions in the poet himself or to divisions between the poet and the imagined subject of his poem, or to both. There are at least two reasons why Wordsworth may have been moved to this. In the first place, the social diversities of his time, turned into conflicts within his self, as Empson describes the process, result in a need for poetic forms which speak to the divided soul of man. Since many people, then and now, have divided souls, it seems a respectable thing for a poet to call them into activity. Secondly, the particular form of such calling into activity required Wordsworth to dramatize poetical composition itself, and this is, in part, effected by a doubled consciousness of metrical language itself, the divesting of 'reality' from the poetic voice which cannot remain singly the voice of 'quelque homme qui parle'. And Wordsworth stands in distinguished company too—Hamlet:

Speake the Speech I pray you, as I pronounc'd it to you trippingly on the Tongue: But if you mouth it, as many of your Players do, I had as live the Town-Cryer had spoke my Lines: Nor do not saw the Ayre too much your hand thus, but use all gently; for in the verie Torrent, Tempest, and (as I may say) the Whirle-winde of Passion, you must acquire and beget a Temperance that may give it Smoothnesse.[151]

Wordsworth essentially attempted, by the metrical intertexture of feeling not strictly connected with the passion, to give to the lyric a dramatic self-consciousness of its own voice such as Hamlet recommends to the players. For example, though he defended 'The Idiot Boy' with grave moral splendour—'. . . I have indeed often looked upon the conduct of fathers and mothers of the lower classes of society towards Idiots as the great triumph of the human heart. It is there that we see the strength, disinterestedness, and grandeur of love, nor have I ever been able to contemplate an object that calls out so many excellent and virtuous sentiments without finding it hallowed thereby and having

[150] *Biographia Literaria*, ii. 16.   [151] III. ii. 1 ff.; Folio, ll. 1849–56.

something in me which bears down before it, like a deluge, every feeble sensation of disgust and aversion'—he also said of it, simply, 'in truth, I never wrote anything with so much glee'.[152] The conjunction of 'strength, disinterestedness and grandeur' with 'glee' provides one instance of the compositional drama which can be played between 'passion' and 'smoothness' by virtue of a break with organic functions of metre, by virtue of rendering the passage from visible to audible rhythmic patterns less secure than Johnson and Coleridge recommended.

There was great prescience in Coleridge's regret that Wordsworth showed an 'undue predilection for the *dramatic* form'.[153] He was thinking particularly of the personative works such as 'The Thorn', but much of Wordsworth's early writing is dramatic in the deeper and more elusive sense that it makes even the most lyrical utterance, utterance that might be thought to spring from the fullness of an 'I', dramatically self-conscious, watched by a 'he'. Coleridge's prescience lies in so exactly anticipating the direction Wordsworth's influence would take—it led to the discovery of the most characteristic and fertile of nineteenth-century poetic forms: the dramatic monologue.

The tendency Wordsworth noted metre had 'to divest language in a certain degree of its reality, and thus to throw a sort of half consciousness of unsubstantial existence over the whole composition' explains in part the particular subtlety of the double nature of the printed voice in poetry. For the 'half consciousness of unsubstantial existence' imparted by metre joins with the retention of voice by the page to enable the poet to fashion, and not merely to suffer from, bafflements of voice, lacks and flusterings in speech, the burdens of address. Poetry in the line of Wordsworth takes shape from, and makes shapes of, what Harry in *The Family Reunion* suffers from as a sheer predicament:

> To be living on several planes at once
> Though one cannot speak with several voices at once.[154]

The sounds of one person's actual voice must happen onc after another (I contrast 'actual voice' with a recorded voice and what is now possible by way of dubbing and multi-track superimposition). This makes it difficult to achieve in fact a simultaneity of different voices to

---

[152] *Wordsworth Letters: Early Years*, i. 357; note dictated to Isabella Fenwick, 1843, in Brett and Jones, op. cit. 292.
[153] *Biographia Literaria*, ii. 135.
[154] (1939). I quote from T. S. Eliot, *The Complete Poems and Plays* (1969), 324.

meet the simultaneity of different demands on your voice. But the
'unsubstantial existence' of poetic voice in print creates the chance of a
polyphony, the chance for a divided soul to speak with something
better than a forked tongue.

Many poets after Wordsworth discovered the practical truth of his
account of metre, and their work shows why metre 'to the eye only'
need not be an inert contrivance. Coleridge's charge was not that the
poems lacked metrical regularity, but that the regularity could not be
made audible without bizarrerie. As he observed of the last three
stanzas of 'The Sailor's Mother', 'If disproportioning the emphasis we
read these stanzas so as to make the rhymes perceptible, even *trisyllable*
rhymes could scarcely produce an equal sense of oddity and
strangeness, as we feel here in finding *rhymes at all* in sentences so
exclusively colloquial.'[155] This is quite true, but it is not true of the
opening three stanzas in which the poet speaks in his own person,
where the diction is comparatively elevated, and there is less
divergence of sense-unit from line-unit than when the Sailor's Mother
speaks. The poem contrasts a traditionally-imagined dignity with an
unexpectedly new form of encountered dignity; there is, then, an
aptness in employing two styles of relation of speech to metrical form.
We would equally need to disproportion the emphasis to make the
metrical regularity of Browning's 'My Last Duchess' perfectly audible,
or of many Hardy lyrics, or of the numerous other poems which take
from Wordsworth's discovery about metre the opportunity for practical
self-consciousness about the conventions of poetry, and about extra-
literary conventions. The most notable development in English poetry
in the nineteenth century, the dramatic monologue, occurs when the
force of Wordsworth's insight is extended to become a principle of
form. The fiction of such poems generally suggests that the imagined
speaker is 'simply speaking' while the poet arranges that his words
happen to fall into verse. In the dramatic monologue, it becomes a
structural principle that metre provides an intertexture of feeling not
strictly connected with the passion, for the consciousnesses of poet and
fictional speaker diverge and coincide in creatively many ways, rather
as diction and subject revolve around each other in the mock-heroic,
and with similarly inquiring effects: where does affectation or
insincerity begin and end? how can one adjust rival senses of value to
each other? might there not at times be more than one thing that needs

[155] *Biographia Literaria*, ii. 70.

to be said? The Wordsworthian poem of encounter with its tales of strange meeting becomes the continuous texture of his successors' work.

His effects continue to make themselves felt, as in Geoffrey Hill's painstaking comment on the actual circumstances of the poetic voice:

The presence of a certain number of social assumptions as to what the poetic voice really is, co-existing . . . with the real existence of a poetic voice, the nature of which is known to poets, certainly creates a situation of exacerbation, or irritation, of misapprehension which I would have thought a good poet could turn to creative ends. This is a way of saying that those old distinctions that you get in the Victorian and Romantic observers and auditors—the kinds of distinction that they are trying to draw all the time between a dramatic and a lyric voice—I think in our time inevitably merge. The lyric voice must exist, or must be heard, or has to be heard, or is constrained to be heard, in a dramatic context.[156]

In this respect, 'our time' is still Wordsworth's.

Nor is it only in the stanzaic poems such as 'Resolution and Independence' that an ill-assortment of eye and ear becomes the medium of that self-consciousness which Dorothy Wordsworth beautifully called 'a sort of restless watchfulness which I know not how to describe, a Tenderness that never sleeps'.[157] In his blank verse too, the half-consciousness of unsubstantial existence in the composition answers to the half-consciousness of the lack of substance in some of those who exist—the soldier in Book IV of *The Prelude*:

> He all the while was in demeanour calm,
> Concise in answer. Solemn and sublime
> He might have seemed, but that in all he said
> There was a strange half-absence, and a tone
> Of weakness and indifference, as of one
> Remembering the importance of his theme
> But feeling it no longer.[158]

or of Wordsworth himself, reading the Paris of the September massacres like a half-open book:

> I crossed—a black and empty area then—
> The square of the Carousel, few weeks back

[156] Transcript of interview, loc. cit.
[157] Letter to Jane Pollard, 16 Feb. 1793, in *Wordsworth Letters: Early Years*, i. 87.
[158] IV. 472–8 (1805 text); I quote from the edition of M. H. Abrams, S. Gill, and J. Wordsworth (1979).

Heaped up with dead and dying, upon these
And other sights looking as doth a man
Upon a volume whose contents he knows
Are memorable but from him locked up,
Being written in a tongue he cannot read,
So that he questions the mute leaves with pain,
And half upbraids their silence.[159]

Only 'half upbraids', though, because at times like these the muteness
of the leaves of a book, their faint tie to the spoken tongue, undertake a
decent reticence, an abstention from the fantasy that one can entirely
comprehend the dimensions of such an evil as is unflinchingly
witnessed here. He questions with pain the mute leaves because it is
natural to wish to understand such unnatural acts as the massacres (it
is not because they are *French* massacres that they are foreign to him,
but because, in their inhumanity, they estrange the human consciousness from itself), but the leaves may also be seen in the written line to
be mute with pain themselves, grief-stricken, self-abashed. This shock
to human nature makes the voice recoil, requires odd lineations which
cut across speech, as at 'looking as doth a man | Upon a volume'. For a
moment, you might wonder how else except 'as doth a man' a man
might look at anything till the simile is completed in the next line, but
the pause at the line-end, 'as doth a man', the tiny weirdness of a
possible pleonasm (more visible than audible), is there to give us pause
for the thought that some other men, those who massacred, must have
looked on these sights not 'as doth a man' but with a horrible capacity
for sightlessness.[160] The effect is comparable to the self-estrangement
implicit in the skewing of the lines about the soldier: 'in all he said |
There was a strange half-absence, and a tone | Of weakness and
indifference, as of one | Remembering . . .'. 'Tone' and 'one' rhyme to
the eye but not to the ear, the eye suggests a compact the ear cannot
fulfil, as the sight of the soldier is the sight of one individual but the
sound of him is doubled, a man half-absent from himself, the 'one |
Remembering' dismembered by the lineation.

Reading Wordsworth, we reflect on and from the limits of poetry's
incorporations of such speech as that of the Sailor's Mother, of the
leech-gatherer and the half-absent soldier, even of Wordsworth's own

[159] Ibid., x. 46–54 (1805 text).
[160] Christopher Ricks has a fine essay on Wordsworth's lineation, 'William
Wordsworth: "A pure organic pleasure from the lines"' (1971), repr. in *The Force of
Poetry*.

shocked voice, to the larger cultural incorporations of which literature is at once the better working and the best image. The labour of 'restless watchfulness' and 'Tenderness that never sleeps' Wordsworth sets forth in his poems is a standard and a guide for those who follow. It may also be their despair.

Keats:.

We heard on passing into Belfast through a most wretched suburb that most disgusting of all noises worse than the Bag pipe, the laugh of a Monkey, the chatter of women *solus* the scream of [a] Macaw—I mean the sound of the Shuttle—What a tremendous difficulty is the improvement of the condition of such people—I cannot conceive how a mind 'with child' of Philanthropy could gra[s]p at possibility—with me it is absolute despair.[161]

Surrounded as he is by this diversity of sounds, and the lives they imply, his imagination both meets the challenge of variety—for example, by the careful gradation of 'laugh', something possibly pleasant to hear, into 'chatter', something probably unpleasant to hear but usually fun to produce, into 'scream', a pain to receive or to produce, into the mechanically abstract 'sound'—and meets the limits of its own vivacity. What issue can be conceived from this jangle and babble? The poet's sense of the acoustic multiplicity of his world turns him from the single speech of 'Philanthropy' with its assurances of what is for the good.

Tennyson equally, though less exuberantly, responded to such fissures in the community of human speech when he failed to make a speech in acceptance of the freedom of the Burgh of Kirkwall, which he was granted, along with Gladstone, in 1883. He let Gladstone speak for him, and explained his silence to his son:

I am never the least shy before great men. Each of them has a personality for which he or she is responsible: but before a crowd, which consists of many personalities, of which I know nothing, I am infinitely shy. The great orator cares nothing about all this. I think of the good man, and the bad man, and the mad man, that may be among them, and can say nothing. *He* takes them all as one man. *He* sways them as one man.[162]

He had been Poet Laureate for thirty-three years, and still could not, and never did, speak in public. The intimacy of his sense of address precluded public speaking, amply and passionately as he wrote for his

---

[161] Letter of 9 July 1818, in Rollins, op. cit. i. 321.
[162] *Memoir*, ii. 280.

countrymen. The 'great orator', Gladstone, said this about him on the
occasion of his silence:

Mr Tennyson's life and labours correspond in point of time as nearly as
possible to my own, but Mr Tennyson's exertions have been on a higher plane
of human action than my own. He has worked in a higher field, and his work
will be more durable. We public men—who play a part which places us much
in view of our countrymen—we are subject to the danger of being momentarily
intoxicated by the kindness, the undue homage of kindness, we may receive. It
is our business to speak, but the words which we speak have wings and fly away
and disappear. The work of Mr Tennyson is of a higher order.[163]

It is magnificent of Gladstone to have said this, and to realize, in the
saying of it, the weakness within that strange power of speech
politicians have—the readiness for intoxication by the present
audience with the consequent ephemerality of what is said. This
moment at Kirkwall, the poet mute and the politician eloquent, is an
emblematic moment in the history of the decline of the bardic, it marks
an emancipation of political speech from the curacy of literary
sensibility, an emancipation which has proceeded, and proceeds,
apace.

The year after Tennyson kept silence at Kirkwall the franchise was
extended once more. (I am not suggesting that Tennyson was
responsible for this). The period from 1832 to 1928, the period in
which most of the poetry with which I am concerned in this book was
written, is one of persistent debate about questions of political
representation. Concurrently and congruously, there ran a more
oblique inquiry into the nature and fairness of artistic representation.
In a parliamentary democracy, political and poetic changes meet at the
level of voice, as John Sterling wrote in his review of Tennyson's 1842
volumes, 'Not merely is there a debate and seeming adjudication in
every country-town on all matters over the whole globe which any
tailor or brazier may choose to argue, but at last the tailor's and the
brazier's voice does really influence the course of human affairs.'[164]
Poetry delineates a political philosophy, responds to political life, in
what Eliot called 'the material of the formulation': '. . . what I mean by
a political philosophy is not merely even the conscious formulation of
the ideal aims of a people, but the substratum of collective tempera-

---

[163] Ibid.
[164] *Quarterly Review*, Sept. 1842, reprinted in J. D. Jump (ed.), *Tennyson: The Critical
Heritage* (1967), 105.

ment, ways of behaviour and unconscious values, which provides the
material of the formulation.'[165] Poetry works in the 'material of the
formulation'. The impress of that material on its work may be both
minutely detailed—in such matters as the relations between the
speech-habits of a community and a poet's versification—and at the
same time abstract, as Eliot's mention of 'unconscious values'
suggests.

That impress appears, for example, in the manner that Walter
Bagehot described in his 1864 essay on Wordsworth, Tennyson, and
Browning. Bagehot distinguishes between a 'pure' classical art and the
'ornate' or 'grotesque' art of Victorian poets. He attributes the grace
and power of the classical to its assured control of salient detail;
classical art marks out the periphery of incident, and finds the centre—
a representative character or action: 'The definition of *pure* literature is
that it describes the type in its simplicity; we mean, with the exact
amount of accessory circumstance which is necessary to bring it before
the mind in finished perfection. . .'.[166] The classical artist is doubly
skilled, then; he identifies the representative instance, and is further
able fully to represent the representative, to bring it 'before the mind in
finished perfection'. Bagehot observes, in an odd phrase, that such
pure work can be produced only by those who are 'under the restraint
of a sensitive talking world'[167]—this is odd because it makes the fluent
society of 'sensitive talking world' sound a bit like a strait-jacket
('under the restraint'), but there may be some truth in the oddity. The
relevance of the talking world to the purity of classical representation is
that it is this world which provides the artist with that vital
determination of the salient; in such a world, it is possible to know,
rather than to dispute, who counts and what matters. Bagehot's sense
of how a 'sensitive talking world' fosters some types of artistic
representation offers us a guide to the connections between political
and artistic representation, connections which exist in the very textures
of nineteenth-century verse, for as the talking world expands, and new
voices come slowly into literary earshot, the perception of saliency
itself is transformed, and the poet is tasked to make himself heard
anew.

---

[165] *The Idea of a Christian Society* (1939), 64.
[166] 'Wordsworth, Tennyson, and Browning; or, Pure, Ornate, and Grotesque Art in
English Poetry' (1864), in *The Collected Works of Walter Bagehot*, ed. Norman St-John
Stevas, 15 vols. (1965–86), ii. 333.
[167] Ibid. ii. 340.

All talk dies, and the communities of speech from which literary creation arises are in consequence only dimly to be guessed. Boswell lamented that, hugely as he recorded what Dr Johnson said, he could not record Johnson speaking; Keats requested the imagination of 'attitude' from his readers. When a readership grows more socially abstract, the jobs of recording and imagining become more urgent and more difficult to perform. Even between friends, as Tennyson felt when in 1836 Richard Monckton Milnes wrote to ask him for a contribution to an 'annual' he was compiling from the work of their circle of friends. Tennyson declined the invitation with what turned out to be offensive jocularity; in particular, he declared his opinion that 'annuals' were vapid. Milnes then accused Tennyson of having slighted the individual pieces rather than the type of collection for which they were intended, a slight that seriously wronged the authors, their mutual friends. In his turn to be nettled, Tennyson took further offence:

But are not annuals vapid? Or could I *possibly* mean that what you or Trench or Darley chose to write therein must be vapid? I thought you knew me better than even to insinuate these things. Had I spoken the same words to you, laughingly, in my chair, and with my own emphasis, you would have seen what they really meant, but coming to read them peradventure, in a fit of indigestion or with a slight matutinal headache . . . you subjected them to such misinterpretation as, if I had not sworn to be true friend to you, till my latest death-ruckle, would have gone far to make me indignant. . .[168]

He doesn't budge from what he wrote (very characteristic, that) but he supplies Keatsian details of 'attitude' to try and make Milnes see what his sense had been—'laughingly, in my chair, and with my own emphasis'. If a friend could read one's writing with such insensitivity to the emphasis it asked, what might one not expect from those who were not exactly one's friends?

Publishers, for instance. Yeats had a quarrel with his publisher, Bullen, who had asked him to write an article on Shakespeare. At that time, in the first decade of this century, Yeats was very keen on dictating his writing to a typist ('dictated 2000 words [of *The Speckled Bird*] in an hour and ten minutes yesterday . . . This dictation is really a discovery . . .'[169]), and he dictated his reply to Bullen:

---

[168] Letter of 8 or 9 Jan. 1837, repr. in *The Letters of Alfred Lord Tennyson*, ed. Cecil Y. Lang and Edgar F. Shannon, Jr., 3 vols. in progress, vol. i (1982), 148.
[169] Letter to Lady Gregory, 10 Apr. 1902, in Wade, op. cit. 370.

My dear Bullen,

I won't do the Shakespeare article for you. My mind is not on Shakespeare at present, and I might very possibly have to upset all my habits to get it there.[170]

Bullen was not surprisingly upset by this letter, though Yeats seems to have been surprised by that. A week after the first note, he is writing to his publisher again:

My dear Bullen,

Certainly I never meant those words 'I won't do the Shakespeare essay' to be rude. I wrote as if I was talking to you, forgetting that what is a friendly petulance when spoken becomes when the tone of voice isn't there a mere brusqueness. I dictated at great speed walking up and down the room and forgot that you weren't there listening to me.[171]

And from the public one might expect at least no better than from a publisher, though a poet's hopes may sensibly exceed his expectations. As Dickens puts it in *Dr Marigold*, his moving story about a man writing a book for his deaf and dumb adopted daughter, and conscious of the fact that he cannot fully convey himself in writing: 'I was aware that I could'nt do myself justice. A man can't write his eye (at least *I* don't know how to), nor yet can a man write his voice, nor the rate of his talk, nor the quickness of his action, nor his general spicy way.'[172] When the public is abstract, these indications become the more useful, but, on the other hand, the gestural and prosodic dimensions of meaning are also the marks of a social particularity which may make the job of mediating between the various groups in the audience all the harder. The deaf and dumb child would only be encumbered with a representation of voice and rate of talk because these elements of communication are foreign to her, and one's 'own emphasis', in Tennyson's phrase, might equally hinder reception from those who are not quite like you and put their stress elsewhere.

The dilemma is theoretically insoluble, which is why it proves so practically fertile. Though poets in this period record a wide gamut of dismay about the prosodic ambiguity of the printed word, the practice of some of them, as I shall show, does not support the belief that the

[170] 28 July 1905, in Wade, 456–7.
[171] Letter (also dictated) of 3 Aug. 1905, Wade, 457.
[172] (1865). I quote the text of 1871 as reprinted in *Christmas Stories* (Oxford, 1956), 452.

absence of vocal indication from the devices of writing is only a loss.
William Allingham reports that

[Tennyson] spoke a good deal about the want of some fixed standard of
English pronunciation, or even some fixed way of indicating a poet's intention
as to the pronunciation of his verses. 'It doesn't matter so much' he said 'in
poetry written for the intellect . . . but in mine it's necessary to know how to
sound it properly.'

I suggested that he might put on record a code for pronouncing his own
poetry, with symbolised examples, and he seemed to think this might be
done.[173]

Those last two clauses seem a bit flat, as if Tennyson's interest in the
project were despondent or sporadic. He may have recognized that,
while matters like pronunciation might be dealt with by a semi-
phonetic script, such as he adopted for his own Lincolnshire poems,
some vital communications of the voice were beyond writing. The
terms Allingham attributes to him are all awry, anyway; a 'fixed
standard of English' pronunciation would have ruled out that
Lincolnshire accent which he kept through his life, and would
certainly have been something draconianly more forceful than a 'fixed
way of indicating a poet's intention as to . . . pronunciation'; the
category 'poetry written for the intellect' (Tennyson gives Browning as
an example of what he means) is either too strict or too loose or,
probably, both.

As Hardy realized, semi-phonetic spelling deformed what it
attempted to represent:

In the printing of standard speech hardly any phonetic principle at all is
observed; and if a writer attempts to exhibit on paper the precise accents of a
rustic speaker he disturbs the proper balance of a true representation by
unduly insisting upon the grotesque element; thus directing attention to a
point of inferior interest, and diverting it from the speaker's meaning, which is
by far the chief concern where the aim is to depict the men and their natures
rather than their dialect forms.[174]

Semi-phonetic spelling does indeed make it look as if speakers with
regional accents mispronounce because it subordinates their speech to
the norms of correct orthography. All accents are partial accents,
whether of region or of class; there is such a thing as correct spelling

[173] A Dairy, ed. H. Allingham and D. Radford (1907), 344.
[174] 'Dialect in Novels', Athenaeum, 30 Nov. 1878, repr. in The Personal Writings of
Thomas Hardy, ed. H. Orel (1967), 91.

but there is no such thing as correct pronunciation. The phoneticists of the nineteenth century keenly recognized the political questions of class relation, or bearing of the metropolis on the regions of the country, which were implicit in the relating of writing to speech. The greatest of them, Henry Sweet, gives a typical account of the rise to dominance of a sectional dialect:

In real life, certain villages would be sure to gain some kind of ascendancy over [others], and thus one or more centres of dialectal influence would be established: till at last, if centralization were strong enough, one dialect would be used as a means of expression all over the territory, as is now the case in England. . . .
   Standard English, like Standard French, is now a class-dialect more than a local dialect: it is the language of the educated all over Great Britain.[175]

His enthusiasm for spelling reform in the direction of phoneticism has a clearly political tinge; phonetic spelling would give 'a genuine and adequate representation of the actual language, not, as is too often the case, of an imaginary language, spoken by imaginary "correct speakers"'.[176] As I pointed out before, though, Sweet believed that an accurate phonetic representation would not be legible, barely decipherable.[177] Those who opposed phonetic spelling reform often did so on even more clearly political grounds—William Archer, for instance, was much against 'that will-o'-the-wisp, phonetic spelling':

If you once make spelling subservient to the inevitable slovenliness of speech, you throw open the floodgates of degeneration. There must be an ideal language, not realised in the speech of any one district, or class, or even of any one man, which shall outlast, and, as it were, keep on absorbing and obliterating, all local and temporary divergences. It is only an ideal language that can resist degradation in the mouths of men. Once admit the right of any man, or set of men, to attempt the representation in literature of his or their actual utterance, and your language will presently lose its individuality in an endless ramification of dialects. In other words, dissolution will have set in.[178]

This is stern stuff—'floodgates of degeneration', 'dissolution'—and not unlike the rhetoric which accompanied each new extension of the franchise. Archer's 'ideal language' is made in the image of a

---

[175] In Henderson, op. cit. 9, 22.
[176] Ibid. 30. The most cogent account recently given of the case against phonetic spelling is in J. Vachek's *Written Language: General Problems and Problems of English* (The Hague, 1973).
[177] See above, p. 18.
[178] *Real Conversations* (1904), 91.

completely equitable polity in which no one 'set of men' commands
dominance. It is therefore extremely distinct from the actuality of
English language and society then as now, and though ideals are not
meant simply to map the realities amidst which they are imagined, they
can perhaps not be as remote as Archer's ideal without losing
coherence and point. It is also in itself a bizarre aspiration he
expresses. An equitable society is not one which absolutely declines to
meet the interests of any of its members, just as a satisfactory written
language is not one which is equally and flatly incapable of
representing any style of speech. It is reasonable to take a more
constructive view of equity, to hope for impartial response rather than
impartial lack of response.

Browning has a phrase for a language which truly responds to
human diversity of need and desire: 'the mediate word'.[179] Such a
word is also an ideal of conduct in the negotiation of competing or
conflicting demands in a society and its language. As such, we can
never guarantee that the ideal will be achieved, or that the possibility of
talking to each other across and through our differences might not one
day break down; it will even seem to some at times that the break-down
has already happened. Yet this 'mediate word' which goes between
people, trying to stave off mutual incomprehension between disparate
groups, trying indeed to compose their disparities in such a way that it
can give pleasure as well as take pains, still seems worth having and
worth hearing. For the Victorian poets, the work of reconciliation in
the 'mediate word' takes place largely in the printed voice, itself
mediate between speaking and writing, able to embody the shifting
relations between normative institutions, whether literary or more
broadly social, and the world they are held to represent.

Reading a Voice

William Empson was short with those who thought we could never
really tell what the intentions in an utterance were and so might as well
give up trying to guess them:

Any speaker, when a baby, wanted to understand what people meant, why
mum was cross for example, and had enough partial success to go on trying;
the effort is usually carried on into adult life, though not always into old age.

---

[179] *The Ring and the Book*, XII. 857, Altick, 628.

Success, it may be argued, is never complete. But it is nearer completeness in a successful piece of literature than in any other use of language.[180]

This is brisk, exasperated with doubters, and noble in its appreciative gratitude to works of literature and the pleasures of intimate communication they can offer us. It is also a welcome emphasis in the current state of criticism. But it is unfortunately not true, or not the whole truth. Many of the methods we have to judge why mum was cross are absent from our relation to works of literature: we can ask a person what the matter is, we can try experiments to see if this or that thing we do will abate or exacerbate the anger, we may share memories on the basis of which we make predictions. Versions of these methods exist in reading too but they are generally more oblique and chancy. This is hidden by the move in this passage from 'any speaker' to an account of where a reader stands and what a reader does. Empson's assurance that understanding is 'nearer completeness in a successful piece of literature than in any other use of language' ignores all those many cues of behaviour and context which assist interpretation of normal speech and which are mostly lacking when we read. (Such cues may not always be helpful, as the case of Iago's deceit shows, but they must more often be helpful than not, otherwise there would have been no point in Iago's abusing them.) The phrase 'successful piece of literature' shows that Empson's argument is circular, for he offers no criterion of literary success other than the implied ability to convey 'what people meant'. It seems that this is not a revealing observation about literature but only a stipulative definition of what is meant by 'successful'.

   Yet Empson's claim touches on something central in our response to literature. It indicates a pleasure we feel in reading, the pleasure of recognized design, something in which we can join and which marks a kinship between us and what we read. We laugh at a book but we more importantly laugh *with* it; we see that we were meant to find this bit funny, and we are pleased that we can and do laugh. As we recognize the appropriateness of our response, we take pleasure in that appropriateness itself; we are sharing a joke with a book, and with its author who may have been dead for years. The same is true when we are touched by writing, or savour its elegance, or take up its indignation. It is difficult to see why someone who never experienced any of these responses should care for literature at all except as, say,

    [180] Preface to *Using Biography* (1984), p. vii.

probably untrustworthy historical documentation or as a source for lexicography. And laughing over Flaubert runs continuously into laughing with a friend (it would be strange if someone saw all the jokes in literature but never understood a joke that somebody made in day-to-day relations). This continuity seems to link, as Empson wished to link, trying to understand why mum was cross and, for example, trying to understand why Flaubert was cross, though it does not justify his assimilation of the ways in which we come at those understandings, nor his confidence that literary mutuality is the more complete.

The young Proust was of one mind with Empson about the perfection of human intimacy in the act of reading, though where Empson characteristically emphasized the way in which literature develops from social relations, Proust ascribed its delights to the way it breaks from social ties:

Sans doute . . . l'amitié qui a égard aux individus, est une chose frivole, et la lecture est une amitié. Mais du moins c'est une amitié sincère, et le fait qu'elle s'address à un mort, à un absent, lui donne quelque chose de désintéressé, de presque touchant. . . . Dans la lecture, l'amitié est soudain ramenée à sa pureté première. Avec des livres, pas d'amabilité. Ces amis-là, si nous passons la soirée avec eux, c'est vraiment que nous en avons envie. Eux, du moins, nous ne les quittons souvent qu'à regret. Et quand nous les avons quittés, aucune de ces pensées qui gâtent l'amitié: Qu'ont-ils pensé de nous?—N'avons-nous pas manqué de tact?—Avons-nous plu?—et la peur d'être oublié pour tel autre. Toutes ces agitations de l'amitié expirent au seuil de cette amitié pure et calme qu'est la lecture. . . . L'atmosphère de cette pure amitié est le silence, plus pur que la parole. Car nous parlons pour les autres, mais nous nous taisons pour nous-mêmes. Aussi le silence ne porte pas, comme la parole, la trace de nos défauts, de nos grimaces. Il est pur, il est vraiment une atmosphère. Entre la pensée de l'auteur et la nôtre il n'interpose pas ces éléments irréductibles, réfractaires à la pensée, de nos égoïsmes différents. Le langage même du livre est pur (si le livre mérite ce nom), rendu transparent par la pensée de l'auteur qui en a retiré tout ce qui n'était pas elle-même jusqu'à le rendre son image fidèle; chaque phrase, au fond, ressemblant aux autres, car toutes sont dites par l'inflexion unique d'une personnalité; de là une sorte de continuité, que les rapports de la vie et ce qu'ils mêlent à la pensée d'éléments qui lui sont étrangers excluent et qui permet très vite de suivre la ligne même de la pensée de l'auteur, les traits de sa physionomie qui se reflètent dans ce calme miroir. Nous savons nous plaire tour à tour aux traits de chacun sans avoir besoin qu'ils soient admirables, car c'est un grand plaisir pour l'esprit de distinguer ces peintures profondes et d'aimer d'une amitié sans égoïsme, sans phrases, comme en soi-même.

[Of course, friendship with particular individuals is a bit frivolous, and reading is a form of friendship. But at least it is a genuine friendship, and the fact that it is felt for a dead person, for someone absent, makes it in a way disinterested, almost moving. . . . When we read, friendship is returned to its initial purity. There's no false sociability with books. If we spend an evening with them, it is because we really want to, and we say good-bye—to them at least—often with true regret. And when we have parted from them we feel none of those anxieties which spoil actual friendships: What did they think of us?—Were we tactless?—Did they like us?—along with the fear of being passed over for somebody else. All such worries of actual friendship die away on the threshold of the pure, calm friendship we experience in reading. . . . This pure friendship breathes the air of silence, a purer medium than speech. We speak for the sake of others, but we keep silence for ourselves. And silence, unlike speech, does not bear the mark of our failings, our antics. Silence is pure, it is a true medium. It does not interpose between the author's thinking and our own those irreducible elements of our diverse selfishnesses which are hostile to thought. The very language of a book is purified (if it really deserves to be called a book), made transparent by the author's thought which removes from the style everything which doesn't belong to his thought, so that the style can be a faithful image of what he thinks; each phrase basically resembles all the others for they are all uttered with the distinctive inflection of a self; a sort of cohesion arises from this resemblance, a cohesion which the dealings of ordinary life (along with the admixture of features alien to an individual's thought which such dealings always produce) prevent us from achieving. This cohesion enables us to follow the author's line of thought swiftly; we trace his physiognomy as it is reflected in the calm mirror of style. We find that we can take pleasure by turns in the aspects of various authors even though some of them are less than admirable, because it is so great a spiritual pleasure to discern the depth at which those aspects have been represented and to love with a selfless love, a love that does without phrase-making, as if an other person were not involved and we were looking into our selves.][181]

This clean air is beautifully described and exquisitely dreamed up. It tells a truth about the activity of reading—its seclusion, generosity, freedom from the twinges and flops of social relations, its opportunities for surrender without fear. These are altitudes on which few read and few write. The silence of the printed voice does indeed suspend the insistence of those 'irreducible elements of our diverse selfishnesses', our will to have our own way and our own emphasis, but it also imperils our selfhood in dispensing with those marks of selfishness. We 'loose the attitude', as Keats wrote, and, though attitude may not be

[181] 'Journées de lecture' from *Pastiches et mélanges* (1919), repr. in *Contre Sainte-Beuve*, ed. P. Clarac and Y. Sandre (Paris, 1971), 187.

'everything', as he also wrote, it is something, and an intrinsic constituent of what is meant in writing. There are two estimations of individuality at odds with each other in the passage, for Proust clears reading of all the 'éléments irréductibles . . . de nos égoïsmes différents' as sedulously as William Archer cleared his ideal language of individual interests. On the other hand, what Proust delights to hear in his reading is 'l' inflexion unique d'une personnalité'; he says little to explain why such an 'inflexion' cannot be yet another of the 'grimaces' of social exchange. The fact that the author is 'un mort . . . un absent' can hardly of itself prove disinterestedness and purity in the relation of writer and reader, for the obvious reason that all authors were alive when they were writing and so were susceptible to the temptation to flatter, themselves or others. From the reader's side, there can also unfortunately be 'amabilité' even towards a dead author—the 'classics' become commodities, the names of authors you have been reading can be dropped as carefully as the names of those you had the privilege to meet at a party. We can, in reading as in other friendships, be obtuse, perverse, or at a loss. The mature Proust, master-analyst of self-deception, would not have relied, as he does at the end of this extract, on being 'en soi-même' as a guarantee of probity.

It seems right to feel, as Proust does here, that reciprocity in imagination to some extent refines the frictions through which we get on with each other from day to day. The feeling is not only Proust's but part of a tradition in which 'impersonality' has assumed a central value. The tradition deserves respect and commands sympathy for many reasons, some of which I gave when discussing the art of cultural negotiation that might be practised in the 'mediate word' by a poet such as Browning. At the extreme pitch of Proust's myth in which reading frees itself from all the trouble of other human activities, it is, however attractive, not to be believed. There is nothing special about literature; we are no better off trying to understand books than when we try to understand other people. In either case, hearing comes through faith. Art has only the doubtful advantage of exemplifying with particular clarity our problems in mutuality and what we can make of them. It may indeed be that the many indications of meaning which we encounter in conversation, all the shy smiles or sudden breaks in the voice, make it difficult for us to understand, just as the comparative paucity of such helps in writing tasks a reader. Proust sentimentalizes the comparison with his sharp antithesis between a superficial and cloudy social world—in which what a person 'really thinks' is hidden amidst self-interest, polite indirection, or an automatic desire to be liked—and the sphere of literature where language is transparent and thought conveyed in its entirety. For, after all, literature too has its obliquities and decorum, and, more importantly, literary convention often has sense only in relation to non-literary practices. The preterite in French, for instance, of which Proust writes so brilliantly in his essay on Flaubert,[182] would scarcely have the character it does were it not a tense which is never heard in ordinary speech; Racine's syntax at times delights, delighted Proust, by being not something apart from social usage but 'la syntaxe vivante en France au XVIIe siècle.[183]

An art devoid of the marks of socialized selfhood, as Proust imagines it in *Journées de lecture*, would mean simply nothing, for meaning is always a derived form of intentionality, and there can be no intention where there is no agent, however attentively impartial that agent strives to be as he creates. This very passage is a case in point. To understand Proust here, to hear the 'inflexion' of this writing, we need to recall that the writer of these words was also the young author of *Les plaisirs et les jours*, a repentant socialite whose ideal of the literary vocation is expressed with such passionate rarefaction partly because he was only inconsistently devoted to it. A remorse for his own *mondanité* speaks through the vehemence with which he banishes the world and the world's business from the realm of imaginative pleasure.

The perfect reader in Proust's great novel does not live in an atmosphere of silence, though she does attune and purify her voice as she reads aloud. That reader is the narrator's mother, reading to him from *François le champi* on the fatal night in Combray with which the book begins:

. . . quand elle lisait la prose de George Sand, qui respire toujours cette bonté, cette distinction morale que maman avait appris de ma grand'mère à tenir pour supérieures à tout dans la vie, et que je ne devais lui apprendre que bien plus tard à ne pas tenir également pour supérieures à tout dans les livres, attentive à bannir de sa voix toute petitesse, toute affectation qui eût pu empêcher le flot puissant d'y être reçu, elle fournissait toute la tendresse naturelle, tout l'ample douceur qu'elles réclamaient à ces phrases qui semblaient écrites pour sa voix et qui pour ainsi dire tenaient tout entières dans le registre de sa sensibilité. Elle retrouvait pour les attaquer dans le ton qu'il faut, l'accent cordial qui leur préexiste et les dicta, mais qui les mots n'indiquent pas; grâce à lui elle amortissait au passage toute crudité dans les temps des verbes, donnait à

---

[182] 'À propos du "style" de Flaubert' (1920), in Clarac and Sandre, op. cit. 586–600.
[183] 'Journées de lecture', ibid. 192.

l'imparfait et au passé défini la douceur qu'il y a dans la bonté, la mélancolie qu'il y a dans la tendresse, dirigeait la phrase qui finissait vers celle qui allait commencer, tantôt pressant, tantôt ralentissant la marche des syllabes pour les faire entrer, quoique leurs quantités fussent différentes, dans un rythme uniforme, elle insufflait à cette prose si commune une sorte de vie sentimentale et continue.

[. . . when she read George Sand's prose, a prose which constantly breathes that kindheartedness and moral distinction which my mother had been taught by my grandmother to consider superior to all other qualities in life, and which I was only much later to show her should not also be considered superior to all other qualities in books, she took care to remove from her voice any pettiness, any affectation which might have hindered the powerful current of the prose from carrying into her voice, she provided all the natural tenderness and broad sweetness which the phrases needed so that they seemed to have been written for her voice and to lie, so to speak, exactly within the range of her temperament. To give the phrases their requisite tone, she rediscovered the accent of the heart, an accent which pre-existed them and dictated them, but which the words do not notate; having found the accent, she was able to soften any roughness in the tenses of the passage, she gave the imperfect and the preterite the sweetness of a kind heart and the melancholy of a tender heart, she steered a phrase which was coming to its end on towards the phrase which was about to begin, sometimes quickening, sometimes halting the pace of the syllables so that, however various their quantities might be, they all joined in a uniform rhythm, and she breathed into this very ordinary prose a sort of continuous life of feeling.][184]

This reading does not remain 'en soi-même'. On the contrary, it has the mark of 'bonté' in being so devoted to others in its apprehension of George Sand's 'accent cordial' and in its care for the troubled child. Though in this respect distinct from the account given in *Journées de lecture*, the two passages are connected by their concern with 'une sorte de continuité' and 'une sorte de vie sentimentale et continue'. Proust's terms in *Du côté de chez Swann* for this continuity are both musical and ethical; the narrator's mother provides 'tendresse naturelle' and 'ample douceur' and also regulates tempo, 'phrases' the passage as we say a violinist 'phrases' a melodic line. She does both things at once because they are in fact at one. For what the mother does is return a voice to George Sand's pages, she gives them back the contours of a characteristic speech—intonation ('le ton qu'il faut'), supra-segmental cohesion and apt collocation ('dirigeait la phrase') in the terms of

---

[184] *Du côté de chez Swann* (1913), repr. from the edition of 1919 in *À la recherche du temps perdu*, 3 vols. ed. P. Clarac and A. Ferré (Paris, 1954), i. 42.

prosodic linguistics. In doing so, she revives George Sand's agency in the writing, that 'accent cordial' which is the unnotated, implicit sound of creative purpose in literary address. 'Revives' is not a metaphor for Proust. His great novel turns on the matter of imaginative resurrection, and this early passage occupies a key position in the whole work through being an instance of such resurrection, as we see when we note the care with which Proust begins his description of the mother's reading by one verb of breath ('respire') and concludes the paragraph with a second ('insufflait . . . une sorte de vie'). The significance of the mother's act for the whole novel appears in Proust's own phrasing of her supreme achievement as a reader: 'Elle *retrouvait* . . . l'accent cordial' (my emphasis)—'retrouver' is the task of the entire work, completed in *Le temps retrouvé*. As Empson said, 'Much of Proust reads like the work of a superb appreciative critic.'[185] It does so because Proust, like a literary critic, is in search of lost accents.

I omitted two words at the beginning of this extract, the words 'De même' ('So too' or 'In the same way'). The description of reading is governed by a comparison of this act of literary response with the mother's social delicacy in her dealings with other people. Proust stresses the continuity between the mother's practical generosity and her virtues as a reader by a syntactic parallel of the two halves of his comparison: 'Même dans la vie . . . elle écartait de sa voix . . .' / 'De même . . . attentive à bannir de sa voix. . .'. To read aloud as the mother does is an act of that apprehensive kindness, of that imaginative anticipation of need, which, under the name of 'bonté', is for him the supreme ethical virtue. The 'accent cordial' gives a better name to what *Journées de lecture* called 'la trace de nos défauts'. Finding that accent is, in the fullest sense, a rediscovery of the illocutionary force of writing. Words do not notate such an accent ('. . . que les mots n'indiquent pas . . .') but they may contain it and do call for it to be rediscovered on the page.

The narrator himself grows to resemble his mother in his appreciation of an author's style as an aspect of character and response to that style as a form of conduct. The 'accent cordial' is heard again in *À l'ombre des jeunes filles en fleurs* when Proust describes the tie between the speech-habits of the novelist, Bergotte, and some features of his prose. Bergotte had a fondness for certain words which he pronounced with special force ('visage', for example); this vocal emphasis

---

[185] *Seven Types of Ambiguity* (1930, 2nd edn., 1947, repr. Harmondsworth, 1961), 287.

corresponded to the syntactical highlighting which he gave to the same
words in his writing and to the requirement thereby created that a
reader should give the words an equivalent prominence if he wished to
'phrase' a passage of Bergotte correctly. The narrator generalizes the
instance:

A cet égard, il y avait plus d'intonations, plus d'accent, dans ses livres que dans
ses propos: accent indépendant de la beauté du style, que l'auteur lui-même
n'a pas perçu sans doute, car il n'est pas séparable de sa personnalité la plus
intime. C'est cet accent qui, aux moments où dans ses livres Bergotte était
entièrement naturel, rythmait les mots souvent alors fort insignifiants qu'il
écrivait. Cet accent n'est pas noté dans le texte, rien ne l'y indique et pourtant
il s'ajoute de lui-même aux phrases, on ne peut pas les dire autrement, il est ce
qu'il y avait de plus éphémère et pourtant de plus profond chez l'écrivain, et
c'est cela qui portera témoignage sur sa nature, qui dira si, malgré toutes les
duretés qu'il a exprimées, il était doux, malgré toutes les sensualités,
sentimental.

[In this sense, there was more intonation, more accent, in his books than in his
conversation: an accent distinct from the beauty of the style, and which the
author himself doubtless had not noticed, for it is inseparable from his most
intimate self. At those moments in his books when Bergotte wrote most
naturally, it was this accent which gave the words their rhythm, lent an
emphasis to the words he wrote which often did not otherwise mean much.
There is no notation for this accent in the text, nothing points it out, and yet it
is joined of its own accord to each phrase, each phrase has to be said this way
rather than that; and this accent is, at the same time, that which is most
ephemeral and that which is most profound in a writer's work, the accent
which bears witness to his moral nature and tells us whether he, after all the
harsh things he has expressed, remains tender, after all the sensual
expressions, remains gentle at heart.][186]

Even as Proust writes of Bergotte's accent, his own accent transpires
through the prose, in the way, for example, that words such as
'indiquer' and 'sentimental' appear in both the *Du côté* . . . and *À
l'ombre* . . . passages, 'sentimental' being syntactically highlighted on
both occasions by coming at the end of a paragraph. More Proustian
still is the cadential emphasis given here to one of his favourite words,
'doux', caught up in the lovely, characteristically mollifying chime of
'toutes' and 'duretés' with 'était doux'. A reader of Proust is now not in
the position of the narrator with regard to Bergotte for he cannot have
an acquaintance with Proust's speech-habits such as the narrator

---

[186] *À l'ombre des jeunes filles en fleurs* (1918), in Clarac and Ferré, op. cit. i. 553.

enjoyed with Bergotte's. Yet a reader of Proust can come to recognize the author's favourites—'doux', 'bonté', 'lilas', 'revenir', insistent caressing pressure on constructions with 'si', chains of three adjectives, and so on. A judgement that these features are characteristic of Proust cannot, of course, be merely statistical; it is not only a matter of how often words or cadences recur but of their prominence in the text. This prominence results as much from the constellation of such aspects into a pattern of significance within an *œuvre* as from the frequency with which they appear. Though we usually lack anything like the narrator's knowledge of Bergotte's physical voice, we can elicit accent from writing and must do so. We can do so by relating artistic compositions to things other than their creators' speech-habits, things which are none the less materials for the instilling of a voice into print, such things as drafts, letters, anecdotes, surrounding and related writings. In what follows, I spend a lot of time on such contextualization in order to elicit as plausibly as I can the accent of the writers studied, and in order to show what sort of witness those accents bear. The lack of much knowledge about a writer's speech-habits does not defeat the search for accent because this literary critical activity is not trying to provide the kind of data which would enable an actor to do an accurate impersonation of a writer, but rather to bring us to understanding and pleasure in a printed voice which remains in print even as we hear it.

It may be asked why I believe that a reader *must* elicit accent. Proust's claim for accent as a witness to the 'nature' of a writer is extreme, and I share it. But perhaps we should not wish to know of a writer whether 'malgré toutes les duretés qu'il a exprimées, il était doux'. Wishing to know such things would be regarded by some aestheticians as showing a biographical curiosity about artists irrelevant to the judgement of artistic work. My own dissent from Empson's assimilation of the way in which we guess speaker-meaning to the interpretation of the illocutionary force of works of literature might look as if it aligned me with proponents of 'impersonality' or with those who celebrate the reader's role as that of dancing on the graves of dead authors. Yet the arguments I gave earlier for the claim that agency, and a conception of the agent, are indispensable from any interpretative activity themselves compel an attention to accent, if accent indeed is a primary witness to the character of the agent who writes. That Empson moves too quickly from wondering why mum was cross to understanding literature does not imply that he was moving in the wrong direction. And though the vocal ambiguities I discuss may add a further type of ambiguity to Empson's list, the addition is not made with the intent of making the possibility of recognizing the character of a writer's design ever more remote. [Ambiguity, in fact, requires the concept of intention, for to detect an ambiguity is to ascribe two or more possible intentions.]Anything less than this amounts only to the empty observation that no sign or set of signs can be its own unequivocal self-definition. 'The niggler is routed here; one has honestly to consider what seems important.'[187] If reading is to be a *response* to a work, then 'what seems important' to a reader must be influenced by what was of moment to the writer of that work. Accent is the sound of what was of moment in writing.

The greatest triumph of understanding art in *À la recherche du temps perdu* comes when Mlle Vinteuil's girl-friend constructs the Vinteuil septet from the drafts the composer left on his death. She is able to evoke this music because she hears the accent of Vinteuil in his previous work and its relation to his unfinished masterpiece. Proust's narrative at this point is dense with meditative cross-reference. The composer Vinteuil's life had been poisoned by his distress over his daughter's lesbian affair; the narrator's glimpse of Mlle Vinteuil and her girl-friend desecrating a photograph of the dead composer is his first major introduction to the 'cruauté' which is the supreme evil in the novel. Yet it is this girl who effects a resurrection that completes the reviving of dead artists which we first encountered in the mother's reading of George Sand. (Like the narrator's mother, Mlle Vinteuil's girl-friend is not named in *À la recherche*.) The exemplar of 'bonté', the mother, and a central figure of 'cruauté', the lesbian lover, meet in this activity, in their ability to recognize accent and, by lending it their voices, to let it sound again.

The narrator makes an astonishing claim for accent at this point in *La prisonnière*:

... c'est bien un accent unique auquel s'élèvent, auquel reviennent malgré eux ces grands chanteurs que sont les musiciens originaux, et qui est une preuve de l'existence irréductiblement individuelle de l'âme.

[. . . for indeed it is to a unique accent which those master-singers, truly original musicians, rise, a unique accent to which they return, and this accent is a proof of the irreducibly individual existence of the soul.][188]

Such a return proves at least the irreducibly individual existence of the

[187] *Some Versions of Pastoral* (1935), 89.
[188] *La prisonnière* (1923), text of Clarac and Ferré, iii. 256.

voice. It is a return which Proust himself, whether he knew it or not, made in this passage, for 'l'existence *irréductiblement* individuelle de l'âme' recalls, while it transfigures, the different view of the self (and whether it is eradicated in art) which Proust had taken in *Journées de lecture* when he deplored 'ces éléments *irréductibles* . . . de nos égoïsmes différents' (my emphases). Proust's writing is of a piece with his self; his inconsistencies are a part of its texture, its consistency. He called his divided soul wholly into writing. It was because he was divided that he had to write, and because he wrote with such integrity he continues to be recognizable in all his divisions. He is not alone in this. So true it is that he who loses his voice shall find it.

2

TENNYSON'S BREATH

Tennyson's Two Voices

TENNYSON on occasion needed not to speak out in order to be the voice of his time. There is an unfinished poem called 'Reticence', which was appropriately not published until after his death, and which begins:

> Not to Silence would I build
> A temple in her naked field;
> Not to her would raise a shrine:
> She no goddess is of mine;
> But to one of finer sense,
> Her half-sister, Reticence.

> Latest of her worshippers,
> I would shrine her in my verse!
> Not like Silence shall she stand,
> Finger-lipt, but with right hand
> Moving toward her lips, and there
> Hovering, thoughtful, poised in air.[1]

Many of his poems have as their special grace this quality of 'Hovering, thoughtful, poised in air', a suspense in communication which fascinated the poet, as when he once stood by a telegraph pole to listen, as he said, 'to the wail of the wires, the souls of dead messages'.[2] Indeed, this slight poem touches for a moment on a source of its creator's virtues, in particular, on his impassioned expertise with the possible messages of dead souls. Where it does so, Tennyson's rhythmic touch is most secure and delicate, altering the resolute trochaics to the irregular

---

[1] First published in *Memoir*, ii. 87–8, probably composed in 1869, reprinted in R. All references to Tennyson's poems are to the text of this edition, given in the form R and a volume and page number. Thus, for this poem: 'Reticence (1869), R iii. 628–9.
[2] *Memoir*, ii. 325.

/ × / × × / · /
Finger-lipt, but with right hand
/ × × × / × /
Moving toward her lips, and there

so that the return of the trochaics at 'Hovering, thoughtful . . .' has a
new poise in its motion given by the small holding of the breath at the
line-end, at 'there', a delay of the impetus of 'Hovering' which makes a
rhythmic reticence. The popular imagination remembers him, as the
popular imagination then received him, mostly as a magniloquent poet,
but he was a shy writer as well as a shy man. R. H. Hutton, whom
Tennyson considered one of his best critics, put the case more fully:
'No poet ever made the dumb speak so effectually', he wrote, but
needed to add that few poets had such a feeling for the 'helplessness
with which the deeper emotions break against the hard and rigid
element of human speech'.[3] This double aptitude—for vocal skill and
for artistic realization of the breaking-points of voice—helped him
imagine throughout his career the drama of speaking, the times of
successful utterance and those other occasions on which words fail us,
or we fail them.

Not all Tennyson's contemporaries managed Hutton's balanced
acceptance of such rhythmic half-measures of revelation. F. W. H.
Myers, the Spiritualist pioneer, had low hopes for the future of poetry
when he read late Tennyson:

It seems sometimes as though poetry, which has always been half art, half
prophecy, must needs abandon her higher mission; must turn only to the
bedecking of things that shall wither and the embalming of things that shall
decay. She will speak, as in the *Earthly Paradise*, to listeners

> laid upon a flowery slope
> 'Twixt inaccessible cliffs and unsailed sea;

and behind all her utterance there will be an awful reticence, an unforgotten
image of the end.[4]

He evidently wanted more, but more of what? 'It seems sometimes . . .'
cannot make up its mind, tonally, between the expression of a casual
mood or of a visionary instant. These tonal indecisions often occur in
Tennyson's own work, but he turns them to an inquiring shape
whereas here Myers just hedges the question of what he wants.

---

[3] 'Tennyson', in his *Literary Essays* (1871, 3rd, enlarged edn., 1888) [hereafter
referred to as Hutton, *Essays*], 372–3.
[4] 'Tennyson as Prophet', *Nineteenth Century* (Mar. 1889), in *Tennyson: The Critical
Heritage*, ed. J. D. Jump (1967) [hereafter referred to as Jump], 412.

Prophecy is clearly poetry's better half, as far as he is concerned, but he also thrills to the idea of a silenced prophetic voice ('*awful reticence*'—my emphasis), and he even draws his words for the 'unforgotten image of the end' from a poetic drama which partly concerns itself with the seductions of prophecy—'Is this the promis'd end? I Or image of that horror.'[5] Myers's 'disappointed ear',[6] disappointed of full-blooded revelations from poetry, can be guessed at plainly enough from the disarray of his style. Both the disappointment, and the failure of his words quite to grasp it, are representative of the demanding hopes for and against which Tennyson had to write. He wrote for such hopes because they were his hopes too—the hope that there would be an afterlife, the hope that an ideal domesticity was possible, the hope that the poet's calling was to serious human responsibilities—and he wrote against them because, as they were relayed back to him from the reading public, they became corroded parodies of what he held dear, requests for edification that could seem at times no more than a clamour for pap, as if he was being asked to write testimonials for, say, Christianity, as he might have been asked to commend a hair-tonic: 'The Poet Laureate writes, "I have tried it and found it works".'

William Allingham remembered his vexation:

T[ennyson]. '. . . I did lately receive a prose book, *Critical Strictures on Great Authors*, "a first hastily scribbled effusion", the writer said. There was this in it, "We exhort Tennyson to abandon the weeping willow with its fragile and earthward-tending twigs, and adopt the poplar, with its one Heaven-pointing finger." ' 'A pop'lar poet,' says I. . . . I went out to the garden, where were Mrs Tennyson with Mrs Patmore and her sister. Returning to the house there was tea, to which Tennyson came in, muttering as he entered the room 'we exhort Tennyson'.—I smiled.[7]

Allingham shows amiable good temper in his bad joke about 'pop'lar' poets and in his smile, but Tennyson shows the temperament of his form of genius in his tenacious grumpiness. He was not one to let things go lightly. He held on to his own past and its roots in the despondent soil of Somersby; he did not want either a popularity or a Heaven out of touch with that dark earth. The two chief subjects of

---

[5] *King Lear*, v. iii. 263–4; Folio, ll. 3224–5.

[6] The phrase is from W. J. Fox's review of *Poems, Chiefly Lyrical*, *Westminster Review* (Jan. 1831), Jump, 32.

[7] *William Allingham: A Diary*, ed. H. Allingham and D. Radford (1907, repr., 1967) [hereafter referred to as Allingham], 62.

his work—the morbid persistences of the past, and the hope for personal immortality—both lead him to refuse to abandon anything, from the past or for the future.

Where Myers feared an awful reticence *behind* all the utterance of poetry, Tennyson intends to celebrate reticence within the words themselves ('I would shrine her *in* my verse!'—my emphasis). We can begin to see the workings of that reticence in an early masterpiece, 'To J.S.':

> I will not say, 'God's ordinance
>   Of Death is blown in every wind;'
> For that is not a common chance
>   That takes away a noble mind.
>
> His memory long will live alone
>   In all our hearts, as mournful light
> That broods above the fallen sun,
>   And dwells in heaven half the night.
>
> Vain solace! Memory standing near
>   Cast down her eyes, and in her throat
> Her voice seemed distant, and a tear
>   Dropt on the letters as I wrote.[8]

'I will not say' is not only an instance of the rhetorical figure of 'apophasis', 'A kind of irony, a denial or refusal to speak . . . when nevertheless we speak and tell all',[9] for Tennyson does not nevertheless speak and tell all here, but writes to Spedding rather than speaking to him. This poem exists primarily in its written form, any vocalization would create problems of tone which the silent writing indicates but does not suffer from.

Consider how, reading this poem aloud, you would distinguish ' "God's ordinance | Of Death is blown in every wind;" ' from the other lines I have quoted. If you are in fact saying all the lines, then these words have to be given a distinct tone to show that they are specially not-said by the poet though many others say such things. Such a tone would tend to emphasize the proverbial quality of these words, make them perhaps a merely formal condolence which conceals behind social propriety something less than kindness, as does Gertrude's 'Thou know'st 'tis common, all that lives must dye, | Passing through

    [8] (1832), R. i. 506.
    [9] See L. A. Sonnino, *A Handbook to Sixteenth-Century Rhetoric* (1968), p. 131, quoting Abraham Fraunce, *The Arcadian Rhetorike* (1588).

Nature, to Eternity.'[10] When Gertrude says that, Hamlet is on stage to rebuke her; in Tennyson's lyric, the drama is within a single person. To give 'God's ordinance . . .' a 'quoting' inflection would tip the balance in weighing up both the help and the insufficiency that may lie in proverbs at such times, as it would also tip the balance if a reader signalled the allusion to Gertrude's ''tis common' by an emphasis of indignant rebuttal—'that is NOT a common chance'. (Hamlet is diffusely present in these lines, 'noble mind' coming from Ophelia's 'O what a Noble minde is heere o're-thrown?') An utterance of the words need not go in this direction, this is only one obvious pitfall of tone, but the poem, being written, and therefore an utterance of absence as well as an utterance about absence, helps us see what speech has to manage at times of bereavement, helps us listen for a true tone of consolation.

Arthur Hallam had felt that letters let us down when we seek intimacy of converse. He wrote to Emily Tennyson:

. . . how wretched it is to be thrown back into the region of letters after treading the giddy heights of existence, in which your dear presence & converse had placed me. Oh it is sad to think how little a letter gives one! Yours today is all precious sweetness; yet it tells but a few moments of your life, a few thoughts of your mind, and it contains no looks, no tones—that is the great, deplorable, alas irremediable loss . . .'[11]

'No looks, no tones—': the body and the voice are withdrawn from writing. This withdrawal is like death, but, being a model of death, it may then enable us to contemplate actual deaths without being led by our own pain and bewilderment to wish for rapid certainties about what has happened to the dead person. If 'To J.S.' were a treatise, it might have a responsibility to argue that people do or do not survive their deaths, but in the poem what is at issue is not the correctness of a general conclusion but a matter of conduct in the face of particular loss. No metaphysical certitude of itself guarantees tactful and apprehensive behaviour, but Tennyson tries to behave just in this way as he writes. Hallam thought the loss of tone an 'alas irremediable loss'; Tennyson turns the loss to a means of humane delicacy, for the voice in the poem is not withdrawn from writing, but into it, preserved by the lines in sight of resuscitation, poised in air. Tennyson's imagination moves from 'I will not say' to 'I wrote', drawing attention to

[10] Hamlet, I. ii. 73–4; Folio, ll. 252–3.
[11] Letter of 7 Apr. 1832, in The Letters of Arthur Henry Hallam, ed. J. Kolb (Columbus, Ohio, 1981) [hereafter referred to as Kolb], 546.

the physical existence of the printed words, onto which a tear can, after that precisely timed delay at the line-end (the brim of the eye), drop. Out of the death of voice, a new body of significance can be made to arise, a loss perpetuated while a grief is consoled. The written words keep James Spedding company in his loss, are close to him by taking into themselves a condition of loss, a Wordsworthian half-absence.

'To J.S.' ends by addressing not James Spedding but his dead brother, Edward:

> Sleep till the end, true soul and sweet.
> Nothing comes to thee new or strange.
> Sleep full of rest from head to feet;
> Lie still, dry dust, secure of change.[12]

Christopher Ricks brings out the intricacy of thought in this lingering melody:

The end of 'To J.S.' is all sleep and no waking. 'Sleep till the end' which may be Judgement Day, or the words may simply mean 'till the end'. 'Nothing comes to thee new or strange'—nothing? 'Lie still, dry dust, secure of change': *still* is delicately dual, and how are we to take 'secure of change'? Does secure mean 'assured of'? In which case the after-life does edge into the poem's conclusion. And yet the whole feeling of these last lines is set against change, is absolute for death. Moreover, the last two lines would seem to recall the innocence in Milton: 'asleep secure of harme' (*Paradise Lost*, IV, 791); and in that case 'secure of' must mean 'secure against'. Sleep, not heaven, would stand as the final solace. For all its mildness, the end of 'To J.S.' is forcefully perplexing. It . . . is a poem of which the ending may not be the ultimate ending, since we cannot speak of sleeping without wondering about waking.[13]

The comment is completely in tune with the lines, especially so in the contending impulses of assurance and tentativeness which speak through the critic's prose as they sing in the verse. The poem closes with 'all sleep and no waking' and 'the whole feeling of these last lines is set against change'. On the other hand, 'we cannot speak of sleeping without wondering about waking' (and the 'we' there must include at least this poet and this critic); the possible pun on 'secure of' to which Professor Ricks alerts us implies that 'the after-life does edge into the poem's conclusion'. 'Edge into' is grudgingly concessive where it should be welcoming. For the stanza, unlike all others but one in the poem, is divided by a full stop at its mid-point. At that stressed mid-point, Tennyson changes the sense of the lines though he does not

[12] R i. 507.   [13] See his *Tennyson* (1972), 98.

change their referent. Edward Spedding is both 'true soul and sweet' and 'dry dust'. These last lines are indeed 'forcefully perplexing' but the force of their perplexity is that of a religious mystery as well as of a personal quandary. As a 'true soul and sweet', Edward Spedding sleeps the sleep of the just until Judgement Day; nothing occurs to the soul in that sleep (it is not, for example, sentient of the strange processes of physical decomposition). As an embodied individual whose body awaits its resurrection, he is 'secure of change' in that whatever changes occur to the matter of his body, it will eventually be recomposed in its integrity—it will be his body still even when it is the body of glory—and he is also 'secure of change' in that, though his body is now become 'dry dust', this is not a dust to which he has finally returned but a dust eventually to be raised and re-united with the 'true soul and sweet'. The stanza is divided as body and soul are for the time divided, but it holds in view, indeed it displays, that resurrection when what has been put asunder will be joined anew. Professor Ricks exactly recognizes what the lines say under their breath about the difficulty of conceiving and desiring what is hoped for in the Christian doctrine of the resurrection of the body. They also speak that doctrine out, plainly and formally. Doing both at once, they are not underhand but open to the stress of living with such a doctrine of the afterlife. It is this unity of faith and tremor which the lines effect in their union of mellifluousness and extreme semantic complexity, something which makes them 'forcefully perplexing' for all their 'mildness', which indeed makes the mildness what is most perplexing about them.

It is possible to say the lines so that the orthodox hope is more evident, and equally possible to voice them so that staying unchanged and not being resurrected is the body's hope, but you cannot have them, vocally, both ways at once. The divergent attitudes would make 'a contradiction on the tongue';[14] the eye can still see the conjunction of meanings in the written words, turn the reciprocal hostility of the opposed senses of hope from bafflement into an authentic richness, a felicitous truth to the complication of our desires for beyond the grave. That is gently intimate, too, with the state of bereavement. It does not belittle loss by making it only temporary (it is not a 'matter of time' before the resurrection of the body at the Last Judgement; 'the end' which the lines mention is the end of time), nor does it bluntly insist that loss is absolute. The whole poem has its attention fixed on James

---

[14] *In Memoriam*, CXXV (1850), R ii. 445.

Spedding, though it turns to Edward at the close. Enclosing address to the dead within a letter to the survivor, Tennyson, for the space of the composition, unites all three of them, and not only in memory or imagination.

Memory is abashed in 'To J.S.', and her self-consciousness comes out when she almost loses her voice: 'Memory standing near | Cast down her eyes, and in her throat | Her voice seemed distant'. This self-conscious vocal thwarting is characteristic of Tennyson. In *The Princess*, Lady Psyche, in the all-female precincts of the University, recognizes her brother through his disguise:

> . . . glowing full-faced welcome, she
> Began to address us, and was moving on
> In gratulation, till as when a boat
> Tacks, and the slackened sail flaps, all her voice
> Faltering and fluttering in her throat, she cried
> 'My brother!'[15]

'Full-faced' shows Psyche at ease with her own social competence (the feeling is that because everything fits for her in her environment she too fits herself snugly, 'full of it' and so 'full of herself'). When she sees her world's rules infringed, she shrinks back, the wind taken out of her sails and out of her voice too, as the rhythmic skewing of the verse away from regular iambics into 'till as when a boat | Tacks, and the slackened sail flaps, all her voice | Faltering and fluttering in her throat' makes an effort to convey both by the jolts of rhythm and by the introduction of internal rhyme and assonance into blank verse ('Tacks'/'slackened'; 'till'/'sail'/'all'/'Faltering'). This meticulous effect in the verse presents a hindrance as much as an opportunity for a reader's delivery. You have to find a way with your voice to preserve the decorum of the verse even while Tennyson skilfully disturbs its poise, just as the cry 'My brother!' has to be made both to break the line in which it occurs and to be contained by that line—you cannot simply cry it aloud, any more than Psyche can, without drawing unwanted attention to it. The rules require, in the narrative situation as in the verse, that the emotional impetus be battened down for the sake of good form. Here Tennyson shows one of the many dramatic opportunities implicit in Wordsworth's understanding of metre— regular iambics can stand, for a moment, as the embodiment of a social system within which the individual and her voice toil.

15 *The Princess* (1847), R ii. 211.

The writing is reticent in evoking a cry without voicing it; we attend in these lines to the desire, and to the dread, of speech, as we do in 'Guinevere':

> Then she stretched out her arms and cried aloud
> 'Oh Arthur!' there her voice brake suddenly,
> Then—as a stream that spouting from a cliff
> Fails in mid air, but gathering at the base
> Re-makes itself, and flashes down the vale—
> Went on in passionate utterance:
>   'Gone—my lord!
> Gone through my sin to slay and to be slain!
> And he forgave me, and I could not speak.
> Farewell? I should have answered his farewell.
> His mercy choked me. . . .'[16]

The epic simile ('as a stream . . . | . . . flashes down the vale'), an old token of rhetorical mastery, gains fresh weight in this interim of the constricted larynx. Tennyson's skill interposes itself in the gaps of Guinevere's utterance, demonstrating a poet's power with the language but making that power instinct with a moral imagination about the speechless needs to which only reticence answers. The lack of strict connection between his composing mind and her passion makes the movement of the verse respond at every point to the thought of what it is for her to speak in these circumstances, as the rhythms of *The Princess* answer to the situation of Psyche's speech. Such imagination stops Tennyson from 'spouting' himself in these lines because his skill inhabits her broken voice; he finds his way eloquently to the dark truth of a sinner's mute resentment of forgiveness, 'His mercy choked me'.

Arthur Hallam said Tennyson had an 'ear of fairy fineness'.[17] The fineness is particularly that which listens in to the interplay of sound and silence, with dramatic feeling for the charged needs in utterance. Across his career, Tennyson finds words for muffled sounds, creates on the page the vestiges of speech, 'the sound of a voice that is still' as 'Break, break, break' calls it.[18] In early poems we catch 'a stifled moan', 'an ancient melody | Of an inward agony' or 'the phantom of a silent

---

[16] (1859), R iii. 545.

[17] 'On some of the characteristics of Modern Poetry, and of the Lyrical Poems of Alfred Tennyson' (1831), *The Writings of A. H. Hallam*, ed. T. H. Vail Motter (New York, 1943) [hereafter referred to as Motter], 191.

[18] (1842), R ii. 24.

song';[19] Gareth tries to speak in the *Idylls of the King* but '(his voice was all ashamed)'; Geraint bites back his words, 'And there he broke the sentence in his heart | Abruptly, as a man upon his tongue | May break it . . .'; and Merlin's elocution leaves something to be desired: 'He spoke in words part heard, in whispers part, | Half-suffocated in the hoary fell | And many-wintered fleece of throat and chin.'[20]

Tennysonian eloquence leads a double life; it invites and repays voicing, it also asks for constant recognition of the quieter life of the words on the page. The experience of a reader, then, resembles that of the poet listening to Christmas bells in *In Memoriam*:

> Four voices of four hamlets round,
> From far and near, on mead and moor,
> Swell out and fail, as if a door
> Were shut between me and the sound . . .[21]

Swelling out and failing are vital motions of the poetry as of Tennyson's life, of anyone's life, the motions of the lungs. Tennyson often said that only he could properly read his own poetry aloud—'the Poet swears no being, existent or possible, can read this but himself' and again 'He will not admit that any one save himself can read aloud his poems properly'.[22] The reason he gave for this was that 'Some of the passages are hard to read because they have to be taken in one breath and require good lungs'.[23] He thought nobody had lungs like his. He may have believed that 'poetry looks better, more convincing, in print'[24] but then had to face the fact that it didn't sound better there, that he couldn't count on its good look to secure the right voicing. He worried about 'Boädicea': 'he "feared that no one could read it except himself, and wanted someone to annotate it musically so that people could understand the rhythm". "If they would only read it straight like prose," he said, "just as it is written, it would come all right." '[25] If nobody could read it except himself, who was this 'someone' who would annotate it for him? And if it would come right when read 'just as it is written' why did he need this unknown 'someone' and the musical annotation?

[19] 'Mariana in the South' (1832), R i. 399; 'Claribel' (1830), R i. 199; 'The Miller's Daughter' (1832), R i. 410.
[20] 'Gareth and Lynette' (1872), R iii. 293; 'Geraint and Enid, (as the second half of 'Enid' 1859), R iii. 351; 'Merlin and Vivien' (as 'Vivien' 1859), R iii. 418.
[21] *In Memoriam*, xxviii, R ii. 346.
[22] Kolb, 385; Allingham, 95.
[23] *Memoir*, i. 395 n.   [24] Ibid. i. 190.   [25] Ibid. i. 459.

The inconsistency of his claims arises from the centrality of respiration, and other physical motor-rhythms, to Tennyson's poetry, a centrality which was the poet's boast and also his quandary. Every body breathes, and poems written in the rhythm of the breath must be of all others those most patent to any body, but no two bodies breathe alike. Two bodies only rarely and ecstatically breathe in synchronicity (which is a source of the pleasure of duets), and so poems written in the rhythm of breathing are of all others the most personal. Hutton commented on 'the lavish strength of what may be called the bodily element in poetry' in Tennyson's work,[26] and other critics have sometimes taken Hutton's insight as the basis of adverse comment. Tennyson is thought to be preoccupied with word-music, with fondling, as it were, the bodies of words, to the exclusion or detriment of responsible thought; he 'is indolent, over-refining, is in danger of neutralizing his earnestness altogether by the scepticism of thought not too strong, but not strong enough to lead or combine, and he runs, or rather reposes, altogether upon feelings (not to speak it offensively) too sensual'.[27] But Tennyson thought *in* melody, and did so because his preoccupation with self-identity over time and beyond time drew him down repeatedly to an encounter with the human body itself as the crucial location of his thinking. His tunes carry the physique of his intellection, as they must, for what he ponders is the body itself, its grain over the years, the tissues he inherits, and where (and if) he would be without it.

He told W. F. Rawnsley, 'I don't think poetry should be *all thought*: there should be some melody.'[28] Loyal to melody as he remained, he was not therefore deaf or dead to thought; acoustics were for him a form of metaphysics. Thinking about the body involves taking account of the fact that our thinking is itself embodied, our concepts sunk in time like a seal in wax, sunk also in the history of a culture's words. This fact tells on and in Tennyson's melodies with their thought immersed in 'matter-moulded forms of speech'.[29] Hallam had thought the 'highest order' of poetry was 'that which deals with the foundations of our being, and never subordinates the thought to the diction'.[30]

[26] Hutton, *Essays*, 364.
[27] Leigh Hunt, *Church of England Quarterly Review* (Apr. 1843), Jump, 128.
[28] 'Personal Recollections of Tennyson', *Nineteenth Century*, (Jan. 1925), repr. in *Tennyson: Interviews and Recollections*, ed. N. Page (1983) [hereafter referred to as Page], 21.
[29] *In Memoriam*, XCV, R ii. 413.
[30] Letter to J. M. Gaskell, 25 June 1828, Kolb, 212.

Tennyson's poetry does not subordinate thought to diction, but it discovers that the 'foundations of our being' lie a deal lower, though not perhaps 'deeper', than Hallam might have liked to think, and that, at those physical foundations, the distinction between 'thought' and 'diction' does not ring as clearly as it did in Hallam's prose. In the context of concern with self-identity, Tennyson's complaint that 'The worst of folk is that they are so unable to understand the poet's mind'[31] turns out to be at one with an anxiety that people did not hear his rhythm or share his lungs.

In his essay on *In Memoriam*, T. S. Eliot suggested a way to sound out Tennyson's mind:

. . . I do not think any poet in English has ever had a finer ear for vowel sound, as well as a subtler feeling for some moods of anguish . . . And this technical gift of Tennyson's is no slight thing. Tennyson lived in a time which was already acutely time-conscious: a great many things seemed to be happening, railways were being built, discoveries were being made, the face of the world was changing. That was a time busy in keeping up to date. It had, for the most part, no hold on permanent things, on permanent truths about man and God and life and death. The surface of Tennyson stirred about with his time; and he had nothing to which to hold fast except his unique and unerring feeling for the sounds of words. But in this he had something that no one else had. Tennyson's surface, his technical accomplishment, is intimate with his depths: what we most quickly see about Tennyson is that which moves between the surface and the depths, that which is of slight importance. By looking innocently at the surface we are most likely to come to the depths, to the abyss of sorrow.[32]

Though the poet who looked for the point of intersection of the timeless with time might have been less severe about the Victorians' search for a union of time-consciousness with 'permanent truths', Eliot's repeated bringing-together of 'surface' with 'depths', of 'vowel sound' with 'moods of anguish' finely senses the nature of Tennyson's accomplishment. The metaphysical debates of Tennyson's time did indeed stir the surface of his verse, as I shall show; his musicality is attuned to the time's questioning remote as that music is from the public manners of intellectual exposition. Most particularly, the music asks, 'what is it to be embodied?' even while the proficiency of the writing celebrates such skills of embodied persons as the having of good lungs. Tennyson's verse sounds as if the body thought.

[31] H. D. Rawnsley, *Memoirs of the Tennysons* (Glasgow, 1900), Page, 68.
[32] 'In Memoriam', in *Selected Essays* (1932, 3rd enlarged edn., 1951, repr. 1976), 337.

It may seem odd to place this emphasis on the physical quality of Tennyson's writing because he so vehemently contended that 'the Spiritual *is* the real: it belongs to one more than the hand and foot'.[33] His keenest concern as regards the spiritual, though, was with the belief in personal immortality. Holding that belief (or clinging to it) requires the believer to ask whether he can distinguish between spirit and matter, between an embodied self and the same self without its body. If immortality is to be truly personal, the defining features of my current empirical self must somehow survive the death of my body, yet my body appears to be one of those defining features. The hope for a resurrection of the body, whatever its philosophical absurdities, marks Christian immortality as radically anti-Cartesian, and distinguishes the Christian conception of immortality from the afterlives of some other religions where survival is at the cost of identity. Coleridge put the orthodox view fervently, but did not mention its difficulties when he wrote: '. . . this last triumphal Crown, the summit and *ne plus ultra* of our immortality, even the union with God is no mystic annihilation of our individuality, no fanciful breaking of the Bottle and blending the contained water with the ocean in which it had been floating, the dreams of oriental Indolence, but on the contrary an *intension*, a perfecting of our Personality.'[34] It is all very well to be scornful about Buddhists ('the dreams of oriental Indolence') but Coleridge says nothing here about how union with the Divine can be compatible with the preservation of a finite identity—though his prose is fervent with unshakeable certainty.

The century after him was to miss his pitch of high confidence. In 1854, H. L. Mansell attacked F. D. Maurice's account of the personal afterlife on the grounds that it was 'a state, of which, because of the nature of human knowledge, we can know nothing. Moreover, he argued, one of the essential conditions for personality to exist would appear to be memory. Since that is linked with temporal notions of succession it was very improbable that, even if the state for which Maurice argued, existed, man would be able to participate in it.'[35] Mansell's argument, as stated here, is incorrect, for memory may not be a necessary attribute of 'man', in the sense of humanity in general,

[33] *Memoir*, ii. 90.
[34] Coleridge, Notebook 36, fo. 65, quoted in J. D. Boulger, *Coleridge as Religious Thinker* (New Haven, 1961).
[35] Geoffrey Rowell, *Hell and the Victorians* (Oxford, 1974), 86, referring to Mansell's *Man's Conception of Eternity* . . . (1854), 22–3.

though it may well be a necessary condition of individuality. The sense of 'memory' is different with reference to personal and to collective or racial memories; a community may well cease to be a community in the degree to which it loses memory of its own past but the members of the community do not by that token cease to be persons, whereas certain severe disorders of memory in an individual may trouble our sense of his self-identity over time. If we re-phrase the argument as 'it was very improbable that, even if the state for which Maurice argued, existed, individuals would be able to participate in it', it is more cogent, and its pertinence to Tennyson becomes clear, for it was just the insistence of the historical personality with its memories whose immortality he desired. John Sterling told Tennyson that he would personally be content with an afterlife in which the self was infinitely developed into higher and higher orders of being; 'I would *not*,' Tennyson replied with an emphasis very much his own, 'I should consider that a liberty had been taken with me if I were made simply a means of ushering in something higher than myself.'[36]

In Memoriam, I, begins by wondering what will happen if the dead do not stop dead but develop endlessly:

> I held it truth, with him who sings
>   To one clear harp in divers tones,
>   That men may rise on stepping-stones
> Of their dead selves to higher things.
>
> But who shall so forecast the years
>   And find in loss a gain to match?
>   Or reach a hand through time to catch
> The far-off interest of tears?
>
> Let Love clasp Grief lest both be drowned,
>   Let darkness keep her raven gloss:
>   Ah, sweeter to be drunk with loss,
> To dance with death, to beat the ground,
>
> Than that the victor Hours should scorn
>   The long result of love, and boast,
>   Behold the man that loved and lost,
> But all he was is overworn.'[37]

The keen intellect of Tennyson's ear shows in the last line of the section where he crowds together 'was' and 'is'. We know that

---

[36] *Memoir*, ii. 129.   [37] R ii. 318–19.

Tennyson was particularly sensitive to the need correctly to manage sibilance in verse. He disdained the opening line of *The Rape of the Lock*:

He quoted
'What dire offence from amorous causes springs.'
' "Amrus causiz springs," horrible! I would sooner die than write such a line!! Archbishop Trench (not then archbishop) was the only critic who said of my first volume, "What a singular absence of the 's'!" . . .'[38]

When his 'Ode Sung at the Opening of the International Exhibition' was printed by *The Times* with errors in the text, he was specially eager to correct 'In this wide hall with earth's intentions stored' to 'In this wide hall with earth's invention stored', complaining to his American publisher of the misrepresentation: 'I should never have written so unmusically as ntions st-'.[39] One reason he revised the fourth line of 'Tithon'—'And after many summers dies the rose'—into the fourth line of 'Tithonus'—'And after many a summer dies the swan'[40]—was to clear the line of the buzzing voiced 's's in the early version. Yet here, at the climax of the first section of a major work, he leaves the jarring 'was is' which sounds as 'horrible' as 'Amrus causiz springs'. Indeed, he accentuated the dissonance. The line previously read, 'Yet all he loves is overworn' in the trial edition.[41] The harshness of the sound to Tennyson's ears stands for his particular horror at the loss of self-identity, and against that blurring of the self, Tennyson insists, at whatever cost of pain, on remaining true to the past, in true with the past, as the third stanza presses on infinitival constructions and rhymes internally on 'be': 'Grief', 'be', 'keep', 'sweeter', 'be', 'beat'.

The 'one who sings' in the section's first stanza is Goethe, and it may be, given Hallam's heavily Germanic intellectual orbit, that Tennyson wished to begin the poem with an emphatic irony: it was Hallam's own death which separated Tennyson from beliefs about death which he had acquired partly from Hallam. Goethian moral evolution, quite as much as Darwin's physical evolution, worried Tennyson because growing thus ever upward, we might outgrow each other, and find that, from the perspective of our higher selves, our past

---

[38] *Memoir*, ii. 286.
[39] See the letters to *The Times*, 25 Apr. 1862 and to J. T. Fields, 11 Feb. 1867, in *The Letters of Alfred, Lord Tennyson*, ed. C. Y. Lang and E. F. Shannon (Oxford, in progress, vol. i, 1982, vol. ii, 1987), ii. 305, 452. All references to Tennyson's letters are to this edition [hereafter referred to as Lang and Shannon].
[40] R ii. 607.   [41] R ii. 319 n.

lives and those we shared them with, had become dead things. The section vehemently repudiates the theory of infinite development because one of the simplest reasons for wishing to believe in immortality is the desire not to be separated from what we have loved, including our own selves. An immortality in which we endlessly change, though, would itself be a process of separation; finite people infinitely developed would eventually become incapable of recognizing each other.

What *In Memoriam* desires above and beneath all is exquisitely and passionately put in the poem's forty-seventh section:

> That each, who seems a separate whole,
>   Should move his rounds and fusing all
>   The skirts of self again, should fall
> Remerging in the general Soul,
>
> Is faith as vague as all unsweet:
>   Eternal form shall still divide
>   The eternal soul from all beside;
> And I shall know him when we meet:
>
> And we shall sit at endless feast,
>   Enjoying each the other's good:
>   What vaster dream can hit the mood
> Of Love on earth? He seeks at least
>
> Upon the last and sharpest height,
>   Before the spirits fade away,
>   Some landing-place, to clasp and say,
> 'Farewell! We lose ourselves in light.'[42]

As a reader learns the relations of individual sections to the whole of *In Memoriam*, he discovers what Tennyson made them to be: not steps in an argument, nor stepping-stones to higher things, but landing-places, to clasp and say farewell, an extraordinary confluence of arrival and departure. Tennyson knew there was no end to such a desire, annotating these lines 'at least one last parting! and always would want it again—of course'.[43]

The philosophical puzzles set by such desire might be swept aside as mere evidence that reason cannot fathom the mysteries of faith, but then the words of Scripture, whatever they meant, would have to be taken wholly on trust and wholly on board. Some of those words,

---

[42] R ii. 364–5.   [43] R ii. 365 n.

though, troubled the age and Tennyson, words such as Christ's reported 'Depart from me, ye cursed, into everlasting fire'.[44] Tennyson was disappointed the translators of the Revised Version had not altered that by changing 'everlasting' to something like 'aeonian'.[45] As Geoffrey Rowell has shown in his *Hell and the Victorians*, the period of Tennyson's writing career coincides with a period in which the orthodox doctrines of eternal punishment were morally and theologically unsettled. H. N. Oxenham, writing in 1878 to Gladstone, felt the faith of centuries had been tampered with; the liberal clergy 'assume that what has certainly been the belief of Christendom from the Apostles' day to our own is an obsolete barbarous anachronism . . . Twenty years ago I doubt if any Anglican clergyman would have embarked on openly averring what had always been looked upon as heretical on this point, and when Maurice first did so, it was in a diffident and tentative tone—not to say without an ambiguity of drift very unlike the confident contempt for the received Catholic doctrine now expressed . . . .'.[46] F. D. Maurice stood godfather to Hallam Tennyson and dedicated his *Theological Essays* to Tennyson, who particularly approved of Maurice's view that the real Hell was the absence of God from the human soul.[47] Both Maurice and Tennyson incline at times to the view that the afterlife permits of moral progression, and that it may therefore be hoped that all will eventually be brought into union with God (how such moral progression is distinguished from the infinite development which Tennyson dreaded was incompatible with self-identity is not clear).

Maurice's opponents were fairly suspicious of the 'ambiguity of drift' in his arguments, and men like Pusey believed that if the traditional doctrine of the eternity of punishment was to be argued away on the grounds of its moral repugnance, other doctrines would have to prepare themselves for the same fate: 'For if the word "eternal" were taken out of its known historical meaning in the Church . . . there would . . . be an end of all teachings.'[48] People feared that it was not only poetry which was losing its magisterial office. Tennyson's own work temporizes about such an end to the teaching authority of the Church. *In Memoriam* can occasionally think that 'all is well, though

[44] Matthew 25: 41.
[45] *Memoir*, i. 322.
[46] Letter of 16 May 1878, quoted Rowell, op. cit. 175–6.
[47] *Theological Essays*, (1853; 4th edn., 1881), 388.
[48] Pusey to W. K. Hamilton, 22 June 1881 or 1882, quoted Rowell, op. cit. 119–20.

faith and form | Be sundered in the night of fear',[49] where 'form' means the institutional disciplines of the Church and 'faith' the beliefs of the individual. The poem is equally moved by the benefits of orthodoxy:

> Leave thou thy sister when she prays,
>   Her early Heaven, her happy views;
>   Nor thou with shadowed hint confuse
> A life that leads melodious days.
>
> Her faith through form is pure as thine . . .[50]

The poem is unclear about the relation of individual religious sentiment to established doctrines; if 'form' has the beneficial effect of purifying 'faith', then something is lost when 'faith' is 'sundered' from form, and if something has been lost then '*all*' (my emphasis) cannot be well—unless, that is, Tennyson's inconsistency here derives from an unexpressed conviction that some people need the assistance of doctrine and religious community whereas others do not. The further possibility then arises that the people who need this assistance are people such as sisters who remain uneducated and immature ('early Heaven') in comparison with their brothers who have become sophisticated at University. We needn't attribute such a view to Tennyson, because neither he nor his poem has a clear view on the matter at all. That may or may not be a failing. The questions of doctrine, and of whether doctrine has a consistent role in the religious life, are themselves subjects of argument in the Victorian period. Broadly speaking, the more Catholic a thinker, the more emphasis he places on the importance of doctrinal consistency; the more modernist (or 'Evangelical', or 'liberal', or 'Low Church'), the lighter the emphasis he gives. At some levels—that of theological reflection, for instance—the need for clarity on such issues is pressing, but this may not be the level at which *In Memoriam* works, nor indeed the level of much human thought. The conceptual difficulties began for Tennyson's age at such a basic stage that they precluded for many anything that could strictly be called 'clear' thought. As Hutton wrote, 'The simple truth is that we are not in a position to say what is body and what is soul, or what is the distinction between them. . . . We cannot say with confidence whether there may not be something essentially material in

---

[49] *In Memoriam*, CXXVII, R ii. 446.
[50] Ibid. XXXIII, R ii. 351.

a finite soul, nor whether there may be something essentially spiritual in a material body.'[51]

There is a reasonable view of philosophical activity according to which it is philosophy's task either to clear away such conceptual confusions or to abstain from using irremediably muddled concepts. Philosophers in this tradition might also urge that human thought should confine itself to propositions that can be evidenced or made coherently articulate. Whether fortunately or not, human thought has for some time been less disciplined than that. The present state of thinking remains much as it was when Arthur Hallam noted that 'the word sympathy . . . like most others in moral science has a fluctuating import'.[52] In these real circumstances, the conceptual hygiene of philosophy may be salutary and yet not equip us for realism about what and how people think. The 'ambiguity of drift' in Maurice's arguments, the 'fluctuating import' in Hallam's terms, indeed bedevil thinking but they may none the less be not a calamity which occurs to human reason from outside but its condition, the air it breathes.

Desirable as timeless truths rid of contradiction are, an exclusive devotion to them may hinder right understanding of some styles people speak and think in—styles that answer to the contradictions of desire, and the thoughts desire prompts. We live in time and so we need the truths of time as well as timeless truths. This is so even with respect to the human desire for personal immortality, because this desire for survival beyond time submits along with other desires to the rhythm of a self's life in time. 'Perhaps what I had experienced,' the narrator writes at the climax of Le temps retrouvé, 'were fragments of a life beyond time, but my contemplation, though a contemplation of eternity, was itself evanescent.'[53] The belief in personal immortality comes and goes within individual Victorians, from the Queen herself asking the Dean of Windsor, 'if there ever come over me (as over her) waves or flashes of doubtfulness whether, after all, it might all be untrue',[54] to Tennyson comically noting that the certainty of an afterlife left him sometimes on the morning after, 'I feel myself to be a centre—can't believe that I shall die. Sometimes I have doubts, of a morning.'[55] The poet generalized these minute swerves of religious

[51] R. H. Hutton, 'The Resurrection of the Body' (1895), in Aspects of Religious and Scientific Thought, ed. E. M. Roscoe (1899), 156.
[52] 'On Sympathy', written 1830, Motter, 133.
[53] À la recherche du temps perdu, iii. 875.
[54] Quoted in Elizabeth Longford, Victoria R.I. (1964), 341.
[55] Allingham, 151.

conviction into the certainty that 'the whole truth is that assurance and
doubt must alternate in the moral world in which we at present live'.[56]
In his poetry, he found a rhythmic shape for such fluctuations so that
their ephemerality might be attended to as well as undergone, while
the permanence of what they appeared to reveal could also be
observed. He relied on the logic of a lyric utterance which is at all
times true to such times as its utterance supposes, but which need not
always be true, for the lyric recognizes that there are other times than
those of which it speaks. Essentially, then, such a lyric voice is
dramatically situated, as when, in the ninety-fifth section of *In
Memoriam*, he chooses to name the much wept-for communion with
the dead Hallam by a form of the word that Queen Victoria chose to
express instants of doubt. Her 'waves or *flashes* of doubtfulness' meet
his

> And all at once it seemed at last
> The living soul was flashed on mine . . .[57]

It has taken Tennyson a prologue, 94 preceding sections, and 34 lines
of this section to get here. *In Memoriam*, xcv marks that time with its
repeated, pausing narrative 'and's'—there are eleven before the 'And'
in these lines, an 'And' of contact as well as of sequence. The drama
and rhythm as Hallam and Tennyson meet again at last sound
throughout the section in its constant melodic gravitation towards
'and'. How perfectly he times this contact with the timeless—'at once'
held permanently in view of 'at last', the time of longing and the
timelessness of fulfilment co-exist in the line. The intonational
bareness of the written word was, for a poet of his imagination, not just
an obstacle to utterance but an opportunity for a new and tacit
eloquence which could speak for a divided soul in its queries about the
division of soul and body.

When Tennyson addresses the yew on Hallam's grave earlier in the
poem, he also makes a poetic virtue of a conflict of actual desires:

> And gazing on thee, sullen tree,
> Sick for thy stubborn hardihood,
> I seem to fail from out my blood
> And grow incorporate into thee.[58]

Does 'Sick for' mean that Tennyson sickens at the sight of the yew's
vigorous indifference to the fact that Hallam has been cut down in his

[56] *Memoir*, i. 304.   [57] R ii. 413.   [58] *In Memoriam*, II, R ii, 320.

prime, or does he long to possess himself such a calm ability just to keep on going? (Two of the *OED*'s examples of 'sick for' meaning 'longing for' come from Tennyson: 'This girl, for whom your heart is sick' and 'Sick as an autumn swallow for a voyage'[59]) The voice can turn the line in either direction and, when the eye recognizes that, a truth about grief comes home: the way the world unperturbedly survives the death of a beloved person affronts the decencies of mourning, and yet, working within the sorrow that cleaves to the disturbing fact of loss, there is in the mourner himself a principle which brings him too to live with his loss. 'Assurance and doubt must alternate in the moral world in which we at present live', but the vocal possibilities of these lines do not alternate, they co-exist. And this is true to the rhythm of assurance and doubt within the experience of grief, for, as Wittgenstein wrote, ' "Grief" describes a pattern which recurs, with different variations, in the weave of our life. If a man's bodily expression of sorrow and of joy alternated, say with the ticking of a clock, here we should not have the characteristic formation of the pattern of sorrow or of the pattern of joy.'[60] The contained ambiguity of voice in such passages of *In Memoriam* preserves a characteristic weave rather than registering the occurrence of distinct sensations, and it does so by taking the events of a voice and holding them on a page whose surface is intimate with depths.

Arthur Hallam in his day had thought 'the doctrine of human immortality . . . [an] excellent . . . theme for the energy of declamation'.[61] Much that Tennyson has to say about immortality cannot be declaimed because it cannot quite be said at all. When 'De Profundis' envisages a future for Tennyson's second Hallam, his son and heir, it does so with vigilance about its own wishes:

> Live, and be happy in thyself, and serve
> This mortal race thy kin so well, that men
> May bless thee as we bless thee, O young life
> Breaking with laughter from the dark; and may
> The fated channel where thy motion lives
> Be prosperously shaped, and sway thy course
> Along the years of haste and random youth
> Unshattered; then full-current through full man;

---

[59] 'The Talking Oak' (1842), R ii. 108; *Harold* (1876), I. i. 55, in *Tennyson: Poems and Plays* (Oxford, 1953, repr. 1965), 609.
[60] *Philosophical Investigations*, 174.
[61] 'Essay on the Philosophical Writings of Cicero', written 1831, Motter, 175.

And last in kindly curves, with gentlest fall,
By quiet fields, a slowly-dying power,
To that last deep where we and thou are still.[62]

Within the plain monosyllables of that last line lies a contrariety of
desire. Tennyson may wish for his son and for himself 'That last deep
where we and thou *are* still' (where we continue to exist) or 'That last
deep where we and thou are *still*' (where we are finally at rest, out of
the turmoil of existing). Metrical considerations cannot settle the issue.
The line is not definitely a regular iambic pentameter; there may be
three consecutive stresses on 'that last deep' (of the other ten lines in
the passage I quote, only five are to my ears regular blank verse). A
stress on 'are' is not metrically impossible. More importantly, metrical
stress and the stress of meaning need not be identical. Stress may fall
metrically on 'still' without this fixing the emphasis of thought, for the
line can be voiced as an iambic pentameter and yet made to mean, with
the intonation contour, 'where we continue to exist'. Begin to allow
that a contrariety of desire may speak in this wish for his son, and the
accent of Tennyson's doubts about survival and inexistence begins to
sound throughout the passage: does the young life break laughingly
from the dark (it laughs because it has escaped, broken from,
something unpleasant) or is the life itself broken by a laughter which
follows it from the dark (compare: 'breaking with fatigue from the
strain')—the second possibility would explain why Tennyson has to
pray that it will survive 'unshattered'? Are the 'quiet fields' in these
lines akin to the 'field' man so happily 'lies beneath', the 'dim fields
about the homes | Of happy men that have the power to die' of
'Tithonus'?[63] And does the 'power to die' of the 1860 poem linger on
in the 'slowly-dying power' of these lines composed some time
between 1852 and 1880? If so, which matters most—that death comes
slowly, or that what comes, however slowly, should be death? The lines
remarkably absorb these many questions in their steady melody, but,
steadily as it goes, the passage stirs with questions.

Human beings sometimes say they want to die but the breath they
use to utter that wish is itself a sign of life, so that, as an early
Tennyson poem puts it, 'No life that breathes with human breath | Has
ever truly longed for death'.[64] But no life that breathes with only *human*
breath has ever wholly longed for immortality. It may be, as those lines

from 'The Two Voices' imply, that the mechanism of breathing strives, as does any other thing which exists, to preserve itself in existence,[65] and that it is therefore 'life, not death, for which we pant'.[66] If indeed it ✗ is our lungs, for example, as much as our selves, which do not want to die, the physical motions wishing on their own account to be a *perpetuum mobile*, then an immortality outside the body will not be quite all which all of a person now desires when he wishes never to die. 'What opposite needs converge to this desire of Immortality!' as an other Victorian exclaimed.[67] Tennyson's genius is to do more than merely exclaim about the fact.

The two voices which speak divergent impulses at explicit length, and in alternation, in his early masterpiece, 'The Two Voices', continue their vital debate in simultaneous vocal ambiguities throughout of the style of his mature work. Wilfrid Ward noted of 'Vastness' that, as Tennyson read it aloud to him, it had 'two distinct voices—the last line being placed in the mouth of a separate speaker who answers the ✗ rest of the poem'.[68] This is excellently said, but is not quite right. 'Vastness' runs for thirty-five lines of torrential lament over the inane cruelties of human life before coming to rest on

> Peace, let it be! for I loved him, and love him for ever:
> the dead are not dead but alive.[69]

'Let it be!' could be an 'amen' to the pointless squandering of creation which the poem details; it may resign to the will of God because, at least, that will has not only permitted the love of Tennyson for Hallam, but assured the eternity such love asks. Or 'let it be!' might decide to leave the whole mess of the world alone, turn from it with a sorrow barely distinguishable from disgust, and concentrate in stead of the world on an exceptional case of decency, so exceptional that it cannot be thought to be part of any scheme which would re-orient all disasters by placing them in the light of a divine plan. The last hemistich also achieves a sublime numbness of tone in print. This is partly because of the logical ambiguity of the colon which introduces it, standing perhaps for a 'because' ('it is because he is an immortal soul that he

---

[65] Spinoza, *Ethica in ordine geometrico demonstrata* (1675), trans. as *Ethics* by W. H. White and A. H. Stirling, ed. J. Gutman (New York, 1949), Part III, proposition vi, p. 135.

[66] 'The Two Voices', R i. 591.

[67] Browning, letter to Julia Wedgwood, 27 June 1864, in *Robert Browning and Julia Wedgwood: A Broken Friendship as Revealed in Their Letters*, ed. R. Curle (1937), 31.

[68] Quoted R iii. 137 n.   [69] R iii. 137.

can be loved as I love him') or perhaps for a 'therefore' ('it is in the perpetuity of love that the dead find their life'). In the former case, the feeling arises from a conviction, in the latter, the conviction is produced by the feeling. The numbness mainly comes from the absence of any indication of attitude in the line, such as might be supplied by 'How frightening!' or 'What joy!'

[Imagining such an indicator of attitude is absurd but we need to make such an act of imagination to see why somebody who confidently opted for one of the possible indicators here would have missed the point of the poem.]This is not because it does not matter which indicator is supplied, but because it is the absence of an indicator which needs to be precisely imagined. Reading the line in the context of the entire poem does not settle the issue, for 'Vastness' is constructed so that the thirty-five lines which detail human wretchedness are countered by this single last line; it is absolutely clear that a great difference is made but not what the difference is. This seems to me honest. If you were convinced that human beings were immortal, it would clearly make a great difference to how the world struck you, but you could not be clear what the difference was unless you also knew fully what being immortal was like, and human beings do not know this. Properly then, the tone with which 'Vastness' ends remains not unsettled but not settled either. We cannot say whether the poem concludes with an achieved serenity or whether the poet gets at last a surprise that may not be entirely pleasant. We can *speak* the line so that it will make up Tennyson's mind and voice for him; indeed, it is impossible to say it without either some tonal indication or an evident and deliberate avoidance of such indication, whereas the written line neither drops hints nor maintains a poker-face. All this configuration of feeling is Tennyson's. Ward was right to say that the line sounded as if 'placed in the mouth of a separate speaker', but Tennyson is also that speaker separate from himself, at odds with his own desires and saved from involution only by the managed detachment of his voice onto the page. (The same is true for any sympathetic reader of the poem.)

He frequently punned around the phrase 'let it be' and its cognates;[70] its range of tones, from patience to an angered curtness,

---

[70] In addition to the examples I discuss below, see 'The First Quarrel' (1880), R iii. 41, especially its cutting lineation of ' "Let bygones be!" | "Bygones! . . ." ' The phrase 'let . . . be' rarely has Tennyson's inquiring richness in other poets of the period. Newman, for example, makes the Soul of Gerontius cry, after the Divine Vision, 'Take me away, and in the lowest deep | There let me be . . .', but the Soul does not wish to be let alone; see *The Dream of Gerontius* in Newman's *Verses on Various Occasions* (1865;

could be made to hold an entire temperament as it looked over its past and looked forward with the troubled hope that such a past might have a permanent future. The speaker of *Maud* rebukes himself for spite against Maud's brother, 'Peace, angry spirit, and let him be! | Has not his sister smiled on me?',[71] but the obstinate workings of his mind against his will cannot leave the thought of the brother alone (let him be) and his angry spirit cannot eventually even leave the brother alive (let him be). After the duel, the speaker himself suffers from not being let be, though, and at the same time, he wishes to die, to be let cease to be:

> But the broad light glares and beats,
> And the shadow flits and fleets
> And will not let me be . . .[72]

The insistent obsession presses in the gruesome echoing of the verse: 'light' becoming 'let' through 'flits', 'beats' and 'fleets' overshadowing 'me' and 'be'. And the echo of 'let me be' back to 'let him be' helps us recognize why the speaker of *Maud* is his own worst enemy—unable to let others be, he cannot himself be let be by himself.

The past maddeningly refuses to die, but then the self obdurately refuses to succumb to death (so that the consciousness in *Maud* seems at once Dracula and his victim). Nor will the past lie down in *The Princess* when Ida repudiates the sentiment of 'Tears, idle tears' with a self-consciously gorgeous tirade of a progressivist colour:

>   nor is it
> Wiser to weep a true occasion lost,
> But trim our sails, and let old bygones be,
> While down the streams that float us each and all
> To the issue, goes, like glittering bergs of ice,
> Throne after throne, and molten on the waste
> Becomes a cloud: for all things serve their time
> Toward that great year of equal mights and rights . . .[73]

She wants to speak against what she thinks is the futile nostalgia of 'Tears, idle tears' but it is as if, try as he might to give Ida the words for

---

Uniform Edition, 1868), 357. On the other hand, Christina Rossetti uses the phrase probably to mean just 'let me alone' in 'Autumn' (1858): 'O love-songs, gurgling from a hundred throats, | O love-pangs, let me be.'—I quote from the variorum edition of Christina Rossetti's poems, ed. R. W. Crump (2 vols., Baton Rouge, 1979), i. 143.

[71] *Maud* (1855), I. xiii, R ii. 548.
[72] Ibid. II. iv, R ii. 575.   [73] R ii. 234.

her projects, Tennyson's own reluctance to let anything die, even the unpleasantness of the past, makes him put in her mouth the ambiguous sentiment of 'Let old bygones be'—'let them continue to exist, don't interfere with them' / 'let them alone, get on to something else'. The way 'let be' can look at once to the past and to the future shows in the printed voice that if a morbid temperament is to be immortal the immortality will itself be morbid.

In 'To the Marquis of Dufferin and Ava', Tennyson gratefully recalls his son Lionel's last letter:

> But ere he left your fatal shore,
>   And lay on that funereal boat,
>   Dying, 'Unspeakable' he wrote
> 'Their kindness,' and he wrote no more;
>
> And sacred is the latest word;
>   And now the Was, the Might-have-been,
>   And those lone rites I have not seen,
> And one drear sound I have not heard,
>
> Are dreams that scarce will let me be,
>   Not there to bid my boy farewell,
>   When That within the coffin fell,
> Fell—and flashed into the Red Sea,
>
> Beneath a hard Arabian moon
>   And alien stars . . .[74]

Tennyson often read the letters of the dead, from 1833 when Hallam died to Lionel's death in 1886, and took those deaths into his own words. This poem, like *In Memoriam*, XCV, turns on the critical word 'flashed', expressing at once the intensity of an attachment and how its object has disappeared. The dead man's words revive in Tennyson's verse, but do so only because the verse consciously fashions for itself a reduction of living voice to meet the dead half-way. The poems do not impersonate dead voices and so do not insert themselves as substitute satisfactions between Tennyson and what he has lost; they do not take the place of the dead, or of his feelings for the dead, though they can stand for the dead and the feelings with which he survives. *In Memoriam*, XCV, reports Hallam's letters indirectly;[75] 'To the Marquis of Dufferin and Ava' displays how Tennyson's writing has changed what Lionel wrote by the evident poetic inversion of the imaginable

---

[74] (1889), R iii. 200–1.   [75] See below, p. 166.

prose syntax of his son's letter. That is, Lionel probably wrote something like 'their kindness has been unspeakable' rather than 'Unspeakable . . . | Their kindness'. No reader versed in the conventions of poetry would believe otherwise (Lionel certainly did not write to his father in *In Memoriam* stanzas anyway). But the conventions are very tender here; they allow the last word Lionel leaves Tennyson to be not 'Unspeakable' but 'kindness'—not inarticulacy, then, but gentle attentions, though, as a reader guesses at the prose within the verse, he hears not only the dumb given effectual speech but also the helpless breakage of deep emotions against the rigid element of print. John Bayley captures the delicacy of the relations between the formal and the colloquial in these lines, when he comments of 'Unspeakable': 'Its social sense, applied to the Viceroy's kindness, is also the literal sense applicable to the grief of the bereaved; and it is touchingly right that the father should find how to fit into his poetry the word his son used.'[76] This hovering reticence which touches social decencies into imaginative decorum weighs thoughtfully too on the capital 'T' in 'When That within the coffin fell'. No vocalization of 'That' can do what the written word does— show that Tennyson's loss was great without saying how great, and at the same time abstain from saying what exactly it was within the coffin even while impressing on the reader how dear whatever it was was. 'The simple truth is that we are not in a position to say what is body and what is soul . . .': but it is true that Tennyson was able to write down the complexity of our not being in such a position.

Breathing Immortally

Hutton admired Tennyson's 'comprehension of grasp, that deliberate rejection of single strands of feeling, which always distinguishes him'.[77] The intonational ambiguities I have noted show one way in which he kept that grasp comprehensive in the verse. Another practical instance of his knowledge that 'all things are double one against another' and his repudiation of 'the swift decision of one who sees only half the

---

[76] 'Tennyson and the Idea of Decadence', in *Studies in Tennyson*, ed. Hallam Tennyson (1981), 204. Compare also Martin Dodsworth's remark that 'the word "unspeakable" receives an almost impossible degree of emphasis', 'Patterns of Morbidity: Repetition in Tennyson's Poetry', in *The Major Victorian Poets: Reconsiderations*, ed. Isobel Armstrong (1969), 26.

[77] Hutton, 'The Modern Poetry of Doubt' (1870), in Roscoe, op. cit. 377.

truth'[78] comes in his thoughtful exploitation of the double timing of a written text and its spoken versions. By which I mean that a vocal utterance may for an instant make clear what a written transcript of that utterance would keep clouded, as, in a text of *Othello*, Iago's 'Indeed?' has a blankness on the page which the voice of an actor can momentarily twitch into suggestive life. Somebody makes a remark which provokes us. If we try to return to what has been said, the tone has gone. Accused of goading, or hinting, he replies, self-righteously, 'What do you mean? I only said . . .' and repeats his words without the offending tone. In written words, the significant tone remains a perpetual possibility but it is evoked only briefly, if evoked at all, on the voice.

Conscious of this existence of writing, an existence with a constant view to speech, we can recognize two timings of a text—its permanent latency on the page, and its instantaneous occurrences as voiced. The second stanza of *In Memoriam*, II reads:

> The seasons bring the flower again,
>   And bring the firstling to the flock;
>   And in the dusk of thee, the clock
> Beats out the little lives of men.[79]

The sentiment looks familiar and uncontentious at first (it seems to say just that time goes on), the versification sounds placidly competent, but the momentary, thick stutter of 'thee, the' in 'the dusk of thee, the clock' clogs the fluent reflections on time. The voice has to pick its way very carefully at that point if it is not to trip over itself. A heavy pause is enforced there, a pause which separates the idyllic platitudes of the first two lines from what, more startlingly, follows. With a swerve of feeling, we seem to collapse from hallowed commonplace into the monotony of obviousness: 'the clock | Beats out the little lives of men'. If these words are read, as they well may be, as regular iambics, they say merely that the clock measures out human lives, that we live, as most of us are perfectly aware, in time. Underneath this placid commonplace the feeling silently swells to a distended menace. Giving a different intonation and stress, we get 'the clock | Beats out the little lives of men', where the timepiece is the agent and not only the witness

[78] William Knight, 'A Reminiscence of Tennyson', *Blackwood's Magazine* (1897), Page, 182; Tennyson, remembered by FitzGerald, *Memoir*, i. 37.

[79] R ii. 319–20.

of our extinction as it beats the living daylight out of us, our life in time made to sound like a continuous, mechanical attrition. The section describes how the passage of time feels to someone who has survived a bereavement, living on in a diminished world, the dwindled light of 'the dusk of thee'. The strain on a grieving imagination as it faces the irreversible push forward of life, away from a time in which the dead were alive, needs to have this quiet, pastoral rote (the first reading) and this squirming under lethal pressure (the second reading) said by the same words in order to compact the sense that these experiences of time come to the same thing, though the lines Tennyson has written also reject each of the single strands of feeling here because you cannot rightly take one view of time and its routines without simultaneously taking the other, as an antidote.

What the voice can make appear when it reads the verse aloud—the sudden horror within the regular seasonal returns—it makes apparent only for a moment, though the writing maintains the permanent possibility of such flashes of sentiment. The vocal ambiguities co-operate with the logic of a dramatized lyric utterance; the potential variety of tone implies a recognition of the changing occasions of speech, and the attendant changes of feeling in the pattern of grief, a 'recognition, implicit in the expression of every experience, of other kinds of experience which are possible'.[80] Able thus to be 'sweet and bitter in a breath',[81] the lines of *In Memoriam* can gauge spasms of emotion against spans of thought, human responses against and within a divine plan. Tennyson works the relation of speech to writing so that his words both express an instant of emotion and take the time to survey their own expression. Doing that, he creates a reply to a question Arthur Hallam had asked: '. . . and what will be the momentary pangs of an atomic existence, when the scheme of that Providential Love, which pervades, sustains, quickens this boundless Universe shall at the last day be unfolded, and adored?'[82] Hallam's piously rhetorical question was perhaps meant to imply that all suffering will at the Last Judgement amount to nothing, be dissolved in beatific comprehension. Tennyson, as usual, is less inclined to let things pass, and whatever the truth about God's plan, the plan of Tennyson's poem does not efface these 'momentary pangs', it grows out of them rather than outgrows them. Gasp and outcry, temporal

---

[80] T. S. Eliot, 'Marvell', in *Selected Essays*, 303.
[81] *In Memoriam*, III, R ii. 320.
[82] Letter to W. E. Gladstone, 14 Sept. 1829, Kolb, 317.

occurrences of the larynx, keep their place in the 'eternity of Print'.[83]

Sometimes it is not a tone of voice which flashes from the writing but a touch of metaphor, as when Tennyson thinks of his tendency to a despair which makes the world empty for him:

> And shall I take a thing so blind,
> Embrace her as my natural good;
> Or crush her, like a vice of blood,
> Upon the threshold of the mind?[84]

Put against 'crush', 'vice' begins to sound like 'a tool composed of two jaws, opened and closed by means of a screw, which firmly grip and hold a piece of work in position' (*OED*, in the language since 1500); it also retains its sense of 'depravity'. 'A vice of blood' would then have a cold, practical tenacity, a bad disposition in your system which has you as well as its victims in its grip—something that conduces to efficiency but which easily deludes you into a self-confidence in areas to which your expertise does not extend. (Something like that happens to Othello, who provides a source of the phrase—'I do confess the vices of my blood'.[85]) 'Embrace her as my natural good' yoked to 'Upon the threshold of the mind' releases briefly a marital image. The lines ask whether he should marry sorrow or whether, carrying sorrow over the threshold of his mind, as a bridegroom carries a bride, he should stop and kill it there. (Tennyson's allusion to *Othello* operates at a great depth.) The suggestion is gruesome—the bridegroom's arms, outstretched to hold his bride, bend up and inwards, crushing her like a piece of wood in a vice turned too tight. The point of this ghastly conceit is to suggest that there is danger in the forces of self-control as well as in the emotions which need to be controlled; the collocation of phrases does not make clear whether it is Tennyson's melancholia or his attempt to control his own temperament which is being compared to 'a vice of blood'. It is then sane to recognize that this is a burst of feeling about his own incapacity to do anything right, a feeling which passes in time, as the metaphor glides under the words, like a shark beneath a keel, and moves off from them.

A brief swell and fail of metaphor in *In Memoriam*, VII touches on the proximity or distance of a past to a present self:

[83] Hallam, letter to Edward Spedding, 23 Aug. 1831, Kolb, 467.
[84] *In Memoriam*, III, R ii. 321.
[85] Ricks points out the allusion, R ii, 321 n.

> Dark house, by which once more I stand
> Here in the long unlovely street,
> Doors, where my heart was used to beat
> So quickly, waiting for a hand . . .[86]

So great was the excitement and keen the anticipation of those old days that the heart thumped so eagerly in the breast as to seem itself beating on the door. That extravagance passed out of his life as it passes out of his lines; it was something he could not hold on to, and the reader should not make of the metaphor here more than a hint, its exquisite aptness being in that glancing quality, though the lines manage to keep it in a state of permanent ephemerality.

The last stanza of this section also drops a vital hint:

> He is not here; but far away
> The noise of life begins again,
> And ghastly through the drizzling rain
> On the bald street breaks the blank day.

It is hard to scan 'He is not here', the words seem to require four even stresses or four even absences of stress. If you spoke the first line as a regular iambic tetrameter to someone who had not seen the printed text of the poem, it might start a hope: 'He *is* not *here*; but *far away*'— not here but, in contrast, far away, and still alive. This had been the implication of the words 'He is not here, but is risen' when the angel spoke them to the women at Christ's tomb (Luke 24: 6).[87] The intimation of a possible immortality plays over the line before the line drops into the blank space of the page, to re-emerge from it as something quite alien to hope: 'But far away | The noise of life begins again'. What revives is not the friend he misses but the beat of regular iambics and the 'noise of life', the daily round, they represent. The metrical impetus disappoints his hopes even as it re-asserts compositional skill. He re-imagines the iambic pulse as an actual pulse, the ictus of heartbeats, which used to beat so quickly but which he would now like to hear the last of, just as the cessation of stress corresponded at the opening of the stanza to Hallam's abeyance in death. A strongly physical intelligence is at work in the verse, sounding out anguish, making anguish out of sounds. W. J. Fox, reviewing *Poems, Chiefly Lyrical* delighted in 'that regularity of measure which is one of the

---

[86] R ii. 325–6.
[87] J. D. Rosenberg draws attention to the scriptural allusion in *The Journal of English and Germanic Philology* (1959), 230.

original elements of poetical enjoyment; made so by the tendency of
the human frame to periodical movements'.[88] Tennyson's insight into
the pleasures of recurrent sound told him that, if this was the case, it
would also be true that, by the tendency of the human frame to weary
of its own periodical movements, regularity of measure could become
something it would be inhuman always to enjoy.

Breathing is the periodical movement he most skilfully frames in his
art. The word 'breath' and its cognates mattered to him (there are 197
entries in Baker's 1914 concordance) because breath intimately brings
together life and poetry in the fact that it is with the breath that keeps
us living that we sing or speak. The root meaning of the word 'spirit' is
'breath', so that breath can be taken as of the essence both of this life
and of any other life we imagine. His verse composes breathing so that
the bare production of the words becomes significant:

> Ah yet, even yet, if this might be,
>   I, falling on his faithful heart,
>   Would breathing through his lips impart
> The life that almost dies in me;
>
> That dies not, but endures with pain,
>   And slowly forms the firmer mind,
>   Treasuring the look it cannot find,
> The words that are not heard again.[89]

The moral drama of wishing to die and wishing to survive death takes
body in the regulated breath we have to take between these stanzas. At
the end of the first stanza quoted, at the end of our breath, the impulse
of self-sacrifice verges on the suicidal, and the exhaustion of our lungs
in speaking the lines endorses that impulse. But the poem is not done,
nor done with us, and requires a firmer resolve which makes itself felt
as an inhalation: 'The life that almost dies in me;' | [the reader
breathes in] 'That dies not . . .'. Absented from the felicity of death
awhile, in this harsh world he draws his breath again in pain to tell
Hallam's story. Tennyson discriminates between his desires, knowing
that the readiness expressed here to do anything to resuscitate Hallam
might, given Tennyson's own desire to die, be only an eagerness to be
given the kiss of death himself. The discrimination is made in these
lines, and particularly in the breathing space between them. The
printed words, of course, do not perform the act of intake they ask a

[88] Op. cit., Jump, 32.   [89] *In Memoriam*, XVIII, R ii. 337.

reader to imagine. That abstention from performance guards against fantasy, the fantasy either of suicide or resuscitation; it also permits a self-conscious contemplation of what it is to feel like this, so that, once again, the words survey what at the same time they express. Henry James, magnificently, though with an edge of complaint, described his impression of such Tennysonian self-awareness when it was prolonged throughout one of the poet's readings to his guests: 'the whole thing was yet *still*, with all the long swing of its motion it yet remained where it was—heaving doubtless grandly enough up and down and beautiful to watch as through the superposed veils of its long self-consciousness'.[90] This stillness is the perpetual immobility of print which James acutely managed to hear even when Tennyson was reading aloud at him. Thus stilled, the gap between 'dies in me' and 'That dies not' represents not only a single breath, one sensation, the tick of a clock, but a pattern of character, something Tennyson lives in every breath he takes.

H. D. Rawnsley remembered how Tennyson read with a drama of pause:

Nor can I forget how, at the intervals or ends of a phrase such as 'And sorrow darkens hamlet and hall', the whole voice which had been mourning forth the impassioned lament suddenly seemed to fail for very grief, to collapse, to drop and die away in silence, but so abruptly that the effect upon one was—'He has come to a full stop; he will not read another line.'[91]

'He has come to a full stop'—Rawnsley may not have intended the pun between 'full stop' as a mark of punctuation and as an absolute halt but it seriously conveys how much existential weight Tennyson could put on a stop, the death in 'he will not read another line' and, equally, the courage for living on at a semicolon: 'The life that almost dies in me; | That dies not'.[92]

Tennyson made the pulse sound like a burden in the returning iambic mechanism of *In Memoriam*, VII, and he can also make the breath reverse its significance and tell of the longing to die in the very fibres and motions which keep us going. *In Memoriam*, LXXXVI:

> Sweet after showers, ambrosial air,
>   That rollest from the gorgeous gloom
>   Of evening over brake and bloom
> And meadow, slowly breathing bare

---

[90] *The Middle Years* (1917), repr. in *Autobiography*, ed. F. W. Dupee (1956), 594.
[91] Page, 63.
[92] Compare also *Maud*, I. xviii, R ii. 555.

The round of space, and rapt below
Through all the dewy-tasselled wood,
And shadowing down the hornèd flood
In ripples, fan my brows and blow

The fever from my cheek, and sigh
The full new life that feeds thy breath
Throughout my frame, till Doubt and Death,
Ill brethren, let the fancy fly

From belt to belt of crimson seas
On leagues of odour streaming far,
To where in yonder orient star
A hundred spirits whisper 'Peace'.[93]

'It all goes together', as Tennyson proudly said of the poem to
Allingham.[94] One implication of the remark is that the section is hard
to read aloud because it has to be taken in one breath and requires
good lungs. Hallam Tennyson wrote that the section 'gives preeminently
[Tennyson's] sense of the joyous peace in Nature',[95] and we can
imagine Tennyson's joy in his own nature, at least in that part of it
which permitted him graceful spans of respiration as he read this poem
out. But is the 'Peace' the poem concludes with that 'peace in Nature'
Hallam speaks of? In so far as the 'spirits' are 'in' a star, however
distant, they can be said to be in Nature, and their peace is in there
with them. Consider, though, the effect of attempting to speak this
poem in a single breath. The exceptionally proficient arch of the verse
blends with the climate it describes, the 'ambrosial air' being both a
song and an atmosphere, so that the human breath and the world's
breeze find a kinship, become 'breath-ren' (the pun is in the poem).
Even the best lungs will be weary at the close of the section, will have
the air left only to whisper the word which is the destination of this
eloquent trajectory, 'Peace'. Said in that way, breathing the reader's
last, the word can sound like the peace that death is, the peace of 'Rest
in peace'. The spirits may be saying 'hush' to the utterance of the
poem ('Peace, peace: | Dost thou not see my Baby at my breast, | That
suckes the Nurse asleepe?'[96]) as well as naming a calm which the poem
attains. The poem comes to completion by telling itself to speak no
more. *In Memoriam*, LXXXVI expresses, in the metaphysical depths of
melody nineteenth-century philosophers often heard in music, the

[93] R ii. 402.   [94] Allingham, 328.   [95] *Memoir*, i. 313.
[96] *Antony and Cleopatra*, v. ii. 307–9, Folio, ll. 3561–3.

longing to be out of Nature, to be dead, and expresses simultaneously a billowing delight in the performing breath as a sign of life.[97] Any vocalization of the poem has the risky privilege of trying to make this balance of absolute impulses audible; on the page, the poise is there to be seen. It creates within the printed voice a feeling for a place 'where beyond these voices there is peace', as the last line of 'Guinevere' has it,[98] by stretching breathing to breaking-point so that utterance of such extremity seems to take one outside of oneself. Gladstone wrote of 'Guinevere': 'No one, we are persuaded, can read this poem without feeling, when it ends, what may be termed the pangs of vacancy—of that void in heart and mind for want of its continuance of which we are conscious when some noble strain of music ceases . . . the withdrawal of it is like the withdrawal of the vital air.'[99] The void in the lungs *In Memoriam*, LXXXVI creates so rapturously does not make us long that we or the poem should be continued, it finishes so perfectly, and the reader who has expended breath to voice this air discovers in the triumph of having done it that he will be content, eventually, himself to have done with life.

Knowles remembered Tennyson admitting that 'O that 'twere possible' was difficult to read aloud because it should be 'read all through without taking breath'.[100] His memory must be at fault, for not even Tennyson could have read ninety-seven lines without taking breath, and the poem is, anyway, clearly split into sections. What Tennyson probably said to Knowles was that the individual sections of 'O that 'twere possible' should be read without taking breath; certainly, the thirteenth section holds up under a single breathing:

> But the broad light glares and beats,
> And the shadow flits and fleets
> And will not let me be;
> And I loathe the squares and streets,
> And the faces that one meets,
> Hearts with no love for me:

[97] Compare, for example, the climactic 'in des Welt-Atems wehendem All '['in the travailing/lamenting totality of the breath of the world'] in the 'Liebestod' of *Tristan und Isolde* (1865). 'Welt-Atems' is sung on a semibreve high G sharp, the highest and the longest note in the sequence; the control of breathing required and celebrated in this magnificent passage is an extended version of that rapturous extinction asked by Tennyson's poem.

[98] R iii. 547.   [99] *Quarterly Review* (Oct. 1859), Jump, 259.
[100] James Knowles, 'Aspects of Tennyson, II: A Personal Reminiscence', *Nineteenth Century* (1893), Page, 94.

> Always I long to creep
> Into some still cavern deep,
> There to weep, and weep, and weep
> My whole soul out to thee.[101]

If you take each 'and' as a cue for fresh intake of breath, you give the verse an itemizing vehemence which invigorates the speaker's despondency beyond what is true to his state when he says this. Of course, you may take pauses, and are required to do so by punctuation at least at the end of the third and sixth lines quoted, but taking a pause and taking a breath are not the same thing. The lines are contrived to sound at their close exactly as if the soul had been wept out with the breath used up in utterance. Hence, their exceptionally strong insistence on the long, expiring breath of the terminal rhymes, the same rhyme-vowel at the end of each line, reinforced by additional internal rhymes on 'me be' and 'weep, and weep, and weep'. The vowel-sounds are a mood of anguish; settled on the page, and made integral to the structure of the passage, they seem not only a mood but a lasting state of torment. Expressive skill assumes a lethal quality in achievements such as this; they help us understand why Tennyson said, 'Poetry is as inexorable as death'.[102]

Inexorable too, as *Maud* presents it, is the insistence of a formed, adult personality, the close weave of temperament which, like *In Memoriam*, LXXXVI but less delightfully, 'all goes together':

> Cold and clear-cut face, why come you so cruelly meek,
> Breaking a slumber in which all spleenful folly was drowned,
> Pale with the golden beam of an eyelash dead on the cheek,
> Passionless, pale, cold face, star-sweet on a gloom profound;
> Womanlike, taking revenge too deep for a transient wrong
> Done but in thought to your beauty, and ever as pale as before
> Growing and fading and growing upon me without a sound,
> Luminous, gemlike, ghostlike, deathlike, half the night long
> Growing and fading and growing, till I could bear it no more,
> But arose, and all by myself in my own dark garden ground,
> Listening now to the tide in its broad-flung shipwrecking roar,
> Now to the scream of a maddened beach dragged down by the wave,
> Walked in a wintry wind by a ghastly glimmer, and found
> The shining daffodil dead, and Orion low in his grave.[103]

---

[101] *Maud*, II. iv, R ii. 575–6.
[102] Reported in Sir Charles Tennyson, *Alfred Tennyson* (1949), 332.
[103] *Maud*, I. iii, R ii. 527–8.

Schoolchildren used to be taught that misrelated participles were a vice of style ('We saw the Alps flying over Italy'); this passage so deranges participial and adjectival relations that it produces a style of vice. 'Pale' refers neither to 'spleenful folly' nor to 'slumber' but arches back over them to the 'face'; 'Passionless' refers not back to 'cheek' but on to 'face'; 'Womanlike' refers not to the face which was beginning to seem to draw everything into its orbit but adverbially modifies 'taking revenge'; 'and ever as pale as before' then reverts to the face which governs the next two lines, till 'half the night long' switches again into an adverbial function to modify 'growing and fading'. It is a lurid demonstration of not being able to get something out of your mind or into focus. Imagine a voice marking these shifts of direction in the sentence, and you hear the self-rebuffs of obsession, but the page turns these swivels of attention into counterpoint, a melodious intricacy. It marks backward- and forward-looking syntactic relations with the same signs: 'Pale' in the third line is retrospectively qualifying, 'Passionless' in the fourth prospectively qualifying, but they are both introduced by commas. This could hardly be otherwise, for English does not have a system of punctuation adequate to show intended collocation, but that things could not be otherwise is here a tragedy of character. The speaker's irresistible propulsion towards thoughts of death moves through the single sentence from the figuratively dead eyelash and the 'deathlike' beauty to the encountered reality of the dead daffodil, and is then lifted up again to the stellar metaphor of 'Orion low in his grave'. The epithets 'shining' and 'low' seem to have swapped places, as if the speaker scarcely differentiated between daffodil and star. The thrill of the new feeling which has broken in on the speaker is 'dragged down by the wave' of the sentence, by the speaker's breathing habits, so that the 'Cold' which began the section meaning 'unresponsive' or 'stand-offish' or 'frigid' begins to sound in the 'wintry wind' with the coldness of the grave. A life's despondency is not lightly shrugged, and a reader can feel the weariness of the speaker growing on him if he speaks the lines, growing in his lungs as they strain to encompass the lines; as the lungs expand and contract, he feels in the tissues of his body, in his speaking substance, how a dreadful lassitude lags on in the fading face of the new. The experience of reading the passage without trying to voice it is an elegant spectator-sport, like cultivating someone else's hysteria; lending it one's own voice, one begins to pay the price of extreme temperamental preciosity.

This, then, is a grim kind of perpetuity, managed by control of the
breath, as were the fond and pious hopes of *In Memoriam* for a
perpetuation of life. The necessity of the page for such achievements
lies in the fact that the page allows us to see at once the triumph and
the toll of thus extending the voice beyond what it can with comfort
span, so that we can envisage the desire to cease in *In Memoriam*,
LXXXVI even while it primarily speaks of eternal peace, and also see the
perverse self-delight of 'Cold and clear-cut face . . .', the speaker's
conscious pride in his own tormented sensibility, though he complains
and longs to be released. Tennyson works out practically in his poetic
technique that if you want your self to survive death, you have to
confront those asperities that come to mind when you think of what
your self has already managed to survive. Immortality, that is, has its
darker congener in that fixity of temperament which Tennyson and his
age called 'morbidity'.

   All the accounts agree that he stubbornly continued to live as Alfred
Tennyson of Somersby even when he was Tennyson Laureate. He
kept the accent of his youth—'His command of the dialect after so
many years' absence from Lincolnshire was very wonderful';[104] he
preserved loneliness in marriage, a blunt social awkwardness when a
peer, and his pints of port against the orders of the best doctors. His
composition of his three greatest enterprises—*In Memoriam*, *Maud*,
and *Idylls of the King*—arched over decades of life in an astonishing
persistence of creative endeavour. His power of poetic incubation was
such that his son, hearing him read *Maud* some sixty years after the
poem began and nearly forty years after he had finished it, was given
the impression 'that he had just written the poem, and that the emotion
which created it was fresh in him'.[105] And

Cotton, the boatman at Freshwater Bay, remembered seeing him one day
stumping along in his cloak and sombrero past two girls who were sitting on
the beach. 'Look!' said one of the girls, 'there goes Tennyson!' 'What,' said the
other, 'did that old man write *Maud*?' Tennyson, whose hearing was always
extraordinarily acute, stopped, turned round to them and said, with
indescribable grimness, 'Yes, THIS OLD MAN wrote *Maud*,' then stumped along
on his way without another word.[106]

This is the best anecdote about Tennyson, and very well told: the poet,
in his eighties wearing still the cloak and sombrero of his twenties, his

---

[104] H. D. Rawnsley, Page, 72.   [105] *Memoir*, ii. 409.
[106] As told in Sir Charles Tennyson, op. cit. 523.

gait a little stiff but his ear supple as ever, walks by girls who might, more than half a century before, have been models for Maud herself. He hears his self called out into the question 'Did that old man write *Maud?*', called out not just by the words but by the assumption, implicit in the volume and pitch of the girl's voice, that he was too deaf to hear it. The answer was perfect, however writing fails to catch the voice in which he gave it ('with indescribable grimness')—'Yes, THIS OLD MAN wrote *Maud*'. Casually taken, it means that the younger man (about forty-five) who wrote *Maud*, re-creating in the poem a relation with an even younger man (about twenty-five), on whose experiences the poem is based, preserves an identity with this stumping codger on the beach. A deeper truth went out in his remark: the man who wrote *Maud*, the man who had the unhappy experiences with Rosa Baring, was already then this old man; the complete integrity of his artistic *œuvre* conferred on him a singleness of personality in all his self-divisions, a singleness like that which Arthur Hallam had thought might be seen by God when he looked at the soul after death: 'For in the Eternal Idea of God a created spirit is perhaps not seen, as a series of successive states, of which some that are evil might be compensated by others that are good, but as one indivisible object of these almost infinitely divisible modes . . .'.[107]

This creative self-identity happens to have been Tennyson's case, but the accident of his life-work yields a more general truth. His son records that 'as a boy he would reel off hundreds of lines such as these . . .

> The quick-wing'd gnat doth make a boat
> Of his old husk wherewith to float
> To a new life! all low things range
> To higher! but I cannot change.'[108]

If a personal life after death can be conceived as well as hoped for, it must be philosophically true that the 'I' cannot change, but then that changelessness, considered not as a logical truth but as a reality to be experienced, does not seem purely desirable. Arthur Hallam set the tone, and glimpsed the need for tonelessness, of Tennyson's writing as the poet of these immaterial matters when he described 'that mood between contentment and despair, in which suffering appears so associated with existence that we would willingly give up one with the

---

[107] 'Theodiciaea Novissima', written probably in 1831, Motter, 210. (Tennyson seems to have attached particular importance to this essay, as Motter notes, 199.)
[108] *Memoir*, i. 18.

other, and look forward with a sort of hope to that silent void where, if there are no smiles, there are at least no tears, and since the heart cannot beat, it will not ever be broken'.[109] Tennyson was very susceptible to such moods; their contour is native to his imagination, particularly in his exceptional sense for the way contentment borders on despair because immortality resembles insomnia. The insomniac's unremitting consciousness is one to which suffering appears so intimately associated with existence that he would willingly give up one with the other. Perhaps the idea that immortality was a form of eternal insomnia came to Tennyson from the way in which we often speak of death as 'rest' or 'sleep'. Whatever the reason, from his earliest poems he expressed the qualm within his deepest hopes in the thought of eternal life as an ecstasy of restlessness: 'Ah! these are thoughts that grieve me | Then, when others rest.'[110]

Remaining above all true to yourself might become a merely stagnant integrity:

> A spot of dull stagnation, without light
>   Or power of movement, seemed my soul,
> 'Mid onward-sloping motions infinite
>   Making for one sure goal.
>
> A still salt pool, locked in with bars of sand,
>   Left on the shore; that hears all night
> The plunging seas draw backward from the land
>   Their moon-led waters white.[111]

Tennyson, along with any other amateur Victorian naturalist, would have seen the pool on the beach, but he uniquely imagined it as orphaned from the sea, protesting its heritage ('still salt'), lying awake to hear all night its relations leaving it behind. The insomniac lives out of step with life's proper rhythms, and so feels his incessant consciousness as a crime: 'Behold me, for I cannot sleep, | And like a guilty thing I creep | At earliest morning to the door.'[112] The speaker of *Maud* promises fidelity for ever—'My dust would hear her and beat, | Had I lain for a century dead; | Would start and tremble under her feet, | And blossom in purple and red'[113]—but he does so as the culmination of a night of accumulated sleeplessness:

---

[109] 'Essay . . . on Cicero', Motter, 176.   [110] 'Memory' (1827), R i. 95.
[111] 'The Palace of Art' (1832), text here as revised in 1842, R i. 454.
[112] *In Memoriam*, VII, R ii. 326.
[113] *Maud*, I. xxii, R ii. 565.

But the rose was awake all night for your sake,
Knowing your promise to me;
The lilies and roses were all awake,
They sighed for the dawn and thee.[114]

Eternal life, it might seem, will only prolong this staying up all night, seeing in the dawn, without a party to pass the time or anybody to see it in with.

That promise to awake for Maud even were he dead returns in a humiliating parody when, at his maddest in Part II, he believes himself dead but still not soundly asleep:

O me, why have they not buried me deep enough?
Is it kind to have made me a grave so rough,
Me, that was never a quiet sleeper?
Maybe still I am but half-dead;
Then I cannot be wholly dumb;
I will cry to the steps above my head
And somebody, surely, some kind heart will come
To bury me, bury me
Deeper, ever so little deeper.[115]

The steps he would cry to now are not Maud's which he had fervently imagined coming to wake him with 'ever so airy a tread'.[116] When the speaker listens for her steps at the end of Part I, he has the enthusiasm of a man who thinks he is coming at last into possession of the social world which is his due; he is going to get his girl. But 'ever so airy', that camp and courtly lilt, alters, as a remark is twisted into a retort, and becomes these words at the end of Part II, where his enthusiasm is all and only for nothingness—'ever so little deeper'. The lilt is not of uppish confidence now; the voice quivers at the close into the cadence of a lullaby sung to himself, a lilt of plea for release from the paradox that all the man who has not wholly died can wish for is to be more absolutely dead.

Tennyson's most acute insomniac is the reluctantly immortal Tithonus. Part of Tithonus' predicament is that he has been awake

---

[114] Ibid., R ii. 564. William Empson's brilliant suggestion that the insomniac and, therefore, talkative flowers of *Through the Looking-Glass* . . . parody the end of Part I of *Maud* is supported by the fact that *Maud* itself recognizes insomnia as a parody of undying love in its own parody of the end of Part I at the end of Part II. See Empson's incomparable pages on nonsense and 'the alarming fierceness of ideal passion' in *Some Versions of Pastoral*, 285–8.

[115] *Maud*, II. v, R ii. 581.   [116] Ibid. I. xxii, R ii. 565.

for innumerable dawns, seeing his wife off to work; the 'gleaming halls
of morn' are the less attractive for him, he thinks with regret of the
'dark world' where he was born, Aurora's 'glimmering thresholds' jar
against his human 'dim fields',[117] because she is a creature whose
element is wakefulness, both in that she is an immortal goddess and in
that she is tutelary deity of the dawn, and he is a creature who has
underslept for aeons. Everything about light in the poem wounds
Tithonus' nerves; he is the antithesis of the Lady of Shalott because
she was half sick of shadows and he is sick for them, for everything
dark and misty. Endless dawns have made his eyes bleary, dulled his
hearing ('the ever-silent spaces of the East'), and perhaps hushed his
voice to dumbness. For 'Tithonus' develops with exceptional moral
sensitivity a question within all print: is this writing ever to be spoken?
He asked Aurora for immortal life, forgetting to stipulate for immortal
youth too. Now he asks for death. But that is a wish perhaps too weak
for words to name, and, immeasurably old as he now is, the words of
the poem may be condemned to roll for ever round inside his skull
because he may have lost his voice along with his dead, past self. It is
one poem if you imagine that he can utter this request (and that, as in
the original myth, she grants it, and turns him into the grasshopper
whose shrivelled voice the myth may be meant to explain); it is an other
poem if you imagine him, as print permits, eternally on the verge of
asking to die. 'Tithonus' began as 'Tithon' in 1833; the earlier version
is vocal and declamatory, it says he wants to die. The completed poem
of 1860 has, over the years, acquired a polished and inquiring
reticence about whether he can ever quite speak that desire. It has the
skill to arrange a meeting of Tithonus' (and Tennyson's) two minds on
the subject of personal immortality.

When it begins

> The woods decay, the woods decay and fall,
> The vapours weep their burthen to the ground,
> Man comes and tills the field and lies beneath,
> And after many a summer dies the swan.[118]

we appear to have a set of beautiful platitudes about mutability; the
verse has a strong propensity to the end of the line, a cadence always
on the way to extinction but drawing back from the end. It seems a
rhythm that wholly sorrows over death. The fifth line changes our
mind and its tune:

[117] R ii. 612.   [118] R ii. 607.

> Me only cruel immortality
> Consumes: I wither slowly in thine arms,
> Here at the quiet limit of the world . . .

What we had heard as the traditional, universal, voice of lyric regret turns out to be the fretful speech of a dramatized individual in a particular state. We hark back to the opening, and hear it again. The lyricism grows taut with a sense of Tithonus' plight, a contrastive intonation veers away from the initially evident sense of the lines. Instead of saying what everybody knows, they now begin to say what everybody knows and what one person knows to his cost: the *woods* decay, the *vapours* come to the ground, but *I* am deathlessly consumed. This contrastive emphasis, in the sound of which we begin to hear that dying may be no bad thing, questions the sentiments about death that welled up in the lyricism we first noticed, though our retrospective understanding does not abolish those sentiments. The page allows a simultaneous perception of both the traditional lament and the individual complaint.

Note how the first four lines come to rest at the line-end, and the way 'immortality' drags on into the 'consumes' of the sixth line, and compare the following four end-stopped lines succeeded by a run-on on the word 'immortality':

> Mighty Prophet! Seer blest!
> On whom those truths do rest,
> Which we are toiling all our lives to find,
> In darkness lost, the darkness of the grave;
> Thou, over whom thy Immortality
> Broods like the Day, a Master o'er a Slave,
> A Presence which is not to be put by . . .[119]

Tennyson retunes the cadence of 'immortality | Broods' into 'immortality | Consumes' as he revalues 'darkness . . . darkness', from which in Wordsworth's ode 'we' long to escape, when Tithonus speaks of the 'dark world', into which he longs to escape. In the Wordsworth, inspirited strength in the human being runs beyond the line-end after four end-stopped lines; in the Tennyson, presumptuous and exhausting error made Tithonus want to go beyond death, beyond 'the goal of ordinance | Where all should pause'.[120] Immortality broods over

[119] Wordsworth, 'Ode: Intimations of Immortality from Recollections of Early Childhood', Hayden, i. 527.
[120] R ii. 609.

Tithonus like the day in a sense more painful than it does over
Wordsworth's babe, because it was the goddess of dawn who gave
Tithonus his withering perpetuity, watching over him like a Mistress
over a Slave, and her presence ('the glow that slowly crimsoned all |
Thy presence and thy portals . . .'[121]) is indeed 'not to be put by'
though for Tithonus that does not express a moral duty to his human
nature, as it does for Wordsworth, but only his lack of nature, of the
natural 'power to die'.

Tradition thinks poems too lack the power to die. They are more
durable than bronze, powerful rhymes which outlive marble and gilded
monuments so that they become themselves monuments of unageing
intellect. We praise verse as deathless, but should we do so if we
recognize that, while we remain only human, the prospect of not dying
cannot be borne? The essence of pleasure in reading 'Tithonus' rests
on the very changelessness which pains its speaker; the poem, like
Aurora herself, renews its beauty reading by reading, or if it does not
do so, we begin to think there is something wrong with it. Our delight
in that permanence surrounds and mutes our pity and dismay at the
human plight it for almost ever contains. The written version is more
durable and happier in its duration than the speech of the individual it
holds; this is a further, and profound, development of Wordsworth's
realization of the potentialities of metre not strictly connected with
passion.

Tennyson himself had no time for the vaunting of artistic
immortality. He wrote to Miss Chapman in 1886 about his opinions on
the matter: 'I should say, as Napoleon is reported to have said. When
someone was urging on him how much more glorious was the
immortality of a great artist, a painter for instance, than that of a great
soldier, he asked how long the best painted and best preserved picture
would last. "About 800 years." "Bah! telle immortalité!" '[122] One way
'Tithonus' and its writer live in time as Tithonus the speaker does not
is through the poem's re-setting of Wordsworth's 'Ode . . .' to new and
dissentient harmonies; another way is in the plain fact that Tennyson
took up the first version of this poem twenty-seven years after writing it
and, with astonishing permanence of temperament, re-worked it into
the perfected poem it now is. This supremely revealing act of
imaginative self-confrontation shows how creativity lives in time and
also shows, in the dubious immortality of the completed work, the

[121] R ii. 611.   [122] *Memoir*, ii. 332–3.

substance of Tennyson's doubts as regards the immortality of persons, even of those who thus create.

Morbidly Speaking

Tennyson, like the nightingale in one of his poems, 'delighteth to prolong'.[123] A central subject of the poetry is identity prolonged back into the past and on into eternity. When he read aloud 'he lengthened out the vowel a in the words "great" and "lamentation" till the words seemed as if they had been spelt "greaat" and "lamentaation" ', and he could be heard 'lingering with solemn sweetness on every vowel sound'.[124] It sounds as if he was describing himself when he revered the sustaining power of St Telemachus's last, martyred words: 'And preachers lingered o'er his dying words, | Which would not die, but echoed on . . .'.[125] He himself could turn such length of line in diverse expressive directions, towards grimness, entrapment, or buoyant self-preservation. Edward FitzGerald looked askance and with a tetchy moralism at the doleful *longueurs* of *In Memoriam*: 'We have surely had enough of men reporting their sorrows: especially when one is aware all the time that the poet wilfully protracts what he complains of . . .'.[126] This is bracing, and Tennyson would have had some sympathy with his old friend's impatience, but FitzGerald, keen as he is that Tennyson 'get on with it', is less clear about what 'it' is—are men to stop having sorrows, or just stop reporting them? The relation between ethical life and imaginative work which FitzGerald's irritation assumes is simple enough: they are alternatives to each other, and there is a moral weight to the choice between them. On the other hand, John Bayley appreciates the way 'Tennyson lingers; indeed it is his genius to linger in a situation in which nothing appears to be done—nothing else, it appears can be done—but to hang around'.[127] What is not clear in this is what the 'situation' is supposed to be; Professor Bayley may mean just 'being alive', and, if that is what he means, that too would be clear about the bearing of imagination on action. With nothing to be done, the imagination serves as an intricate pastime, a luxury on the

---

[123] 'The Palace of Art', R i. 449.
[124] H. D. Rawnsley, Page, 63; Allingham, 158.
[125] 'St. Telemachus' (1892), R iii. 227.
[126] Letter to W. B. Donne, 28 Feb. 1845, in *The Letters of Edward FitzGerald*, Terhune, i. 486.
[127] Bayley, op. cit. 193.

desert island of ethics. Tennyson's work is situated at the junction of these views. Early poems such as 'The Palace of Art' explicitly debate the conflicting demands; the later self-consciousness of the mature style constantly implies them. And it is a particular characteristic of the dramatized lyric that it presents imaginative activity as a form of conduct, so that, for example, literary conventions appear as answers to needs or impulses, and those values which might narrowly be thought 'aesthetic', such as prolongation of melody or spinning a yarn or vowel-music, resound in imagined contexts where they are judged as behaviour. Martin Dodsworth gives an exemplary instance of criticism responding to such aesthetic conduct when he says of the patterns of verbal repetition in Tennyson's poetry that they show 'the difficulty with which this poet turned his mind from the subjects of real concern to him'.[128] The critic recognizes the texture of composition, its wrought and appreciable facets, as a witness to the moral nature (not always perhaps a model to be imitated) of the writer; he is asked to do so by the writing, for such dramatizations of the lyric manifest self-consciously the process by which, in Proust's terms, an accent is formed.

Within the skill of Tennyson's writing the persistence of a self can be heard. He was long governed by a sense of the idiosyncratic, indeed, even the socially shameful, needs that drove him to the creation of poems: overwhelmed despondency, suicidal desires, erotic anxiety—the poems conserve such obscure longings in their lyricism. His material conservatism works not by obscuring these rooted states but by bringing them to the light of a day shareable in eloquence. Writing saves Tennyson from himself but also (more obdurately determined against the wishes of the public culture the poems lay claim to) saves him for himself. The poems convert intolerable material into objects of delight so that Tennyson shall not be altogether lost to himself when he is at last countenanced by his age. If a writer seeks an audience, he looks for a responsive community, and one which cannot be identified as a group solely by its relation to his work, for such a community would not stand to the writer as a conscience, but only as an echo. The writer who mediates between groups noticeably diverse, as Empson describes the task, and especially a writer as popular as Tennyson was, will inevitably find partial much of what these independent groups offer by way of response to his work,

[128] Dodsworth, 'Patterns of Morbidity' in I. Armstrong (ed.), *The Major Victorian Poets*, 7.

partial even to the point of parody. His public self, the self received in his work, may strike him as alien, and difficult to recognize, but that self-alienation too must find its way into his writing. After having published, he is doubly aware of himself as, in part, taken up into the community (even as one of its 'standards') and as, in part, deflected from it, his own character a residue left out of public account. Tennyson creates his reticences as a measure of and for his community.

It was not petulance only or touchiness which made him miserable about people's comments on his work. Meredith asked him, of one carper who had said that Tennyson was not a great poet, 'Why should you mind what such a man says?' and Tennyson answered, 'I mind what *everybody* says.'[129] Meredith was too confident that a poet can insulate himself linguistically from what is said around and about him; Tennyson knew that his creative life went on so entirely *in* the language that he could not but mind what '*everybody* says' because what everybody said was the air he breathed, the material he composed. This '*everybody*' constantly throws a shadow over Tennyson's trust—in religious doctrines, in his own abilities, in the public—for his poetry strives to be the perfection of the surface, but the shadow threatens to make it only perfectly superficial. That is, natural languages make some desires sound simple and capable of satisfaction, as when the narrator's mother in *A la recherche* . . . 'longed that the dead might return to life, so that she could have her mother near her sometimes'.[130] The words are familiar, and the longing sounds simple, though reflection on the language tells us that this longing is deeply paradoxical—for two people to be 'near' each other they must both have bodies, but what body has a dead person? Similar processes operate in any shared vocabulary; they affect ethical and political as well as religious discourse. Words pass through many mouths and lives, and, as they pass, they become rich and solid but also curiously elusive and hollow, as Baudelaire noted in his journal—'Commonplaces. Holes dug by generations of ants.' The deeper a hole, the deeper its emptiness. Tennyson works for trust in the surfaces of language, and in the hopes of those surfaces, while he is aware of what reflection teaches about the deceptions of the surface, and, even more acutely aware, of how the surface of language belongs to that '*everybody*' who

---

[129] Reported by T. H. Warren, *The Centenary of Tennyson, 1809–1909* (Oxford, 1909), Page, 157.
[130] *Le côté de Guermantes, À la recherche* . . . , ii. 343.

will say almost anything, thereby producing the risk that a poet's
reliance on the language might turn into a shallow trust in illusory
depths, his own hopes being only the refraction of corrupt idiom.

Dilemmas of trust had been with Tennyson from early days. He
experienced them not only in face of the abstract diversity of his
audience but in the bosom of his family. He was wary of those 'who
may be almost termed callous to all the kindlier sympathies, however
they may cloak their deficiencies with the vesture of *form*',[131] wary to
the point of disliking the often-used subscription to letters, 'thine
affectly' because he feared that it might mean 'affectedly' quite as
easily as 'affectionately'.[132] He suspected poetic forms as well as social
ones of being sometimes mere 'vesture'. Indeed, *In Memoriam* worries
for a long time that elegiac expression amounts to no more than a
customary donning of black—'In words, like weeds, I'll wrap me
o'er'.[133] The poem tries to trust in God; it also has to make an effort to
trust in itself. This is one source of its structural looseness, because
Tennyson searched for a compositional flexibility which would
authentically match the rhythms of the experiential theodicy in the
poem. God's plan in the event which prompted *In Memoriam*, Hallam's
precocious death, was hard to understand, and an argued neatness of
structure would not have conveyed the strenuous exercise of trust that
event required of Tennyson as he composed the poem. So too, the
word 'trust' itself ranges through the work from stiff, semi-formal
pronouncements like those of an official ('I trust I have not wasted
breath'[134]) to intimate expressions of a struggle with perplexity:

> . . . the songs, the stirring air,
> The life re-orient out of dust,
> Cry through the sense to hearten trust
> In that which made the world so fair.[135]

He places 'stirring' close to 'dust', though it qualifies 'air', and the
slight hint of stirring up dust touches the image of the spirit
resurrecting the dead body which the lines contain with a delicate but
full sense of the complexities of imagining a body's resurrection. The
trust in the lines is all the stronger for having so little to go on; the
creator of the world is here no more than a 'that which', as again in the

---

[131] Letter to Elizabeth Russell, 10 Mar. 1833, in Lang and Shannon, i. 89.
[132] Letter to R. M. Milnes, 8 or 9 Jan. 1837, in Lang and Shannon, i. 149.
[133] *In Memoriam*, v, R ii. 322.
[134] Ibid. cxx, R ii. 440.   [135] Ibid. cxvi, R ii. 437.

awed bareness of *In Memoriam*, XCV, 'And came on that which is'.[136]

This troubled and creative relation to the surface of language, these vagaries of trust in form, could be called Tennyson's sense of 'shame'. Shame, for Tennyson, is an awareness of the base motives which can create aspiration towards the Good, as in 'St Simeon Stylites', of the manner in which skill may convert itself into a hollow knack, of the selfish, circling energies which inhabit even an artist as honest and diligent as he also knew himself to be. The first book he published had as its epigraph 'haec nos novimus esse nihil' ('we know these works of ours are nothing worth').[137] Hallam Tennyson recalled that when he and Lionel were children 'he would never teach us his own poems, or allow us to get them by heart',[138] perhaps because Tennyson thought his poems were not fit for children, even (or especially) his own. The epigrams he wrote in his notebook in the late 1860s turn on the conjunction of trust and shame, the swell and fail, of Tennyson's feeling for his own work:

> What I most am shamed about,
> That I least am blamed about;
> What I least am loud about,
> That I most am praised about.[139]

And *Idylls of the King* is an epic of shame, shame in front of a culture's standards, shame at those standards, shame in those standards.

The interim between *Poems by Two Brothers* (1827) and the completed *Idylls* (1874) is full of shame that nobody thought shameful:

> Better thou and I were lying, hidden from the heart's disgrace,
> Rolled in one another's arms, and silent in a last embrace.[140]

A close acquaintance with the state of shame speaks in the ambiguous caesura at 'Better thou and I were lying' where it seems, briefly, that deceit and burial come to the same thing, because shame would equally rather lie or die than be brought to face itself—which is why the classic expression of shame is that it cannot face an other person or itself, shamefacedness:

> but I,
> So much a kind of shame within me wrought,
> Not yet endured to meet her opening eyes . . .[141]

---

[136] Ibid. XCV, R ii. 413.   [137] *Poems, By Two Brothers* (1827), title-page.
[138] *Memoir*, i. 371.   [139] R iii. 7.
[140] 'Locksley Hall' (1842), R ii. 123.   [141] *The Princess*, R ii. 238.

'Guinevere' pivots on that critical, Tennysonian moment when 'the withering look | Of men and angels . . . will turn | Their dreadful gaze on me alone'.[142] It imagines the climax of judgement as an aversion of face, an evasion of voice:

> But when the Queen immersed in such a trance,
> And moving through the past unconsciously,
> Came to that point where first she saw the King
> Ride toward her from the city, sighed to find
> Her journey done, glanced at him, thought him cold,
> High, self-contained, and passionless, not like him,
> 'Not like my Lancelot'—while she brooded thus
> And grew half-guilty in her thoughts again,
> There rode an armèd warrior to the doors.
> A murmuring whisper through the nunnery ran,
> Then on a sudden a cry, 'The King.' She sat
> Stiff-stricken, listening; but when armèd feet
> Through the long gallery from the outer doors
> Rang coming, prone from off her seat she fell,
> And grovelled with her face against the floor:
> There with her milkwhite arms and shadowy hair
> She made her face a darkness from the King:
> And in the darkness heard his armèd feet
> Pause by her; then came silence, then a voice,
> Monotonous and hollow like a Ghost's
> Denouncing judgment, but though changed, the King's:

> 'Liest thou here so low, the child of one
> I honoured, happy, dead before thy shame?
> Well is it that no child is born of thee. . . .'[143]

The self-encirclement of shame is embodied in the line 'There with her milkwhite arms and shadowy hair', because the rhyme of 'There' and 'hair' at the beginning and the end turns the line in on itself, as Guinevere's consciousness turns in on itself in a motion of self-protection. She hides her face by making it 'a darkness' but then she is *in* her own darkness when she hears his voice. The slight formal difference of 'There with her milkwhite arms and shadowy hair' from the blank verse which surrounds it corresponds to her own difficult adjustment to the Arthurian world. Such differences come again in the varying levels of homeliness and epic dignity in this passage with its small locutions petrified in a grand style. 'She sat' coming at a line-end

---

[142] 'Remorse' (1827), R i. 100.   [143] R iii. 540–1.

leaves her for a moment sitting just as any woman might, before the continuation of the phrase into 'Stiff-stricken' freezes her with the decorum of the epic, recalls her to her status as a Queen. 'High, self-contained, and passionless, not like him' works the shift from Queen to woman in the reverse direction. She thinks of Arthur first in terms that are apt to him, apt to the level of linguistic elevation he demands; we might imagine the lines continuing 'not like him, | When first with burning greaves he strode the field' or some such fabricated dignity, but what comes is a break into a childishly fond and vulnerable snatch of Guinevere's own words athwart the tenor of the verse—'not like him, | "Not like my Lancelot" '. '*My* Lancelot': it is daring and right of Tennyson to let such an accent of lovers' babble into the verse. The cat's-cradle of self-inculpation and self-exculpation which occupies Guinevere, as she simultaneously blames herself for not living up to Arthur's ideals and blames the ideals for their lifelessness, demands such a double loyalty of the writing to the divergent styles of her duty and her desires. We hear the correctness of the style as the embodiment of Arthur, and across that, in the phrase 'My Lancelot', in the dip of the voice it suggests, the lovable and illicit body of Lancelot. When Arthur at last speaks to her, his voice too shares in these styles at variance with each other, alive with conflicting impulses in the words 'the child of one | I honoured, happy, dead, before thy shame'. These words stand on the dignity of a Latinate syntax by trailing the adjectives so far behind what they qualify, by the ellipsis of 'happy, dead before thy shame' (probably an English version of an ablative absolute, meaning 'fortunate in that she died before you were disgraced'), but the rhythm of his speech conveys the gulp and rush of his judgement, his struggle even now to be fair. That rhythm, far from the measured fluency of iambics, makes the epic elevation sound like psychologically realistic and fractured speech. To Guinevere in her shame, the verdict comes 'monotonous and hollow', but the page allows us to see that the words may not have sounded only like that; the skill of the verse gives us dispassionately culprit and judge and the air between them.

His most shameful poem is *Maud*. Hutton, as usual in sympathetic accord with the poet, remarked of *In Memoriam* that 'Tennyson always shows a certain tendency to over-express any morbid thought or feeling he wishes to resist' but Bagehot simply feared that the story of *Maud* was 'evidently very likely to bring into prominence the exaggerated feelings and distorted notions which we call unhealthy'.

George Brimley, eager to vindicate Tennyson against the chorus of 'morbid', 'unhealthy', 'perverse', 'infantile', and so on and on was prompted to complain after the reception of *Maud* that literary critics worked 'morbid' too hard. He regretted that it had 'acquired a perfectly new meaning of late years, and is made to include all works of art, and all views of life that are coloured by other than comfortable feelings'.[144]

This 'perfectly new meaning' had come to be in the hundred years up to the publication of *Maud*. Dr Johnson's *Dictionary* gives only 'Diseased; in a state contrary to health' for 'morbid' and his only example is from Dr Arbuthnot; 'Though every human constitution is morbid, yet there are diseases consistent with the common functions of life.' His dictionary also gives 'morbidness', 'morbifical' and 'morbifick', and gives all of them in the same physiological sense. 'Morbid' was a firmly medical term, but in the course of the succeeding century was ever more pressed from its physiological home into the service of the new needs of a developing psychology. Johnson himself resisted the attempt to translate the notion of morbidity from the physical into the mental or temperamental sphere in his forty-third *Rambler* essay:

It is observed by those who have written on the constitution of the human body, and the original of those diseases by which it is afflicted, that every man comes into the world morbid, that there is no temperature so exactly regulated but that some humour is fatally predominant, and that we are generally impregnated, in our first entrance upon life, with the seeds of that malady, which, in time, shall bring us to the grave.

This remark has been extended by others to the intellectual faculties. Some that imagine themselves to have looked with more than common penetration into human nature, have endeavoured to persuade us, that each man is born with a mind formed peculiarly for certain purposes, and with desires unalterably determined to particular objects, from which the attention cannot long be diverted, and which alone, as they are well or ill persued, must produce the praise or blame, the happiness or misery, of his future life.

This position has not, indeed, been hitherto proved with strength proportionate to the assurance with which it has been advanced, and, perhaps, never will gain much prevalence by a close examination.[145]

---

[144] Hutton, *Essays*, 392; Bagehot, *National Review* (Oct. 1859), Jump, 218—see also the phrases 'mental malaria', 'semi-diseased feeling', and 'diseased moodiness of feeling' in the same review; Brimley, 'Alfred Tennyson's Poems', in his *Cambridge Essays* (1855), Jump, 191. Some of the background to this matter is discussed by A. C. Colley in her *Tennyson and Madness* (Athens, Ga, 1983).

[145] *The Rambler*, No. 43 (14 Aug. 1750). I quote from the Yale edition, vol. iii, ed. W. J. Bate and A. B. Strauss (New Haven, 1969), 231-2.

Dr Johnson refers with fine scepticism to eighteenth-century psycho-
logists in his 'some that imagine themselves to have looked with more
than common penetration into human nature' but his scepticism
depends on the conviction that any pretence to 'more than common
penetration' into human nature is bound to turn out bogus because
human nature is just the region of the common, and so no specialized
technique of inquiry suits it. The era of experts in human nature was
dawning round Johnson even as he held out against it; the notion of
'close examination' of the human mind and character altered.
Johnson's resistance to the theory stems from his need to reassert the
moral primacy of resolve in human affairs, from his characteristic
allegiance to freedom of the will even though he had no great
expectations of what the exercise of that freedom could achieve. He
knew from experience that to feel yourself inhabited by impulses you
do not quite control was a common lot; his own backslidings and
shiftless lassitudes bore witness to the possible truth of a theory of the
constitutional morbidity of personality. Such self-embattled ethical
striving in Johnson remains an imperative for the Victorians too, but
for them it shares the scene of possible attitudes to one's own mind
with versions of the medico-psychological account which Johnson
repudiated. This change of scene also gives a new setting and role to
the concept of will as it comes in the English nineteenth century to be
tugged about between theological, metaphysical, and socio-psycho-
logical senses (a change in the word FitzGerald ignored when he
complained so 'straightforwardly' of Tennyson's wilful lack of proper
resolve in taking so long over *In Memoriam*).

In the Preface to *Julian and Maddalo*, Shelley describes himself
with urbane irony as he describes Julian:

Julian is an Englishman of good family, passionately attached to those
philosophical notions which assert the power of man over his own mind, and
the immense improvements of which, by the extinction of certain moral
superstitions, human society may yet be susceptible. Without concealing the
evil in the world, he is for ever speculating how good may be made superior . . .
Julian is rather serious.[146]

'The power of man over his own mind' does not for Shelley lie in a
Johnsonian resolve but in implementing that technical knowledge of
the self which psychology promised to yield; it depends in fact on the

---

[146] 'Preface' to *Julian and Maddalo: A Conversation* (1819); I quote from T.
Hutchinson's edition of the *Poetical Works* (1905, repr. 1943), 190.

very theories of 'more than common penetration' which Johnson resisted. (Shelley uses 'philosophical' in the old sense of 'related to an empirical science'). The passage betrays an enthusiastic cast of mind, showing an unphilosophical passion in its attachment to philosophical notions. 'Certain moral superstitions' is a suave phrase for Christianity and the other world-religions; '*immense* improvements' (my emphasis) flushes with a less subtle excitement, and the immensity of the improvements expected permits Shelley to write of the 'extinction' of those superstitions without bothering to make it clear whether the superstitions will become extinct quietly and inevitably by natural wastage or whether they, and those who hold them, will have to be extinguished. Again, to be 'for ever speculating' is here a good thing; it shows a mind tirelessly alert and challenging. There seems to be no worry in the phrase that eternal speculation could be a condition of permanent intellectual disquiet.

The nature of Shelley's enthusiasm is political, and the belief that the supposed discoveries of psychology had a directly political significance is common at his time. Thus, Mill writes of his father's work in his *Analysis of the Phenomena of the Human Mind* (1822–9): 'In psychology, his fundamental doctrine was the formation of all human character by circumstances, through the universal Principle of Association, and the consequent unlimited possibility of improving the moral and intellectual condition of mankind by education.'[147] Shelley's 'immense improvements' have grown up to become James Mill's 'unlimited possibility of improving the moral and intellectual condition of mankind by education'. The swelling of the political aspirations is in direct proportion to the degree in which the psychological hypotheses are regarded as indubitable—Johnson's 'position' about the determination of human character becomes Shelley's 'philosophical notions which assert' and then Mill's 'fundamental doctrine'. Power over one's own mind extends now to power over the minds of others 'by education'. It did not sufficiently trouble Mill or his father that if 'all human chracter' is formed by circumstances, and if our notions of good and bad are part of our character, then our view of what constitutes an 'improvement' in the human condition is itself a function of determining circumstances and, therefore, as we are formed under very various circumstances of prosperity, of cult, of culture, and so on, we may have correspondingly various, and perhaps

---

[147] *Autobiography* (1873; repr. from MS, 1971), ed. J. Stillinger, 65–6.

irreconcilable, plans for each other's improvement. The ethical issue of what makes an 'improvement', and how to reconcile divergent views on that issue, is not recognized as such, being considered uncontentious, a matter only of technical description.

Wordsworth was more daunted by these prospects than Shelley or the Mills. The 1805 *Prelude* reads:

> Hard task to analyse a soul, in which
> Not only general habits and desires,
> But each most obvious and particular thought—
> Not in a mystical and idle sense,
> But in the words of reason deeply weighed—
> Hath no beginning.[148]

The 1850 version revises this to 'Hard task, vain hope, to analyse the mind . . .', perhaps in response to claims such as those of Mill's *Analysis of the . . . Mind.* The revision also makes the lines more exact in terminology; the aim of psychological science is not the soul but the mind; there are, then, new problems about the relation between 'soul' and 'mind' to add to the classic questions about the relation of 'soul' and 'body'. Arnold was also impressed by the vanity of psychological hopes. He may have thought the Romantic poets did not know enough, but the doom of his own generation, he sometimes felt, was to know too much:

> We shut our eyes, and muse
> How our own minds are made.
> What springs of thought they use,
> How rightened, how betrayed—
> And spend our wit to name what most employ unnamed.[149]

When the speaker of 'Resolution and Independence' submits to 'Dim sadness—and blind thoughts, I knew not, nor could name',[150] he feels some loss in his lack of a terminology for his own experience; the loss is calmly noted and endured, it does not prompt him to systematic curiosity. Arnold's attitude to psychological naming is more harried and less clearly purposive. He romanticizes empirical study of the mind when he presents it as introspective musing, but then he has a conversely bleak view of the results of study—they merely hamper us

---

[148] *Prelude*, II. 232 ff.
[149] *Empedocles on Etna*, I. ii. 327 ff. I quote from the edition of Kenneth Allott (1965), 169.
[150] Hayden, i. 552.

with unwieldy knowledge, an inner lexicon which never descriptively exhausts the mind though it tires the mind out to compile it, making the mind unfit for the proper business of right conduct. Arnold's recourse to a Johnsonian call for resolve, for employing of one's faculties rather than worrying how they are to be described, doesn't carry Johnson's conviction or convincingness. Specifically, Arnold lacks Johnson's assumption that, being in a minority, the speculators on the mind are deluding themselves about their own 'more than common penetration'. Arnold's few 'we' may not be a happy few but it is impossible not to feel they are sensitive and distinguished beyond the gainfully employing 'most'. Johnson doesn't believe in the psychological knowledge; Arnold believes he is possessed of it, but wishes he weren't. It looks as if Arnold believed that one thing it was not desirable to have a free play of the mind about was the mind.

Introspection, psychological self-contemplation, has become a symptom of morbidity, not because the Victorians wanted to get on with the business of life in a state of robustly unexamined self-confidence, but because the ethical imperative, 'know thyself', changes its meaning as a result of the development of the institutional discipline of empirical psychology which thinks it has supra-individual, ethically neutral terms for the individual, and in whose hands the self is, in the words of F. H. Bradley, 'invaded by another, broken up into selfless elements, put together again, mastered and handled, just as a poor dead thing is mastered by man'.[151] Jowett wrote to soothe Tennyson about accusations of morbidity: 'A certain man on a particular day has his stomach out of order and the stomach "getteth him up into the brain", and he calls another man "morbid". He is morbid himself and wants soothing words, and the whole world is morbid with dissecting and analysing itself and wants to be comforted and put together again.'[152] By the middle of the nineteenth century, awareness of the circumstances which form the self was not only a key to unlimited human self-improvement but also a possible cause of mental unease, perhaps even disease; the 'analysis' of mind which was to combat morbidity and other failings then came to be suspected of promoting what it was supposed to remove, as people became, in Jowett's phrase, 'morbid with dissecting and analysing' their selves. In a world 'crammed with self-consciousness',[153] the power of man over his own

---

[151] *Ethical Studies* (1876, 2nd edn., 1927), 20.
[152] *Memoir*, i. 426.
[153] Ibid.

mind turned out to be self-thwarting, and left its possessors impotently knowledgeable—morbid, in fact.

Such is the crippled state Arnold described in 1853 when he gave his reasons for not reprinting *Empedocles on Etna*:

What then are the situations, from the representation of which, though accurate, no poetical enjoyment can be derived? They are those in which the suffering finds no vent in action; in which a continuous state of mental distress is prolonged, unrelieved by incident, hope, or resistance; in which there is everything to be endured, nothing to be done. In such situations there is inevitably something morbid, in the description of them something monotonous. When they occur in actual life, they are painful, not tragic; the representation of them in poetry is painful also.[154]

Arnold gave up too soon on the potential prospects of such a subject for poetry. He might have guessed from Tennyson's earlier work and the capacity it showed in poems like 'Mariana' for creatively prolonged distress that two years after he wrote the 1853 'Preface' he would be proved wrong by *Maud*. Tennyson called the hero of that poem 'a morbid poetic soul, under the blighting influence of a recklessly speculative age'[155] and, remembering Shelley's 'he is for ever speculating', we realize he was not thinking only of the financial speculation which figures prominently in the story. The proximity of 'morbid' to 'poetic' in 'a morbid poetic soul' both distinguishes morbidity from poetry and suggests their possible interlinking. As the word 'morbid' had come to mean something like 'hyper-sensitive' or 'extremely self-conscious', it is not hard to see why a poet might have been thought professionally morbid. It was, for instance, reported of Carlyle that 'He gravely thought poetry a sort of disease . . a sort of fungus of the brain—& held as a serious opinion, that nobody could be properly well who exercised it as an art . . . .'.[156] The poets shared their time's mistrust of poets and poeticality: Browning wrote of Shelley that 'sympathy with his kind was evidently developed in him to an extraordinary and even morbid degree';[157] Tennyson trounced the historian, Lecky, who was a fan of Tennyson's 'Lucretius', when Lecky 'quoted with admiration the lines: "Poor little life that toddles half an

[154] Preface to *Poems* (1853), repr. in Allott's edition, 592.

[155] *Memoir*, i. 396.

[156] Elizabeth Barrett Barrett, letter of 11 Aug. 1845 to Robert Browning, in *The Letters of Robert Browning and Elizabeth Barrett Barrett, 1845–1846*, ed. E. Kintner, 2 vols. (Cambridge, Mass., 1969), 151. The ellipsis marked by two dots is in the original.

[157] 'Introductory Essay' (1852), to *Letters of Percy Bysshe Shelley*, P. ii. 1009.

hour | Crown'd with a flower or two, and there an end—" ', the poet
turned on his admirer and 'said that my liking for them only showed
the morbidness of my nature'.[158]

What makes Tennyson in this matter adept at composition of his
age's contradictory attitudes to poetry—prophecy and disease—is his
double sense of poetry both as an expression of morbidity and as an
expressive power which releases from morbidity. In *Julian and
Maddalo* and *Empedocles on Etna*, Shelley and Arnold contrast the
deranged pulses of madness or morbidity to the measured pulsations
of music; they count on the belief that music hath charms to soothe the
ravaged mind. So Arnold has Empedocles' friend lament the fact that
Empedocles no longer takes comfort from music: '. . . he has laid the
use of music by, | And all which might relax his settled gloom'.[159] In
Maud, the music is often the deepest madness. What counts for us as a
literary value—intensity of expression or acuteness of observation or
skilful patterning of speech, for example—transmutes itself into a
possible symptom of the morbidity of the speaker as the poem
dramatizes the roots of lyricism in the pathological.

Arthur Hallam had praised Keats and Shelley for being 'poets of
sensation rather than reflection':

Susceptible of the slightest impulse from external nature, their fine organs
trembled into emotion at colours, and sounds, and movements, unperceived or
unregarded by duller temperaments. Rich and clear were their perceptions of
visible forms; full and deep their feelings of music. So vivid was the delight
attending the simple exertions of eye and ear that it became mingled more and
more with their trains of active thought, and tended to absorb their whole
being into the energy of sense. Other poets *seek* for images to illustrate their
conceptions; these men had no need to seek; they lived in a world of images . . .[160]

It does not take much to turn such poetic souls into victims of morbidity
with their tremulous vitality, absorption in the senses. Like the Lady of
Shalott, they occupy a 'world of images', and may be 'half sick' or
'semi-diseased', though Hallam does not here register the risk
involved in their abilities. It is that risk *Maud* takes as its subject.

Consider the moment when the morbid poetic speaker concentrates
on Maud's hand:

[158] *Memoir*, ii. 206.
[159] *Empedocles* . . . , I. i. 83–4, Allott's ed., 152.
[160] 'On some of the characteristics . . .', Motter, 186.

O heart of stone, are you flesh, and caught
By that you swore to withstand?
For what was it else within me wrought
But, I fear, the new strong wine of love,
That made my tongue so stammer and trip
When I saw the treasured splendour, her hand,
Come sliding out of her sacred glove,
And the sunlight broke from her lip?[161]

In so far as this is poetry we may be pleased with the sharpness of
visual focus, the slow motion and close-up which 'sliding' in 'her hand
| Come sliding out of her sacred glove' suggests, for these offer
testimony to the delightful exertion of the eye, absorbing thought into
the energy of sense. Yet the fiction of this extended dramatic
monologue is that the speaker is not trying to produce poetry, though
what he says may be poetical; the fiction of the monodrama requires us
to believe that the speaker does not aim at 'vivid imagery' for the sake
of his readers because he has no readers. He betrays rather in this
exaggerated language the threatening avidity with which he looks at
Maud's hand, a passion charged up by the long waits and physical
abstemiousness of Victorian courtship. We may rightly fear for Maud
with his stare on her, and fear for the speaker who calls a glove 'sacred'
and means it. The speaker fears too, though his 'I fear' seems
colloquially casual. Indeed the problem of how to take that 'I fear'—
whether with the passing lightness which suits conversation, or with
the scrutiny and brooding which we give to texts—structures the
poem, for we are constantly having to decide how seriously to take
what we read him saying, to decide whether he is 'only' being poetical.
    The dramatic context in which Tennyson makes us hear the lyric
voice opens up questions about the motive of the patterns in the lines,
as when the speaker turns his nailing stare onto Maud's brother:

His face, as I grant, in spite of spite,
Has a broad-blown comeliness, red and white,
And six feet two, as I think, he stands;
But his essences turned the live air sick,
And barbarous opulence jewel-thick
Sunned itself on his breast and his hands.[162]

The texture of the verse thickens without warning at the 'But'.
Considered as poetry, such unforeseen alterations may be accepted

[161] *Maud*, I. vi, R ii. 536–7.   [162] I. xiii, R ii. 546.

purely as having the value of a composed surprise—they show a
nimble attention to the demands of variety. Coming from the speaker,
the shift of tone here might indicate rather a jumpiness in his account,
a switch of focus which may be not deliberate but scatty. On the one
hand, 'his essences turned the live air sick' simply means that the man
was wearing too much after-shave; on the other hand, the sickly
redolence of his cologne is swiftly transformed by the pun on
'essences' into the very aroma of his soul. The effect of the writing is
brilliant, but we are to imagine the speaker is not speaking or writing
for effect. The expressive intensity of the writing may be heard, in the
dramatic situation, as the speech of a morbidly hyper-sensitive
individual. Or take 'barbarous opulence jewel-thick | Sunned itself on
his breast and his hands'. 'Barbarous opulence' may come from the
description of Satan at the start of *Paradise Lost*, Book II—'where the
gorgeous East with richest hand | Showers on her kings barbaric pearl
and gold'.[163] While Tennyson might be said to allude to Milton, the
speaker does not so certainly allude. He may, of course, remember
Milton's lines, but, as he is not engaged in literary composition, he may
not be invoking the device of literary allusion. There could be rather
an insinuating pressure of the thought 'Maud's brother is the devil', a
thought that comes near to being a delusion. Again, the fantastic
metaphorical vivacity of 'Sunned itself on his breast and his hands',
where the jewels bask on his waistcoat and hands as a lizard basks in
the sun, comes from the speaker with a hatred which fuels his lurid
sight. Speech-act and writing-act are not the same, coming in the
fiction from different people.   IE, dramatic monolog

   Tennyson insisted on the distinction between the composition and
the utterance of *Maud*: '. . . there are two or three points in your
comment to which I should take exception, e.g. [you refer to] "The
writer of the fragments, etc.,", surely the speaker or the thinker rather
than the writer . . .'.[164] The distinction needs to be held on to if we are
to recognize the inquiring achievement of *Maud*. I have pointed out
already instances where Tennyson exploits locally the ambiguous
existence of his text as written and spoken, to produce intonational
puns which delineate a conflict of attitudes, or to adjust a flash of
feeling in the voice to the calmer thoughts also expressed in the written
words. Here the possibilities of creation in the region between writing

---

[163] *Paradise Lost*, II. 3–4; I quote from the text of Alastair Fowler (1968, corrected
edn., 1980), 509.
[164] *Memoir*, i. 408.

and speech are realized across an entire poem in the attribution of its
metre and general formal arrangement to its author (Tennyson) and its
passions to its speaker (who is nameless), so as to draw attention to the
drama of why he needs to speak like this, why Tennyson writes this
way. Tennyson's further distinction between 'the speaker' and 'the
thinker' of *Maud* reveals another opportunity he took from the silence
of writing, the opportunity to give the force of speech withheld, of
inward brooding, as in

> For by the hearth the children sit
>   Cold in that atmosphere of Death,
>   And scarce endure to draw the breath,
> Or like to noiseless phantoms flit:
>
> But open converse is there none,
> So much the vital spirits sink
>   To see the vacant chair and think,
> 'How good! how kind! and he is gone.'[165]

The silent oppression all about these children broods in their capture
by print; they do not say 'How good! . . .', something seals their lips, as
Tithonus perhaps never quite brings himself to say 'Release me, and
restore me to the ground', but the page shows them thinking it, and
suggests the constriction which denies them speech. Any reading out
of the poem will obviously have to say that line aloud, and so lose the
working of the written word as a sophisticated kind of 'thought-
bubble'. Compare the last lines of 'Pelleas and Etarre':

>   The Queen
> Looked hard upon her lover, he on her;
> And each foresaw the dolorous day to be:
> And all talk died, as in a grove all song
> Beneath the shadow of some bird of prey;
> Then a long silence came upon the hall,
> And Modred thought, 'The time is hard at hand.'[166]

'All talk died'—the phrase makes us despair of community, as *Idylls of
the King* often does. 'All talk' is something less than the 'open converse'
which the children miss in *In Memoriam*; it sounds like the mere clack
of tongues. Yet when even talk dies, something significant is lost,
people shrink back into their isolate, shamed selves, and we are left
only with the grim thought of Modred voicelessly conspiring.

---

[165] *In Memoriam*, xx, R ii. 339.   [166] (1869), R iii. 508.

There is no open converse in *Maud*, for the speaker is speaking to nobody in particular, except perhaps himself. The imaginative richness of a Browning dramatic monologue often derives from the subtle tensions implicit between speaker and suggested interlocutor, but Tennyson's monodrama plays rather on what it is to have nobody to talk to. The speaker has an odd way of, as it were, button-holing the absence of his interlocutor. The opening of the poem is very frank and communicative, too much so, in fact, for no conversation could comfortably start as this poem starts:

> I hate the dreadful hollow behind the little wood,
> Its lips in the field above are dabbled with blood-red heath,
> The red-ribbed ledges drip with a silent horror of blood,
> And Echo there, whatever is asked her, answers 'Death'.[167]

Nothing can be answered to these lines. (What could you say—'Oh, really, how fascinating' or 'Yes, yes, so do I, so do I'?) '*The* dreadful hollow', '*the* little wood': the definite articles press on us the obsessive predominance of these places in the speaker's imagination but as yet the reader, not knowing the story, has no idea which hollow and wood are of such concern, and so even less idea why they matter. Conversely:

And my pulses closed on their gates with a shock on my heart as I heard
The shrill-edged shriek of a mother divide the shuddering night.[168]

Here it is the indefinite article, 'a mother', that is odd. In the context of the story he is telling about his father's death, it is probably his own mother he means by this vague reference which could sound desultory, coolly generalizing or appalled in its shying from the particular. He changes his mind and his meaning of his words faster than a reader can follow, though the reader eventually latches on to his unhinged vagaries of thought:

Villainy somewhere! whose? One says, we are villains all.
Not he: his honest fame should at least by me be maintained:
But that old man, now lord of the broad estate and the Hall,
Dropt off gorged from a scheme that had left us flaccid and drained.

Why do they prate of the blessings of Peace? We have made them a curse . . .[169]

In 'we are villains all', 'we' has the universal reference of a proverb; the speaker rejects this claim—'Not he', that is, 'Not my father, he at least

---

[167] *Maud*, I. i, R ii. 519.   [168] Ibid., R ii. 520.   [169] Ibid.

was not a villain'. The 'us' of 'left us flaccid and drained' indicates only the limited family circle of the speaker. The speaker seems carefully to watch the extension of his first persons plural so as to avoid the entrapment of generalities such as 'we are villains all', but at 'we have made them a curse' the reference of the 'we' has become extremely general, it cannot mean only the speaker's family. This instablity in the scope of the 'we' pervasively affects the prophetic voice denouncing a society's ills which the hero tries to summon in the opening section of _Maud_, so that the tirade sounds on occasion more like a catalogue of his family's wrongs. The reader wonders whether he would be so righteously indignant if he were not domestically so upset.

Sometimes he seems not to know quite who Maud is either:

> Perhaps the smile and tender tone
> Came out of her pitying womanhood,
> For am I not, am I not, here alone
> So many a summer since she died,
> My mother, who was so gentle and good?[170]

Maud is the woman who pitied him, the woman behind the third person feminine 'her', but two lines later the third person feminine 'she' means the speaker's mother, without signal of the shift of reference. He puts it awkwardly—'since she died, | My mother'—as if he recognized the grammatical unclarity of his own use of the pronoun and had to add an explanation in the next line. The reader has been puzzled for a moment and is left with the feeling that he mistakes Maud for his mother. Indeed, he does not securely know who he is himself:

> . . . I know her own rose-garden,
> And mean to linger in it
> Till the dancing will be over;
> And then, oh then, come out to me
> For a minute, but for a minute,
> Come out to your own true lover . . .[171]

He shifts the mood of his verbs from indicative ('know . . . mean') to imperative ('come'), and it can sound as if the speaker is saying 'I intend to linger in the garden till the ball is over and then come out to myself', as if he were both himself and the Maud for whom he waits, to whom he pleads. Tennyson has marked the move from indicative to

---

[170] Ibid. I. vi, R ii. 536. [171] Ibid. I. xx, R ii. 561.

imperative with a semicolon between 'over' and 'then', a pause re-inforced by the lineation. We can eventually *see* what he means but that makes us only the more wonder quite whether he clearly hears what he says. The lines continue:

> That your true lover may see
> Your glory also, and render
> All homage to his own darling,
> Queen Maud in all her splendour.

The second-personal emphases ('your . . . your . . . your') bring him closer to the Maud who first appeared in this passage in the third person ('her own rose-garden'); this approach to her, along with the imperatives, might lead us to expect that 'Queen Maud . . .' is a vocative, but then the syntax reverts to referring to her in the third person ('all her splendour') and the phrase turns out to be in apposition to 'his own darling', as if shyness had overcome him and he could not quite address her directly. As she retreats into grammatical distance, so does he; he begins as a first person ('I know'), becomes a second person, placed from Maud's position, and fades into the third person of 'his own darling'.

Later, he seems to be both himself and Maud's brother. The second part of the poem begins, ' "The fault was mine, the fault was mine" ', and we learn that the speaker has killed Maud's brother in a duel. But we do not learn for thirty lines that these words were said by the dying brother and that the speaker is only quoting them. The effect then is to divide the guilt between them, and share too the consequences of that guilt, as the speaker seems to suggest in the odd line 'Was it he lay there with a fading eye?'[172] It has the thrill of a melodrama's rhetoric when he says it, but as Part II of *Maud* proceeds, and we find that Maud has died and that the speaker is under the delusion that he himself is dead, then the line asks a graver question, '*Was* it only the brother who died?' The three of them come closer together in death than they ever were in life. The unstable pronominal referents in *Maud* could often be made clear in advance to an auditor of the poem by a careful, guiding vocalization but in the reading of the printed text such a voice is not supplied but looked out for.

Tone and concern veer:

> Morning arises stormy and pale,
> No sun, but a wannish glare

[172] Ibid. II. i, R ii. 566–7.

> In fold upon fold of hueless cloud,
> And the budded peaks of the wood are bowed
> Caught and cuffed by the gale:
> I had fancied it would be fair.[173]

The description sets out with literary assurance and shows its grasp on what is being said in the agile, improvisatory rhythms which fold with the contours of what is seen—'bowed | Caught and cuffed by the gale'. This is reassuringly adroit in a recognizably literary manner, but then we get a strange, invasive sense of pique, or self-bemusement, at 'I had fancied it would be fair'. Out of nowhere the verse goes slack as the speaker outsteps the proprieties of poetic description to append this note about himself. Such swivels of attention pass nearly unnoticed in a conversation (they are endemic in self-communion) but stuck down *writing* like this in writing they trouble communication by their egotistically casual manner, like someone who has deliberately under-dressed for a formal gathering. Or take the glint when the speaker imagines the proprietorial satisfaction of Maud's brother:

> Seeing his gewgaw castle shine,
> New as his title, built last year,
> There amid perky larches and pine,
> And over the sullen-purple moor
> (Look at it) pricking a cockney ear.[174]

The imperative '(Look at it)' is placed in brackets to add to the phantom confidentiality of the passage, as if the speaker nudged his interlocutor in the happy expectation of an answering and deprecatory smirk only to find his elbow meet nothing but air (for nobody is listening to him). The command reaches out for an imagined obedience but only he is there to share his joke, to follow his own instructions.

The shock to reason comes again when the maddened speaker imagines that his dead fiancée visits his bedside:

> Do I hear her sing as of old,
> My bird with the shining head,
> My own dove with the tender eye?
> But there rings on a sudden a passionate cry,
> There is some one dying or dead,
> And a sullen thunder is rolled;

---

[173] Ibid. I. vi, R ii. 534.   [174] Ibid. I. x, R ii. 540.

> For a tumult shakes the city,
> And I wake, my dream is fled;
> In the shuddering dawn, behold,
> Without knowledge, without pity,
> By the curtains of my bed
> That abiding phantom cold.[175]

'The shuddering dawn' exquisitely transfers the epithet from the speaker, perhaps waking to find the blankets have slipped off, to the dawn itself, but in the context of his mania the transference of his shudders to the dawn implies less a rhetorical skill than a crazed inability to distinguish between his own feelings and actuality (making dramatic the fallaciousness of the so-called pathetic fallacy as Browning does in his poems of erotic madness such as 'Porphyria's Lover'). Madness peaks at 'behold' because we have no idea who he imagines is listening to him now; the command comes over as self-addressed, but surely he needs no encouragement to attend to what wholly preoccupies him. Even were we to understand the address of the 'behold', we would not know how to obey it because of the ambiguous function of 'Without knowledge, without pity', which may adverbially modify 'behold' or qualify 'that abiding phantom'. How sharp too the demonstrative 'that', like a hand held out in a seizure of adjuration towards the beloved, at the same time holding her off and holding her to him. When Lear says 'Looke on her? Looke her lips, | Looke there, looke there',[176] he has those about him who would fain call him master and do what he says, if they could. Not so the speaker of *Maud*.

Tennyson supremely creates here the contour of a mind obsessed and in recoil from its own obsession. He creates it 'without knowledge, without pity'—without recourse to the explanatory categories of empirical psychology or to the plangencies of *bel canto* mad-scenes or the gothic ravings of the Maniac in *Julian and Maddalo*. The clarity of his realization would have been impossible without the dramatic understanding of lyricism which the printed voice permits him, for only that drama of the lyrical enables the poem's incomparable surface to be intimate with its depths, neither solely as a symptom nor solely as an assuagement but as the very predicament of music. To the distinction between the speaker or thinker of *Maud* must correspond a distinction between the listener and the reader of the poem. If we

---

[175] Ibid. II. iv, R ii. 573–4.
[176] *King Lear*, v. iii. 310–11; Folio ll. 3282–3.

vocalize the text, we may listen to it, but we are then listening to our own voice, even if that voice is speaking for the poem's fictional speaker. We take on ourselves the task of being fair representatives of his idiosyncrasy but, just because it is an idiosyncrasy, we cannot be too sure we grasp it. If we do not vocalize it, we can be privy to his thoughts but only at the price of witnessing the fracture of his personality as he becomes his own interlocutor. *Maud* has imagined terrible implications for Mill's unworried dictum that 'eloquence is *heard*, poetry is *over*heard'.[177] Overhearing the piece as poetry, we supervise a plight, with delight or perhaps clinically. It shames us to overhear this record of shame which was not exactly meant for us. The poem's swaying between its status as speech and as writing causes responses to the poem to diverge. A modern critic states that '. . . the hero's feeling prevents the reader from identifying with him fully and sympathetically, despite the fact that such identification is almost required by the nature of the poem'[178] but Hallam Tennyson felt that when you heard Tennyson read *Maud* 'You were at once put in sympathy with the hero.'[179] These two views respond to different though connected versions of the poem—Hallam's responds to the text read aloud in company, when there is visibly an audience for what the poem says though not quite for the person who speaks in the poem; the modern critic comments rather on the text read alone and privately voiced. The difference between these two versions, especially when Tennyson was his own reader, is 'truly that between looking on a black and white engraving and the coloured picture from which it had been taken',[180] though we should be careful about this description for it implies that the written text is derivative and depleted, whereas, like an engraving, it has a sharp point of its own. Naturally, we regret the loss of the poet's guiding and reviving voice, but the poem, though created with that voice, was not created only for it; indeed, the exceptional delicacy of its sense for loneliness of utterance lives best in the written text as that text, like a person in need, solicits and shies from what we can do for it, from our voicings. It is both an attempt at kindness and a form of theft to say to the afflicted, 'I know *exactly* how you feel'. Created

[177] 'Thoughts on Poetry and its Varieties' (1833), repr. in *Dissertations and Discussions, Political, Philosophical and Historical*, 4 vols. (1859), i. 71.

[178] R. W. Rader, *Tennyson's 'Maud': The Biographical Genesis* (Berkeley and Los Angeles, 1963), 116.

[179] *Memoir*, i. 396.

[180] T. A. Trollope, *What I Remember* (2 vols., 1887), ii. 293–4, quoted in P. Collins, *Reading Aloud* (Lincoln, 1972), 4.

dilemmas of voice in *Maud* keep us alive to the fact that, because we
don't know exactly how he sounds, we can't know exactly how he feels.
The confidence of empirical psychology in the indubitability which
attaches to observations of the mind because 'our perception and
belief' of '*the Mind itself*' 'are necessarily more immediate than of
anything else'[181] gets duly shaken as the poem's double status as
written and spoken puts in question the notion of immediacy, of a
transparence of the personality either to itself or to expert observers.
Its other double status as fictionally uncomposed and actually
composed permits Tennyson to stage, in the texture of the verse itself,
the bearings on each other of lyrical convention and pathological
symptom and, by doing so, to bring to light the ethical life of our terms
both for the mad and for our selves.

The convention he most alters to bring out its substantial life is that
of the lyric address, in which a poet speaks to non-human or inanimate
creation. A poet writes, 'I said to the lily, "There is but one . . ." ' and 'I
said to the rose, "The brief night goes . . ." '[182] and it is read as a mild
and traditional pretence. When the speaker of *Maud* says these things,
we cannot rely on their being only elegant, for there is an accumulation
of crazed speech harboured behind the lines. He talks to flowers
because Maud has still not come into the garden to talk to him. His
lyricism has disorder at its roots, a fear that Maud and everybody else
in the unresponsive world will fail to give him a hearing. Tennyson's
accomplished composition is intimate with distress here, as it is in *In
Memoriam* which equally addresses itself to things which cannot speak
(wild bells, witchelms, dim dawn, sweet new-year delaying long) as if
they could hear the poet, and does so because the person who once
best listened to him has now become a speechless thing. He was wary
of conventionalities in speaking of the dead (about dead selves or dead
others), and this can make his private letters of condolence sound curt.
But the poems give new life to wariness in a proper vigilance as they
feel for the actual needs to which poetic conventions may respond. As
he writes and imagines, the world of 'all talk' in these matters, of ' 'Tis
common' and Goethe's truth, remains the dark world where his
utterance was born, and to which it answers in return. The potential
threat of a corruption in surface-idiom which would make his lines
merely continuous with evasive chatter is absorbed into this self-

---

[181] See the discussion of introspective immediacy in A. Kenny, *Action, Emotion and
Will*, esp. chs. 1 and 2.
[182] *Maud*, I. xxii, R ii. 563.

conscious eloquence which knows when and why it has to keep quiet in order to keep faith. 'Faith is the substance of things hoped for';[183] hope might rush towards its object, want it grasped as soon as glimpsed—so too a reader might wish to fill a printed voice at once with speech, and lack the patience to falter over its reticent ambiguities. Faith, however, is an activity of recognition and waiting: it must find that its hopes are already fulfilled (believing that is what makes it 'faith'), though it still so lives in time that it is sensible to speak of faith as continuing to have 'hope'.[184] Speech is both only potential and already evident within a printed voice. As such, it is for a poet the 'substance of things hoped for'. Tennyson's faith is a form of trust, trust both in the objects of its belief and in the significance of its own capacity to hold these beliefs and hold on to hopes. It is, then, in a paused voice, a reviving convention, that Tennyson can do more than worry at how others will receive his works or fret over their failure to hear. He can make the workings of his apprehension the element of his desires. The lyric invocation of inhuman objects is filled with a dramatic life; a longing for actual conversation permeates the conventional address, and, what is more, the conventional entirely accords with the actuality of longing.

Both *Maud* and *In Memoriam* feel most of all what it is to lose a loved voice. The speaker of *Maud* is moved first to 'adore, | Not her, who is neither courtly nor kind, | Not her, not her, but a voice',[185] a voice we never hear speak directly in the poem. Hallam Tennyson notes that *In Memoriam* grew from the question 'Where is the voice I loved?'[186] Tennyson had many recollections in the poem but none dearer than recalling Hallam's irrevocable voice, as is shown by the fact that the text of *In Memoriam* doubles the word 'dear' only once—in section CXVI, line 11: 'The dear, dear voice that I have known'.[187] Death did its worst when it interrupted their conversation:

> For this alone on Death I wreak
> The wrath that garners in my heart:
> He put our lives so far apart
> We cannot hear each other speak.[188]

---

[183] Epistle to the Hebrews 11: 1.

[184] Commentators have disagreed about the correct translation of 'hypostasis', which I give here in the King James version as 'substance'; others prefer 'assurance' or 'conviction'. See F. F. Bruce, *The Epistle to the Hebrews* (1964), 277–9.

[185] *Maud*, I. v, R ii. 533.   [186] *Memoir*, i. 107.

[187] The line read so in 1850–1; Tennyson revised to 'And that dear voice I once have known' in 1855 and subsequent editions, R ii. 437.

[188] *In Memoriam*, LXXXII. R ii. 394.

The poem's corresponding hope is to 'lift from out the dust | A voice as unto him that hears'[189]—a hope for resurrected speech which takes the condition of the printed voice in which Tennyson wrote as the emblem of his aspirations as well as their perfect embodiment.

Hence, the imaginative centrality to his *œuvre* of the ninety-fifth section of *In Memoriam* in which the writer becomes a reader, as Tennyson goes back to the letters of the dead Hallam and comes close to his friend though he does not feel, as Hallam had felt, 'how wretched it is to be thrown back into the region of letters after treading the giddy heights of existence, in which your dear presence & converse had placed me'.[190]

> we sang old songs that pealed
> From knoll to knoll, where, couched at ease,
> The white kine glimmered, and the trees
> Laid their dark arms about the field.
>
> But when those others, one by one,
> Withdrew themselves from me and night,
> And in the house light after light
> Went out, and I was all alone,
>
> A hunger seized my heart; I read
> Of that glad year which once had been,
> In those fallen leaves which kept their green,
> The noble letters of the dead:
>
> And strangely on the silence broke
> The silent-speaking words, and strange
> Was love's dumb cry defying change
> To test his worth; and strangely spoke
>
> The faith, the vigour, bold to dwell
> On doubts that drive the coward back,
> And keen through wordy snares to track
> Suggestion to her inmost cell.
>
> So word by word, and line by line,
> The dead man touched me from the past . . .[191]

Tennyson begins to read as the community singing is hushed; he stays awake, as so often, while others sleep. When he writes of what

[189] CXXXI. R ii. 451.
[190] Letter to Emily Tennyson, 7 Apr. 1832, Kolb, 546.
[191] *In Memoriam*, XCV. R ii. 412–3.

happened then, the lines tug against the knowledge they still maintain, the knowledge that Hallam, and the past self he corresponded with, are gone, perhaps for good. He puts the emphasis repeatedly, as if in doubt that he could ever secure it in writing. The stress is semantic— two oxymorons, 'silent-speaking' and 'love's dumb cry', coupled with a 'strangely' underlined by 'strange' then double-underlined by a further 'strangely'. He breaks the 'periodical movement' of that human frame, the stanza, as he breaks it nowhere else in the section; for the first time in this section, he runs over two successive stanza-breaks ('the dead: | And strangely', 'spoke | The faith'), and, also for the first time, he runs on over the line-end for five lines in succession (from 'the silence broke' to 'drive the coward back'). It is apt that the rhyme which frames this climactic stanza of written utterance should be 'broke' | 'spoke' because his speech comes near to breaking when he reads what Hallam wrote. It is this broken voice which allows us to hear, through it, the accents of the dead. Rhythm, sense and rhyme collaborate to make that break.

Through it, the dead man reaches from the past. The fading-out of a sense of regular line and stanza produces an avid syntax, pressing on past all previously-fixed boundaries, so that a speaker of the lines gets vocally involved to the point where, the energy of declamation becoming most present to the voice, it can seem that it is Tennyson who declares his constancy in these words, the poetic composition sounding as the utterance of someone still alive rather than as the record of a document from the past. It is with effort that a voicing remembers these are Hallam's vows—there is an exceptional feeling that Hallam here speaks for Tennyson, much though Tennyson lends him the voice he needs so to speak.

But the page has not forgotten loss in its own atmosphere of death. This passage of absolute touch is introduced by an elegant pun on paper. Hallam's correspondence now reads as 'those fallen leaves which kept their green'; his letters are deciduous ('fallen') as well as evergreen to match the way in which he both died and has become immortal in his death. Those letters are 'leaves', bits of paper, but also living parts of a tree—the section later observes 'The large leaves of the sycamore' moved by a breath of air, as these dead words move in Tennyson's breath. When we read Hallam's writing through Tennyson's printed voice, the poetic composition crosses and brings to life the passion originally and long ago expressed. 'Love's dumb cry defying change' sounds for a moment as if love showed itself as love by withstanding time; it could be imagined that Hallam had written something like 'Nothing will ever change me, or my feelings for you'. (It can only be imagined as Hallam's letters to Tennyson have in fact been destroyed.) The lineation momentarily halts the words at this point, and the intense, proximate internal rhyme of 'cry defying' makes the line pivot on itself, as if it refused to move forward or change. The verse, though, moves on over the line, and changes into 'love's dumb cry defying change | To test his worth'. This means that Hallam had not refused to contemplate changing any more than his revived words refuse change; it must now be imagined that he wrote something like, 'As we grow older and change, I shall through all such changes stay true to you'. Hallam when alive found in letters only 'a few moments of your life . . . no looks, no tones'. The exchange of words between Tennyson and Hallam in this section contains no looks, no tones, but this ambiguity of voice and gesture manages to give a single moment of Tennyson's life, one night out on the lawn, the weight and shine of something he took to be more than this life. In the light of such writing, it may matter less that Tennyson or Hallam (or any one else) is immortal than that we have our moments.

In Memoriam, XCV lends itself welcomingly to a reader's voice. It is melodious in the extreme. Yet at heart (and its heart is in its mouth), it is born from and lives with 'silent-speaking words', a reticence in trust. A Proustian 'accent cordial'[192] makes itself out, though nothing in the text supplies a notation for that accent. The accent can be elicited, for instance, from the cadential handling of monosyllabic chains and disyllabic breaks in those chains:

> I read
> Of that glad year which once had been
> In those fallen leaves which kept their green,
> The noble letters of the dead . . .

Nineteen successive monosyllables ('fallen' pronounced 'fall'n' not 'fallen') swell up to 'noble letters'. The word 'noble' demands to be heard in its full insistence against the previous run of monosyllables, heard with the long-nurtured esteem for Hallam which speaks in the word. Tennyson may have wilfully protracted his sorrows—according to FitzGerald—but the protraction rings in 'noble' with respect and fidelity, not with repinings. Or again:

---

[192] *Du côté de chez Swann, À la recherche . . .* , I. 42.

> So word by word, and line by line,
>   The dead man touched me from the past,
>   And all at once it seemed at last
> The living soul was flashed on mine . . .

Twenty-five monosyllables conduct a reading to the only disyllable in the stanza, the flash of vocal emphasis on 'living' with all its vehemence of asseveration that the 'dead man' *is* a 'living soul', just as Edward Spedding was both 'dry dust' and a 'true soul and sweet', just as 'Vastness' ends in the knowledge that 'the dead are not dead but alive'. Here, as there, though, the writing asks the voice to persevere amidst the perplexities of assurance, sensing how doubt is through-composed even with such passionate conviction. 'Noble' in 'noble letters of the dead' requires a rhythmic contrast to what has gone before, a resonant allegiance to a past value, but Hallam's worth, even as Tennyson declares it, undergoes change, for the juxtaposition of 'noble letters' asks a pause to stop the consonants from running together, and so imposes a halt to vocal afflatus. That pause is for thought, thought about how Hallam's nobility can be sensed now only in his letters, and, more generally, about the very literary status of the concept of 'the noble' at the time of Tennyson's writing. What has gone before and what comes after 'noble letters' perfectly places the phrase as a key-scene of lyric declaration.

So too, at the most abstract reach of *In Memoriam*, XCV:

> The living soul was flashed on mine,
>
> And mine in this was wound, and whirled
>   About empyreal heights of thought,
>   And came on that which is, and caught
> The deep pulsations of the world . . .

The section often relies on closely-packed stresses to point its urgent concerns, packed usually in groups of three ('that glad year', 'those fallen leaves', 'love's dumb cry', 'dead man touched'). Right here, Tennyson places five stresses consecutively—'came on that which is'—an emphasis of absolute being, made rhythmically, which answers back to the absent emphasis of *In Memoriam*, VII's 'He is not here'. The voice is given free rein to convey a fullness recognized, but is then at once reined in at 'deep pulsations', where the forced neighbourhood of terminal and initial 'p' asks a dip of voice, an abeyance between the two surges of stress, a silence in which to hear the very depth of the pulse it suspends. The verse swells and fails, and, as it does so, gains a rhythmic composure to meet the alternations of assurance and doubt in the moral world where Tennyson lived. He gives up his voice to find it, and this is not, as Hallam had thought it, an 'alas irremediable loss'.[193]

In the eighty-ninth section of *In Memoriam*, Tennyson remembers, not Hallam writing, but Hallam speaking:

> But if I praised the busy town,
>   He loved to rail against it still,
>   For 'ground in yonder social mill
> We rub each other's angles down,
>
> 'And merge' he said 'in form and gloss
>   The picturesque of man and man.'
>   We talked: the stream beneath us ran,
> The wine-flask lying couched in moss,
>
> Or cooled within the glooming wave . . .[194]

Hallam speaks four rhyme-words ('mill', 'down', 'gloss', 'man) but none of his rhyme-words rhymes with any other word he speaks; Tennyson writes four rhyme-words ('town', 'still', 'ran', 'moss') and none of his rhyme-words rhymes with the rest of what he has written. Tennyson's words and Hallam's words yet rhyme completely with each other here, keeping each other perfect company.[195] Hallam's voice comes to life in Tennyson's verse, and the change it undergoes there only tests its worth. The same is true of Tennyson's own voice as he wrote it. Speaking the stanzas aloud, we must decide whether or not to distinguish Hallam's quoted words from Tennyson's verse, and we have only a voice to make that distinction. The printed voice is shared between them, without distinction (though the inverted commas preserve their distinct identities); they are composed as one, become one now in the fact that each equally needs to draw on a reader's breath for continued life. It is not a fantasy, an impersonation, or a ventriloquial show that Tennyson achieves and undergoes in these lines but rather the measured fullness of his reticent art, alive with the depths of surface. Death might have put Hallam and Tennyson so far apart they could not hear each other speak but not so far apart that Hallam might not still be able to read what Tennyson had become able to do with their conversation, as we still can.

[193] Kolb, 546.
[194] *In Memoriam*, LXXXIX. R ii. 407.
[195] Charlie Meredith pointed this out to me.

3

COMPANIONABLE FORMS

Ideals of Marriage

TENNYSON died and was buried with his copy of *Cymbeline* open at
Act v, Scene v, line 263:

POSTHUMUS.　　　Hang there like fruit, my soul,
　Till the tree die.

which 'he always called among the tenderest lines in Shakespeare'.[1]
The lines have an ambiguity which makes their tenderness more
whole-hearted, though less fully expressed. Reading them, we cannot
be sure who or what is Posthumus's 'soul'. He embraces Imogen after
the trial of their long separation, and may either command himself in
future to keep more perfect faith with her—'Cling to Imogen while
your body lives, my soul'—or assure her that from now on she will be
able to depend on him—'Cling to me, Imogen, I will be true to death'.
He may address his soul as an Imogen or Imogen as his soul. The
living Imogen is an emblem, or at least a promise, for Posthumus.
George Bernard Shaw attributed the co-existence of an actual and an
emblematic Imogen in the play to the fuddled state of the playwright's
mind and leanings; Imogen presented 'a double image—a real woman
divined by Shakespear without his knowing it clearly . . . and an idiotic
paragon of virtue produced by Shakespear's *views* of what a woman
ought to be'.[2] 'Paragon' is Iachimo's word for Imogen[3] but the phrase
'idiotic paragon' belongs wholly to Shaw, as does the exasperated force
behind it which wants to make it a pleonasm. For Shaw, idiocy lies in
any conversion of people into paragons, not in the conversion of this
fictional person into a paragon. In his ethical geometry, 'idiotic
paragon' expresses an analytic truth as 'three-sided triangle' does in

---

[1] *Memoir*, ii. 428. I quote *Cymbeline* here in the text given in *Memoir*.
[2] *Shaw on Shakespeare*, ed. Edwin Wilson (1961), 40.
[3] *Cymbeline*, v. v. 147; Folio, l. 3426.

Euclidean geometry, and so the question whether Shakespeare could have handled Imogen's transformation better does not arise because the point is that he should not have touched such stuff. There's an impatient pressure in Shaw's comment which pushes him to that italicized antithesis—'*divined*' as against '*views*'; the antithesis itself glamourizes an artist's power to create beyond what he can articulately desire. Shakespeare's views were less obtuse than Shaw allows, and his vision less preternaturally acute. For one thing, a 'paragon' in Shakespeare's English is a point of comparison, a competitor, and a marriage-partner as well as something supremely excellent (see *OED*, senses 2 and 3—senses which appear to have died out of Shaw's English). The word ranges thoughtfully from discernment to extreme admiration, moves between rivalry and true consorting. Shakespeare's plays cover the same range; they show, and many characters in them know, that the human inclination to idealize the beloved may be freakish, or disastrous, or ennobling. Shakespeare in his writing so variously demonstrates the role of 'views' of love in loving, so intimately delineates fact, aspiration to an ideal, and the fact of such aspiration, that he demonstrates a knowledge of the place of love in his world more accurate and reliable than any divination could afford. Shaw's tetchiness suggests that, beneath a humane acquiescence in foible and inconsistency which seems to be the substance of his protest against Shakespearian paragons, there may lie a desire, not to save human beings from the tyranny of idealization, but to rescue ideals from contamination by the human.

But Shaw may have been annoyed less by Shakespeare than by what he thought the nineteenth century had made of Shakespeare. Certainly, Imogen was amongst those pressed into service for that confection of *das Ewig-Weibliche* which Shaw was only one of the earliest to find distasteful. Indeed, it might be said that my suggestion of a pun on 'my soul' is redolent of the nineteenth-century attitudes and hankerings Shaw so disliked, reading Shakespeare through Victorian ears, bestowing on him an ardour not his own but, say, Browning's:

> I changed for you the very laws of life:
> Made you the standard of all right, all fair.[4]

---

[4] *The Inn Album* (1875), ll. 1868–9. All quotations from Browning's poems, with the exception of *The Ring and the Book*, are from the edition of John Pettigrew, completed by Thomas J. Collins, referred to as P with a volume and page number. Thus, for this reference: P ii. 283.

This sounds like the accent of the Victorians, this is what calling your wife your soul involves. Yet a 'standard of all right, all fair' would be a good definition of the root-sense of 'paragon' in Shakespeare's day. Anyway, Browning's lines figure in a context as doubtful about romantic idealization as anything in the fifth act of *Cymbeline*, and then this is not the only place Shakespeare uses 'my soul' to mean 'my beloved',[5] nor was he alone in his time in this usage. 'Better half' and 'better part', the *OED* shows, have long in the language referred both to a spouse and to the soul. People have long been fond of each other in England and have expressed their fondness in such conjugal phrases where a wife might be a soul and the soul a wife. These phrases have a tradition in the Church too. The Victorians may not have invented the idealization of love but it could be said that they brought it to a pitch at which it was defiled and defiling, injuriously divorced from an actuality which it rather occluded than informed. When Geoffrey Hill writes of Imogen that 'Shakespeare has shaped his play to procure the reality of the woman from the romance of her setting',[6] the 'from' is well judged to mean both 'out of' and 'out of the clutches of'; the achieved life of Imogen is both set off against the context of romance as against a foil and also pitted against that context as in combat. Imogen acts to procure her own reality, to vindicate herself against Iachimo's fictions and to free herself from the Queen's plots; Shakespeare and Imogen collaborate with each other. A figure such as Pompilia in *The Ring and* ~~transition~~ *the Book*, though, remains more passive in the setting where her creator has placed her; her excellent qualities are more conferred on her by Browning's arrangement of matters than by initiatives we imagine as her own. Correspondingly, Browning appears to 'procure' her reality for us in a less respectable sense, pandering to the desire of the reader for spiritual gratifications. Pompilia might have addressed the poet as Elizabeth Barrett Barrett addressed her lover: 'Have pity on me, my own dearest, & consider how I must feel to see myself idealized away, little by little, like Ossian's spirits into the mist . . .'.[7] Still, even Pompilia has her own moments. In her monologue, there are triumphs of created *naïveté* which convey with beautiful authenticity her attempt

---

[5] See *A Midsummer Night's Dream*, III. ii. 246 (with a hint of disbelief) 'My love, my life, my soule, faire *Helena*!'; Folio, l. 1273.

[6] ' "The True Conduct of Human Judgment": Some Observations on *Cymbeline*' (1969), collected in *The Lords of Limit: Essays on Literature and Ideas* (1984), 57. Throughout this chapter I am indebted to Hill's subtle discussion of *Cymbeline*.

[7] EBB to RB, 21 Apr. 1846, in *RB/EBB* ii. 640.

to grasp the reality of her setting, as when, as she dies, she thinks of the priest, Caponsacchi, who tried to save her life:

> He is a priest;
> He cannot marry therefore, which is right:
> I think he would not marry if he could.
> Marriage on earth seems such a counterfeit,
> Mere imitation of the inimitable:
> In heaven we have the real and true and sure.
> 'T is there they neither marry nor are given
> In marriage but are as the angels: right,
> Oh how right that is, how like Jesus Christ
> To say that![8]

These reaches towards the hard saying about marriage in heaven are very touching as they flutter, like last breaths, around the words of her concern to know the rights of marriage ('marry . . . right . . . marry . . . marriage . . . marry . . . marriage . . . right . . . right'). Browning intimates most delicately that perhaps Pompilia might have wished to marry Caponsacchi but that she moves herself with great purity to embrace not him, but his faith; the difficult steps by which she comes to that sharing of belief offer the most decisive testimony against the rumours, with which *The Ring and the Book* is rife, that she and Caponsacchi were lovers according to this world. And that extraordinarily small cry of her delight, 'how like Jesus Christ | To say that', so idiomatic, personal, and un-awed, is itself a sign that she is achieving an imitation of the inimitable, not a 'mere imitation' either but an *imitatio Christi*, for you can say 'how like so-and so!' only of someone you have come to know well.

⌈Whether we feel that people are whittled down or worked up into paragons, the conduct of such transformations is central for a study of the Victorian poetry of love and marriage.⌋ The very fact that the Victorians idealized men and women, and their feelings for each other, has preoccupied some writers on the period to the exclusion of the different manners of apotheosis. Shaw's short temper with Victorian ideals has been inherited by many students of the period. While scholars preserve an unsurprised calm in their discoveries that previous ideals of love, such as Petrarchanism, fail precisely to match their ages' recorded practice of love and marriage, the tone of

---

[8] *The Ring and the Book* (1868–9), ed. R. D. Altick (Harmondsworth, 1971), VII, 1821–30, 375; all references to the poem are to this edition, hereafter referred to as Altick.

comment on the disparity between Victorian actuality and Victorian ideal usually registers shock, whether dismayed or gleeful, that life in those days came short of what those days imagined as exemplary. Sometimes, the shock assumes a style of knowledgeable cool: '[the] Victorians, who sentimentally worshipped and unsentimentally neglected their wives'.[9] Or the writer may be so irked by what is taken to be the age's duplicity that evidence is not weighed but weighted, as when Katharine Moore writes in her *Victorian Wives*:

A certain Dr Acton pronounced that any attribution of sexual pleasure to women was a 'vile aspersion'. The average husband not only took this ignorance for granted but desired it, as in his mind ignorance was equated with innocence:

> Pure as a bride's blush
> When she says 'I will' unto she knows not what . . .[10]

It is odd to quote (or, in fact, *mis*quote) Coventry Patmore as an instance of what 'the average husband' took for granted, and equally odd, though a hallowed gambit, to quote Acton as a representative Victorian thinker about sexuality.[11] It is certainly wrong to believe that Patmore meant by the lines quoted just that a bride undertook to have sex with her husband without knowing what she was letting herself in for. The words 'I will' promise a good deal more than that in the marriage service, so much more that a bride might well blush at the daring of her vow.

Jane Welsh Carlyle, writing to a friend in 1859, comes closer to what Patmore had in mind:

And you are actually going to get married! you! already! And you expect me to congratulate you! or 'perhaps not'. I admire the judiciousness of that 'perhaps not'. Frankly, my dear, I wish you all happiness in the new life that is opening to you; and you are marrying under good auspices, since your father approves of the marriage. But congratulation on such occasions seems to me a tempting of Providence. The triumphal-procession-air which, in our manners and customs, is given to marriage at the outset—that singing of *Te Deum* before the battle has begun—has, ever since I could reflect, struck me as somewhat senseless and somewhat impious. If ever one is to pray—if ever one is to feel grave and anxious—if ever one is to shrink from vain show and vain babble— surely it is just on the occasion of two human beings binding themselves to one

[9] W. Irvine and P. Honan, *The Book, the Ring and the Poet* (1975), 459.

[10] *Victorian Wives* (1974), p. xvi.

[11] See Peter Gay's remarks on Acton's limitations as a source in his *The Bourgeois Experience*, vol. i, *Education of the Senses* (New York, 1984), 481.

another, for better and for worse, till death part them; just on that occasion which it is customary to celebrate only with rejoicings, and congratulations, and *trousseaux*, and white ribbon! Good God![12]

Does 'ever since I could reflect' mean 'ever since I reached the age of reason' or 'ever since I was in a position to judge what being married is like'? The ambiguity of the phrase manages dextrously to avoid an admonitory 'If only you knew . . .' while still conveying to Miss Barnes that Mrs Carlyle has not herself found marriage an unmixed blessing. Consider her 'two human beings binding themselves to one another, for better and for worse'. The 'and' there hints at a gloom which is not in the Book of Common Prayer's 'for better for worse, for richer for poorer, in sickness and in health', the point of which, I think, is that, through sickness *and* health are both to be expected, the couple may find things better *or* worse just as they may become richer *or* poorer.

If we set Patmore's lines back where they belong, they reveal—in their more placid way—something kin to Jane Welsh Carlyle's nervy tact:

> There's nothing happier than the days
>   In which young Love makes every thought
> Pure as a bride's blush, when she says
>   'I will' unto she knows not what;
> And lovers, on the love-lit globe,
>   For love's sweet sake, walk yet aloof,
> And hear Time weave the marriage-robe,
>   Attraction warp and reverence woof![13]

A blunt paraphrase would run: 'The happiest time in the life of men and women is when they are engaged in a chaste courtship. They have such high ideals of love that they are constantly embarrassing themselves by their own desires, desires which haven't been satisfied yet and which they therefore don't fully understand. Yet they practise pre-marital chastity ('walk yet aloof') because of their ideals, and they have the patience to allow their relation to develop into maturity before they get married. This patience is a blend of sexual desire ('attraction') and respect for the other person ('reverence').' But the lines clearly do not refer only to women's pre-marital ignorance; the plural 'lovers'

---

[12] Letter of 24 Aug. 1859, in *Jane Welsh Carlyle: A New Selection of her Letters* ed. T. Bliss (1949), 288.
[13] *The Angel in the House* (1854–6), II. v. Quotations from Patmore's poems are from *The Poems of Coventry Patmore* ed. F. Page (Oxford, 1949), hereafter referred to as Patmore, *Poems*, 167–8.

settles that. 'Attraction' and 'reverence' may not always have consorted so happily together as Patmore's weaving metaphor hopes (rates of pre-marital pregnancy were rising during the first half of the nineteenth century[14]). There is a trickiness, too, in its remaining unclear whether the globe only appears 'love-lit' to lovers or whether it is actually so illuminated. The passage describes and lyrically endorses a world of pre-nuptial enthusiasm, a 'singing of Te Deum before the battle has begun', a world blissfully ignorant of many things human beings can do to each other, out of bed as well as in it. That ignorance is quite as much the man's as the woman's.

In what follows, it may seem that I am ignorant of, or wilfully ignore, the wife-beating, the prostitution, the inequality of the divorce laws as between men's and women's rights during the period in which these poets worked, and other such realities. My silence about these matters does not mean that I think they are not important. It is rather that the kind of understanding of marital lyricism in Victorian poetry which I hope to achieve expects that Victorian lyricism will resemble the lyricisms of other periods about this and other matters, for the lyric utterance consciously idealizes the facts of the dark world it issues from as it idealizes that world's speech. It doesn't come as a surprise when we find that people actually behaved towards each other in ways quite different from what we might imagine if we read works of the imagination as if they weren't at all imaginative.

Particularly in the case of love and marriage, we should expect to find that imagination has been quite vigorously operative on actualities. Coventry Patmore might have been presciently defending his own work against modern critics and historians who have detected his failure, his age's failure, to love up to the height of the aspirations he expressed, when he spoke up for the imagination:

The vulgar cynic, blessing when he only means to bray, declares that love between the sexes is 'all imagination'. What can be truer? What baser thing is there than such love, when it is not of imagination all compact? or what more nearly divine, when it is? Why? Because the imagination deals with the spiritual realities to which the material realities correspond, and of which they are only, as it were, the ultimate and sensible expressions.[15]

---

[14] See P. Laslett, K. Oosterveen, and R. M. Smith (eds.), *Bastardy and its Comparative History* (1980), table 1.3, p. 23.

[15] 'Imagination', in *Principle in Art, Religio Poetae and other Essays* (1889; one vol. edn., 1913), 306 [hereafter referred to as *Principle . . .*].

This perverts the Shakespeare lines it feeds off. When Theseus remarks that 'The Lunaticke, the Lover, and the Poet, | Are of imagination all compact',[16] he does not mean to rehabilitate the lunatic and the lover with those words, though Patmore is not alone in feeling that Shakespeare offers through the lines a defence of the imagination beyond what Theseus in his own right would probably have wanted to say. Yet the passage shows, in its compacting of love and imagination, the dense embodiment of ideals in the actual which faces us in Victorian poetry and culture, an embodiment which confidences about the demarcation between 'illusion' and 'reality' are destined to misapprehend.[17]

Isobel Armstrong, in her sympathetic essay on 'Browning and Victorian Poetry of Sexual Love', puts Patmore's views in words which reach us now more easily than do his fervid neo-Platonisms:

Browning explores another way of limiting the autonomy of the person you love, which is to create an image which is offered to the loved one as a possible way of being, a possible role. Of course, all love requires and lives upon this act of imagination. Browning looks mostly at the vulnerability of this creative act of will, but he acknowledges the human need to create and recreate a person in the imagination, the need to make consistent wholes out of people.[18]

The excellence of this lies in the way it is completely of its own time (1974) while managing to be responsive to the Victorian writing. 'Autonomy' demands quite as much allegiance nowadays as 'spiritual realities' were granted by Patmore. And Professor Armstrong's words are momentarily out of true at points, as Patmore's are; it is bleak to think that any offer of a role is inevitably a 'limiting' of autonomy, and conversely, it is roseate to imagine that lovers only *offer* each other roles. The acquisition of a role may afford the beloved unforeseen horizons, and yet be felt as an imposition rather than as something offered for free acceptance. Such small falsities of phrase attest to the true engagement of both Professor Armstrong and Coventry Patmore

---

[16] *A Midsummer Night's Dream*, v. i. 7 ff.; Folio, ll. 1799–1800.

[17] A characteristically Victorian sense of the relation of ideal to real appears in the title of J. R. Miller's *Home Making of the Ideal Family Life* (n.d.) with its sense of the ideal as something which is built up in real circumstances, like knitting by the fire-side. A more recent sociologist who draws on Miller's book assumes a different relation of ideals and actualities when he entitles one of his chapters 'The Victorian Family: Illusion and Reality'. Here the author relies on hard facts to show plain truths about ideals, ideals which are either documentable realities or mere illusions. See O. R. McGregor, *Divorce in England: A Centenary Study* (1957).

[18] *Writers and their Background: Robert Browning*, ed. I. Armstrong (1974), 288.

in their writing about writing about love, for they display what she calls 'the complicatedness and vulnerability of any *beliefs* about loving, any formulations about feeling'[19] and substantiate the truth of what he claimed for love's history:

> Here they speak best who best express
> Their inability to speak,
> And none are strong, but who confess
> With happy skill that they are weak.[20]

The twining of literary idealization and actual practice is particularly intimate as regards love and marriage, for so many of the conceptions through which people come to love, which form their notion of what it is to love, have been derived from works of the imagination. The great nineteenth-century study of the potential for hurt in that fact is *Madame Bovary*; but the aspect of tormented heroism in Emma's parching and delusive belief in an imaginary 'love' does not make us think of her as simply 'lacking common sense'. Her story satirically diagnoses amorous and other illusions but it also protests, with that distinctive Flaubertian blend of anger and absolute despondency, against a world which comes so short of her capacity to imagine. Literature may indeed be an accomplice in her tragedy but we do not wish that she had contented herself with something less than she had learned from literature to desire. There are peculiar responsibilities for the imagination in such matters. Particularly, in that a post-Kantian notion of imaginative activity as autonomous, having itself as its own end, and expressive of individuality, breaks down here. This doubt about the autonomy of the loving imagination puts itself to writers in terms of the very conventionality of their own lyricism. Early in his correspondence with EBB, Browning despairs of finding words sufficiently 'his' to convey, as he wishes to do, his self to her:

.. if these words were but my own, and fresh-minted for this moment's use! ..

> Yours ever faithfully,
>
> R. Browning[21]

It is not easy to give your self to an other person when you have not even your own terms in which to speak the deed of gift. The impulsive urgency of his 'if these words were but my own' comes up against the conventional subscription 'Yours ever faithfully' and the formal signature, 'R. Browning'. 'Fresh-minted', though, begins to tell, even against Browning's wishes at the moment of writing, why his hopes cannot be fulfilled, for if words, as the metaphor implies, are like coins, then they cannot be minted by any individual, and any individual who tried to mint his own words would be no better than a forger—his words would have not even face value. If words are to go between people, they cannot belong to any one person in particular.

Similarly, Felix resolves early in *The Angel in the House* to have his own way with words:

> I will not hearken blame or praise;
>   For so should I dishonour do
> To that sweet Power by which these Lays
>   Alone are lovely, good, and true;
> Nor credence to the world's cries give,
>   Which ever preach and still prevent
>   Pure passion's high prerogative
> To make, not follow, precedent.[22]

Sheer *fiat* cannot secure the desired individuality of utterance. The poet might try disregarding the 'world's cries' but he cannot but speak in the world's hearing, a hearing much influenced by an accumulation of those cries. So, even the octosyllabics of this defiance sound within the metrical culture of English, in that world of acquaintance with this and other verse forms which shapes our sense of the character of those forms. Patmore knew this, as the enthusiastic amateur poet, Felix, who speaks *The Angel in the House* perhaps did not, for he found that his readers accorded greater intellectual respect to *To The Unknown Eros* . . . than to *The Angel in the House* simply because its metre sounded more imposing: 'I have immensely gained in reputation with these ninnies by mounting a "mail-phaeton". I have even had thoughts of re-writing the "Angel" for them, in the metre of the "Unknown Eros".'[23] The 'sweet Power' which supplies Patmore's 'Lays' with what there is of good in them is intended to be love, and the specific beloved, Emily Augusta Patmore ('Honoria' in the poem), but the world crowds itself betwixt poet and the unique addressee, lover and sole beloved, yielding up to public judgement lines which the fiction pretends are an intimate exchange. This would have happened even had Patmore not published

---

[22] *The Angel in the House*, I. i, Patmore, *Poems*, 65.
[23] Basil Champneys, *Memoirs and Correspondence of Coventry Patmore*, 2 vols. (1900) [hereafter referred to as Champneys], i. 161.

the poem, for the words are not his own, and carry in them the potential judgements of others who might use them. That he did publish it, of course, shows his determination to surrender to that very judgement his words defy, and it requires him to attend to 'precedent' in the very fabric of his writing. In this, a poet's necessary encounter with the past of his material provides an image and a means of love's equally necessary encounter with the past of feelings, the fact that people have felt like this before, thereby setting precedents of the heart.

Literature both contributes to such a patterned body of anticipations and works with it. Thus Patmore, planning his poem on the marriage of the Blessed Virgin, recognizes such precedent as part of the material for his composition: 'The whole difficulty of the subject will be in getting rid of the vulgar ideas of 'greatness' and seeing the matter in its essential smallness and homeliness and sweet warmheartedness. But this is an enormous difficulty, and one which I don't see how I can get over.'[24] Vulgar sounds as well as vulgar ideas of greatness, such as those sounds which made some people hear the octosyllabics of *The Angel in the House* as unserious. The many attitudes in the world, to domesticity, to the demands of marriage, to the glamour of romance, are distilled into the propensities and weights of the words people use about these things, the cadences that go with longing or disappointment, the diction of attachments. Here too there is a complexity of acoustic texture, like that dinning variety which surrounded Keats on entering Belfast,[25] which the printed voice is tasked to represent and compose.

Literature, to a large extent, supplies that very 'greatness' which Patmore struggles with in writing. As John Bayley has observed of Hardy, 'Art demands—and with emphasis in Hardy's time—appearances of greatness where the nature of things in life is smallness, and the relation of the two in Hardy is never anything else than equivocal.'[26] This is overgenerous to Hardy in making it seem that he, alone or specially, in his time withstood the demands of art and its magnitudes, and unfair to him in making out that his withstanding never got further than an equivocal shifting from foot to foot. Yet Professor Bayley acutely brings together what 'art demands' and 'the nature of things in life'. His disjunction brings us back, from another angle, to the turns and turns about of 'ideal' and 'real'. Both words, and their cognates, put out many new offshoots in the nineteenth

[24] Champneys, i. 255.   [25] See above, p. 78.
[26] *An Essay on Hardy* (Cambridge, 1978), 15.

century, new senses which branched with irregular luxuriance in many directions. [To draw all the relevant Empsonian equations for the interpenetrating structure of these two complex words would be as big a job as charting the double helix of DNA.] Two developments in their relations as regards artistic practice show that the Victorian sense of the ways in which art's demands may meet or mar the nature of things in life was both richer and more muddled than Professor Bayley's terms recognize. The *OED* gives as sense 2 of 'idealist': 'One who idealizes; an artist or writer who treats a subject imaginatively' (1805); 'idealism' acquires the parallel sense as an abstract noun in 1829, and 'idealize' as a verb in about 1834. 'Realism', on the other hand, becomes a term for a type of artistic practice in 1856; Emerson uses 'realist' to mean 'one who is devoted to what is real, as opposed to what is fictitious or imaginary' in 1847 whereas Leslie Stephen praises Fielding as a 'realist' in 1874 and H. C. Robinson in 1829 had contrasted Goethe's 'realistic' poetry to the 'idealism' of Wordsworth's. Thus, allowing for variations of standpoint, by the mid-century in English it would have been possible to say (and to mean something by saying) 'The realistic writer must idealize his subjects' and 'Only idealists can realize the real' and 'No novelist is a realist'. These abundant opportunities for formulae were not the prerogative of the English. Indeed, the history of aesthetics in Europe in the nineteenth century shows how international was the privilege of employing such elastic terms for interminable debate.[27] And the addition to 'ideal' in about 1849 of the sense 'an actual thing or person regarded as realizing [a standard of conceived perfection]' meant that the permutations for argument and desire grew even more numerous and extended to less abstract matters, as Joyce wittily indicated in the question, at once philosophical, political, domestic, amorous, and simply moony: '*Art thou real, my ideal?*'[28]

These shifting semantic grounds correspond to the variety, indeed, at times, the self-division of the practice of Victorian poets. An arch-idealizer like Patmore is wary of art's idealisms, and looks for smallness not greatness in his work, even while he writes of the beloved 'She is

[27] See Croce's description of such debates in chs. 9–16 of his *Estetica* (Bari, 1902), trans. Douglas Ainslie as *Aesthetic as Science of Expression and General Linguistic* (1909, rev. edn., 1922).

[28] *Ulysses* (Paris, 1922). I quote from the edition of H. W. Gabler *et al.* (3 vols., New York and London, 1984), ii. 782–3. Joyce attributes the question, apparently correctly, to Louis J. Walsh of Magherafelt.

both heaven and the way'.[29] Hardy retorted on Patmore's zeal for treating love as redemptive when he looked down on 'that large and happy section of the reading public which has not yet reached ripeness of years; those to whom marriage is the pilgrim's Eternal City, and not a milestone on the way'.[30] But Hardy shares Patmore's wariness; neither of them relies on an exceedingly sharp distinction between the demands of art and the nature of things in life. As poets, they could hardly have felt any such reliance, for poetry has as its material the language of the world; the demands of the art require the treatment of the nature of words in life, and so for a poet (at least, and probably for any artist not in the grip of enthusiasm or despair) the line between life and art looks like a bridge and not like a gulf. Art is more modest, and life more avid of grandeurs, than Professor Bayley allows. Consider, for example, the permeation with hyperbole of our casual idioms for affection—'angel', 'sweetest', and 'I'll love you for ever'. Literature nourishes, cherishes, but also checks the eagerness of appetite which speaks in such terms, as Hazlitt did when he noted, 'What idle sounds the common phrases, *adorable creature, angel, divinity*, are!',[31] or as Flaubert did in his passionate rebuke to Louise Colet: 'Tu crois que tu m'aimeras toujours, enfant. Toujours, quelle présomption dans une bouche humaine!'[32] Afflatus is not the prerogative of art, as even a quite ordinary love-life can teach us. This may have been in Henry Patmore's mind when he wrote to his father:

I don't know what to think of poetry, whether to think it is all empty sham or that it is a relic of man's greatness left after the Fall. For it makes you feel so elevated for the moment, and then you see it was all an illusion, and fall back into unpoetical reality or unreality, whichever it is . . .[33]

That last, unexpected and thoughtful swerve, 'or unreality', casts a wary eye, like his father's, on the precariousness and variety of the appeal of greatness, on the pitfalls of momentary elevation, in life and in art.

The cynical villain in Browning's *The Inn Album* finds the young

---

[29] *The Angel in the House*, II. ix, Patmore, *Poems*, 189.

[30] Preface (1912) to *A Laodicean*, reprinted in *Thomas Hardy's Personal Writings*, ed. H. Orel (1967) [hereafter referred to as Orel], 16.

[31] *Liber Amoris* (1823), in P. P. Howe (ed.), *The Complete Works of William Hazlitt*, 21 vols., (1930–4), ix. 134.

[32] Letter of 6 or 7 Aug. 1846, from Flaubert, *Correspondance*, 2 vols., ed. J. Bruneau (Paris, 1973), i. 276. ['And you believe you will love me for ever, you poor child. "For ever", audacious words in the mouth of a human being!']

[33] Letter of 14 Feb. 1878, in Champneys, i. 311.

hero's loyalty to a disappointed love not too typical for words but typical just of the wordiness of love. The young man 'had loved, and vainly loved: | Whence blight and blackness, just for all the world | As Byron used to teach us boys.'[34] The closer a pattern of life comes to a literary formula, the emptier of reality it looks to this man; he says so, and Browning arranges for the words to appear to back him up in the alliterating chain of 'blight and blackness . . . Byron . . . boys' which archly indicates the interconnectedness of these notions in the speaker's mind and mouth. That alliteration is meant as a worldly aplomb by the speaker, but he reads as protesting too much and being too fond of the sound of his own voice. The dual control which operates in Browning's dramatic poems often asks us to suspect linguistic showing-off on the part of the speaker, and trust instead Browning's own concealed artistry behind that displayed virtuosity. The villain of *The Inn Album* descends directly from the mendaciously self-deprecating Duke of 'My Last Duchess' with his 'Even had you skill | In speech—(which I have not) . . .'.[35] Both worldlings look to us steeped in artifice though they wish to sound like plain speakers, because their 'skill in speech' is circumscribed by Browning's greater, less self-interested skill, the skill, for example, which arranges that Ferrara's 'Even had you skill | In speech . . .' should unbeknownst to him rhyme with 'to make your will | Quite clear'. The pretence that there is nothing 'literary' or 'conventional' about *your* life and attitudes turns out to be a very literary gambit in both poems.

Conversely, when Browning wants to show someone who stands for life, both as representing and as championing it, we find the verse dense with literary allusions, though they are not allusions we think the fictional speaker makes. The gypsy in 'The Flight of the Duchess' urges the Duchess to abandon the vacant formality of her marriage and take up a new life as an 'enfant de Bohème':

> It is our life at thy feet we throw
> To step with into light and joy;
> Not a power of life but we employ
> To satisfy thy nature's want;
> Art thou the tree that props the plant,
> Or the climbing plant that seeks the tree—
> Canst thou help us, must we help thee?

---

[34] ll. 2474–6, P ii. 398–9.
[35] 'My Last Duchess' (1842), P i. 350.

If any two creatures grew into one,
They would do more than the world has done:
Though each apart were never so weak,
Ye vainly through the world should seek
For the knowledge and the might
Which in such union grew their right:
So, to approach at least that end,
And blend,—as much as may be, blend
Thee with us or us with thee,—
As climbing plant or propping tree,
Shall some one deck thee, over and down
   Up and about, with blossoms and leaves?
Fix his heart's fruit for thy garland-crown,
   Cling with his soul as the gourd-vine cleaves,
Die on thy boughs and disappear
While not a leaf of thine is sere?[36]

Shakespeare breathes through this magnificent passage: the trochaics of 'For the knowledge and the might | Which in such union grew their right' recall 'The Phoenix and the Turtle'; 'While not a leaf of thine is sere' remembers Macbeth's 'my way of life | Is falne into the Seare, the yellow Leafe';[37] and 'Fix his heart's fruit for thy garland-crown, | Cling with his soul as the gourd-vine cleaves, | Die on thy boughs . . .' stems from Posthumus's 'Hang there like fruit, my soul, | Till the tree die', the lines Tennyson was fond of till his own death. Because we do not suppose the gypsy alludes to Shakespeare, the effect is to make Shakespeare's lines take on the character of folk-wisdom. The extraordinary coincidence of her words and Shakespeare's makes the 'classics' of English poetry sound at one with, say, a Romany knack with horses.

The Duchess leaves her husband in the poem, and sets off for the reality of a life of romance with the gypsies, but the poem's speaker, an old retainer, stays on at court to serve his despised master, the Duke, and, with a more endeared fidelity, sticks by his own wife till death parts them. He could seem a merely repining conformist in contrast to the woman who rode away, were it not for the vigour Browning lends to his garrulity by the bravura of the versification. Consider the rich doubleness of aspect in

---

[36] 'The Flight of the Duchess' (1845), ll. 619–41, P i. 440.
[37] *Macbeth*, v. iii. 22–3; Folio, ll. 2239–40.

> Well, early in autumn, at first winter-warning,
> When the stag had to break with his foot, of a morning,
> A drinking-hole out of the fresh tender ice
> That covered the pond, till the sun, in a trice,
> Loosening it, let out a ripple of gold,
>   And another and another, and faster and faster,
> Till, dimpling to blindness, the wide water rolled:
>   Then it so chanced that . . .[38]

Coming from the fictional narrator, this syntax ludicrously suspends getting to the narrative point: 'Well, early in autumn . . . | . . . it chanced that . . .'—the interim seems from one point of view like an interminable and gratuitous specification of chronology ('Now, let me see, was it Thursday . . . no, hang on, it was Wednesday, I think, because . . . or was it Tuesday, now I come to think of it, anyway . . .'). Yet the lines delicately prefigure the Duchess's own release from the harsh frigidity of her married life, and do so with a perfect rhythmic responsiveness to the 'loosening' they image. The prefiguring could not be so delicate, nor the responsiveness of the verse so agile, did not the speaker's character provide each of them with, as it were, an alibi which allows them to be less deliberatedly there. Browning creates the lines as dramatically characterizing their fictional speaker; their rhythm answers to his imaginable way of speaking. At the same time, coming from Browning, the verse enacts with great subtlety a major theme of the poem. From the speaker's point of view, that enactment is an accident; equally, if we abstract the poem from the fictional character who utters the words which form it, the depiction of his speech-habits is only a by-product. But the old retainer and the poet are actually at one in this exceptionally flexible style, as the gypsy and Shakespeare grew together; each has his point and his dignity, and the poem comprehends both. That condition of extrinsic relation between words and the pattern into which they poetically fall which Coleridge deplored in some of Wordsworth once again shows its potential as the ground of a poetic amplitude, of a reciprocity of lives.[39]

The stay-at-home narrator is best placed to convey the appeal of the Duchess's vagabondage just because he didn't himself answer that call of the wild. He knows what she has missed—a tolerable domesticity —and misses her with her exuberant courage for romance as the price of his knowledge. The choice of this figure to tell the story makes

[38] 'The Flight of the Duchess', ll. 216–23, P i. 430.
[39] See above, p. 71.

possible the poem's adjudication of the claims of spontaneity and commitment, the romance of love and the responsibilities of being married. (It was written between April 1845, when the poet was 'R. Browning' at the end of his letters to EBB, and November of that year, by which time he was confidently hoping to marry her, 'to feel you in my very heart and hold you there for ever, thro' all chance and earthly changes . . .'; he had begun to sign himself 'Ever your own R.' or 'Your own R.'.[40] After all, the life into which the Duchess escapes is the life of an old ballad, a taste for which in the 1840s could be protected against the charge of being either *faux-naïf* or else wantonly unrealistic only by some method of locating the literary self-consciousness of the poem dramatically within the piece itself. Browning finds such a place in the person of the narrator with his good-hearted longings for a gypsy life on which he yet knows it is better for him not to embark. By permitting the reader to see and hear the gypsy and her promises of a fuller existence only through the narrator's voice, a voice marked everywhere with the inflections of a life of honest maintenance, Browning admits the element of fantasy in presenting the gypsy life as a substantial alternative to the shrivelled, institutional life of the court. The stay-at-home most feels the allure of the open road but then that is because he feels nothing but its allure. Browning did not want Elizabeth Barrett Barrett to stay in her home, nor did he want to expose her to the harsh words and harsher conditions that would have been her lot had she done in fact what the Duchess does in the poem. At the point in their relation when it was written, 'The Flight of the Duchess' was a lyrical narrative whose point was to persuade her to leave Wimpole Street; yet it is one of the most responsible poems of seduction in the language, responsible in the vigilance with which it sets its literary fervour for escape from obligations in the voice of a very humble and ever-obliging servant. The many dramatic dimensions of the poem do justice both to his sense of the value of what he offered her and to her sense of the value of what he asked her to give up. More generally, and more permanently, the poem weighs in the double aspect of its style the claims of loyalties which can set each other on edge—the loyalty to a family, to a social frame, and to an individual, one's self or an other. Doing that, the poem creates in imagination what Robert Browning and Elizabeth Barrett Barrett actually achieved— a respectable elopement, one ideal of Victorian marriage.

[40] *RB/EBB*, respectively, i. 283; i. 999; ii. 436.

The pivotal moment of the poem's exquisite balance comes when the gypsy's voice breaks down into the narrator's report. She is concluding her forecast of the Duchess's career if she leaves her husband:

> So, at the last shall come old age,
> Decrepit as befits that stage;
> How else wouldst thou retire apart
> With the hoarded memories of thy heart,
> And gather all to the very least
> Of the fragments of life's earlier feast,
> Let fall through eagerness to find
> The crowning dainties yet behind?
> Ponder on the entire past
> Laid together thus at last,
> When the twilight helps to fuse
> The first fresh with the faded hues,
> And the outline of the whole,
> As round eve's shades their framework roll,
> Grandly fronts for once thy soul.
> And then as, 'mid the dark, a gleam
>   Of yet another morning breaks,
> And like the hand which ends a dream,
> Death, with the might of his sunbeam,
>   Touches the flesh and the soul awakes,
> Then—
>
>   Ay, then indeed something would happen!
>   But what? For here her voice changed like a bird's;
>   There grew more of the music and less of the words;
> Had Jacynth only been by me to clap pen
> To paper and put you down every syllable
>   With those clever clerkly fingers,
>   All I've forgotten as well as what lingers
> In this old brain of mine that's but ill able
> To give you even this poor version
>   Of the speech I spoil, as it were, with stammering
>   —More fault of those who had the hammering
>   Of prosody into me and syntax,
>   And did it, not with hobnails but tintacks!
> But to return from this excursion,—
> Just, do you mark, when the song was sweetest,
> The peace most deep and the charm completest,

> There came, shall I say, a snap—
> And the charm vanished![41]

This is a crucial passage for Browning's imagination. Its importance to him can be indicated by noting how 'Ponder on the entire past | Laid together thus at last, | When the twilight helps to fuse | The first fresh with the faded hues . . .' contains the germ of his greatest poem, 'Andrea del Sarto', but its centrality lies deeper than in such anticipations, for what it makes explicit is the more general significance which he finds in the difficulty with which even 'clever clerkly fingers' capture a 'voice . . . like a bird's'. Browning sets himself a hard task. How can he cap the gypsy's revelations with some account of the afterlife? By interrupting her 'Then—' with the antanaclasis of the narrator's 'Ay, then', a rebuff if not of the meaning of the second 'then' from the first, then at least of the tone of the two occurrences, he does something other than spare himself a deal of trouble and accord his poem a portentousness about eternity (as if to say 'I could tell you, if I wanted to . . .'). Breaking off here does not cheat the poem into a cryptic grandeur, because the interruption brings home to us the unimaginability of the gypsy life to the narrator, an unimaginability which matches and conveys the unimaginability of an afterlife to anybody living. It is this actual disparity between lives in this world which is Browning's chief subject in the poem, a disparity which makes a 'choice of life' difficult but also pressing. And such actual disparity taxes human thinking with a severity which may in part prepare us for being teased out of thought by metaphysical problems about other worlds. To contemplate a very different way of life, to meet a *complete* stranger, someone who makes us ask, 'Are we both living in the same world?', is one way to face the question, 'Could *I* live in a wholly other world?' Transcendence as a theological problem confronts us across the gap between worlds, or between the world and something which is not the world, but the experience of finding one's way between worlds is one that may be had on even one small planet. The narrator's difficulty of transcribing gypsy speech in 'The Flight of the Duchess' is a model of the pleasures and the work of the imagination in many Browning poems. It keeps in mind and ear the motto Browning supplied from the Psalms for 'Caliban upon Setebos', 'Thou thoughtest that I was altogether such a one as thyself', and it provides the reader with means to guard against such presumptions of identity.

---

[41] 'The Flight of the Duchess', ll. 669–706, P i. 441–2.

In his work, the main way he asks us to find out who we are as we discover the people he has created is by tasking us to speak the poems aloud, and to see when and why we cannot quite do that. The task is clearest and most seriously delightful at those moments of intonational ambiguity which arise from the character of the printed voice. So here, 'Ay, then indeed something would happen!' has a jokey vocal pun concealed in it: we may read the exclamation as 'Ay, *then* indeed *something* would happen!' (meaning 'Oh yes, some important and unsearchable thing would occur at the supreme moment of death') or as 'Ay, then indeed something *would* happen!' (as when the television goes on the blink at a key moment, 'Oh, wouldn't you just know it *would* happen then!'). Just when he seemed about to learn, the transmission faltered but that cut-off invests the gypsy's speech with a substantial reality. The incompleteness of what she says attests to its autonomy from the narrator, and thereby helps us infer the reality of the life in whose behalf the gypsy speaks, a life which Browning's imagination brings to birth and then triumphs by making seem independent of what fostered it, as a child grows independent of its mother. As Hardy said, 'It is the incompleteness that is loved, when love is sterling and true. This is what differentiates the real one from the imaginary, the practicable from the impossible, the Love who returns the kiss from the Vision that melts away.'[42] The 'incompleteness' generally integral to writing when it conveys speech acquires a particular drama of respect here.

Even had Jacynth been there, and been a good stenographer, she could not have 'put you down every syllable', any more than Browning has put down every syllable of his narrator's speech. The written poem contains at this point, exactly where it imagines the lost chance of pinning down speech permanently, an emblem of the impossibility of such ambitions in the rhymes 'happen' | 'clap pen' and 'syllable' | 'ill able', which cannot be voiced without bizarrerie in English, though we can see in the phantom evocation of a partial voice that they are meant as rhymes. This is one of Browning's developments of the possibilities inherent in 'metre to the eye only', slightly unusual in this case because the speaker consciously tries to make verse out of the 'speech I spoil' where, more usually, the protagonist of the dramatic monologue is

---

[42] Journal entry for 28 Oct. 1891, in *The Life and Work of Thomas Hardy*, ed. Michael Millgate (1984) from *The Early Life of Thomas Hardy* (1928) and *The Later Years of Thomas Hardy* (1930), 251. All quotations from Hardy's (auto)biography are from Millgate's edition, hereafter referred to as Hardy, *Life*.

unaware that his speech falls into the odd rhymes Browning contrives. The purpose of the rhymes, here as elsewhere in Browning, when they have a purpose rather than merely signalling his own exuberance, is to set the voice on edge with the demands of the page, indicating the distinct existences of the written text and vocal renditions of it, so that we shall not think that they are altogether such as each other, nor that we are altogether such ones as the speaker whose speech the text informs us of.

Coleridge had complained of Wordsworth that a disproportionate emphasis (an unusual stressing of the words) was required for some of his rhymes to be made audible.[43] Coleridge had conservative principles, and some arguments for them, but no argued principle about artistic practice—especially not a conservative one—can withstand the pressure of achieved creative instances. Byron made the world of excessive and defective rhymes in *Don Juan*, and the nineteenth century profusely invented poised expressiveness from a source Coleridge had thought merely disproportionate. Some of these rhyming pole-vaults or pratfalls exploit comic opportunities which could have been familiar to Coleridge, and would not have worried him in Wordsworth. So, for example, Browning in 'The Pied Piper . . .' makes light verse as W. S. Gilbert made light verse, by pretending to slight the rules of verse. It is in the gravity of implication which Browning (and other poets of the period) can build into these skewed rhymes that the distinct, creative potential of Wordsworth's supposed fault lies. Especially when a double aspect can be heard in a rhyme, heard as sounding one way from the fictional speaker of a monologue, and heard again and differently from the actual writer of that speech, such rhymes help the dramatic monologue to be a form which enquires into social and historical disparities, not a form which conduces to impersonation or colonizing in the imagination. The gap between writing and speech, straddled by the conflicting demands of these straining rhymes, is also the ground on which a poet of Browning's genius practically elaborates that consciousness of a disparity between ideals and actualities which confronts us as the challenge in responding to Victorian poetry of love and marriage.

Looking at the relations of ideal and actuality in the Victorian imagination of love, we find them swapping roles with a swift zest which is bewildering or exhilarating, depending on your stamina and

---

[43] See *Biographia Literaria*, eds. Engell and Bate, ii. 70.

pliancy. This is also true of the relations of writing and speech: writing
may be regarded by the Victorians as an idealization of the actuality of
speech, its refinement and correction, or speech may be the longed-for
ideal of full and vivid communion on the condition of which writing
continually only verges. The changes in the life of a poem, as it moves
from spoken utterance of the poet to written text and then to printed
book conduct a courtship in and of the language, from the first
impulses of still unformulated desire to the intimate declarations of
lovers to each other and then to a restricted circle of family and
friends, and finally to the public avowal of an attachment that is to last
in the ceremony of marriage.

The Poetry of Being Married: The Brownings

Victorian poets vary in their attitudes to the writing out and printing of
their poems. For Elizabeth Barrett Browning and Coventry Patmore,
reading yourself in print renders your work alien from the pleasurable
impulses which you put into it, which put 'you' into it. The individual
gets lost in an alien and institutional medium. EBB:

> Like to write? Of course, of course I do. I seem to live while I write—it is life,
> for me. Why what is to live? Not to eat & drink & breathe, . . but to feel the life
> in you down all the fibres of being, passionately & joyfully. And thus, one lives
> in composition surely . . not always . . but when the wheel goes round & the
> procession is uninterrupted. Is it not so with you? oh—it must be so. For the
> rest, there will be necessarily a reaction; & in my own particular case,
> whenever I see a poem of mine in print, or even smoothly transcribed, the
> reaction is most painful. The pleasure, . . the sense of power, . . without which
> I could not write a line, is gone in a moment; & nothing remains but
> disappointment and humiliation.[44]

Coventry Patmore:

> . . . as usual, seeing the poems in print has entirely destroyed my own pleasure
> in them. All the meaning and beauty I fancied I saw in them seems to have
> vanished, and it requires a great exercise of faith not to conclude that, after all,
> I am nothing but a miserable self-deluded Poetaster.[45]

EBB's desire to believe that Browning felt as she did about print—'Is
it not so with you? oh—it must be so'—seems to have been one of the

---

[44] Letter of 20 Mar. 1845, *RB/EBB* i. 42.
[45] Champneys, ii. 117.

several occasions in their correspondence when a hope that they could not possibly disagree about anything was affectionate rather than correct, for, at least later in his career, Browning felt quite the reverse. He wrote to Julia Wedgwood, '. . . so, pray you to observe that it has been a particularly weary business to write this whole long work by my dear self—I who used always to be helped by an amanuensis—for, I cannot clearly see what is done, or undone, so long as it is thru' the medium of my own handwriting—about which there is nothing *sacred*—imperative for, or repellent of—change: in print, or alien charactery, I *see* tolerably well . . .'.[46] He had told EBB even before her 'Is it not so with you?', 'I have no pleasure in writing myself—none, in the mere act . . .'.[47] Browning's need for 'alien charactery' ranges from such minute and literal dependence on print or another's script to his preference for speaking through dramatic characters noticeably distinct from himself.

He was very outgoing as a writer, unlike Hardy who felt lost in print, perhaps as a result of having been brought up to mix 'the printed tongue' of standard English with 'the unwritten, dying, Wessex English'.[48] Mabel Robinson records a sad example of Hardy trying to recall his own voice from the 'beautiful manuscript' of *Tess*, a manuscript written with that finical calligraphy so characteristic of him and indicative of his long effort to transform the pained correctness of a man not wholly at ease with 'the printed tongue' into the lineaments of a personal style:

. . . after dinner Emma lit a bright fire in the drawingroom and he read aloud bits from the novel he was engaged on. He read very badly & was suddenly overwhelmed with a sense of the inadequacy of his words 'No: No. Its not at all what I thought!' much turning of pages 'Lets try here this is—' etc etc, but neither was *that* what he expected, & he dipped elsewhere in the vain hope of touching his own heart.[49]

As he tries to get back the accents of his heart from the mute leaves of his manuscript, Hardy encounters writing's power to preserve and generalize utterance; he finds that the power comes at the price of rendering vestigial, out of reach, that warmth of feeling with which he

---

[46] Letter of 1 Feb. 1869, in *Robert Browning and Julia Wedgwood: A Broken Friendship as revealed in their Letters*, ed. R. Curle (1937), 175 [hereafter referred to as Curle].

[47] Letter of 1 Mar. 1845, *RB/EBB* i. 39.

[48] 'The Dorsetshire Labourer' (1883), Orel, 170.

[49] Letter of 17 Dec. 1937, quoted in M. Millgate, *Thomas Hardy: A Biography* (1982), 299.

had written. The first, fine, careful rapture now stands apart from the writer's emotion; it is achieved and, as such, remote. The experience of conning again what has been composed with enthusiasm need not always be as dismaying as it was for Hardy on this occasion (think of Swift's delight on re-reading *The Tale of a Tub*), but any writer will go through some form of this strange meeting between an experiencing and a recorded self on the page to which he has consigned his voice. In his poetry, such wry or regretful lacking of a past is one of Hardy's main subjects; he treats the way words fade in time, and particularly the words of romantic delight as they dwindle into the familiar routines of married life.

Hardy is not the only poet of the period to invest the movement between speech and print with the expression of attitudes to the survival or failing of individual passion in the long institution of a marriage. The Victorians make such frequent connection between loving and literary formalities that their practice suggests a formula: 'speech is to writing as romance is to marriage'. That is, the relations created between the imagined utterance of a poem and its existence on the printed page model the relations the poet is depicting between amorous impulse and conjugal life. In Hardy, the attenuation of speech within poetic forms, its awkwardness and distance, stand often for his sense of the hollowness of some marriages. In Patmore, conversely, the exuberance of idiom and the mellifluousness of speech-rhythm stylistically embody his sense of a permanent freshness within married life, its continuity with the heady days of first love. The ingenious variety of Victorian practice in this matter exceeds the formula, but even the blunt ratio points us towards the intricacy and the closely detailed creativity with which these poets wrote of their ideals, the sharp ear they had to measure the reality of their desires.

Writing on 'The Philosophy of Shelley's Poetry', Yeats claimed that 'All the machineries of poetry are parts of the convictions of antiquity and readily become again convictions in minds that brood over them with visionary intensity.'[50] If it is true that the form and the content of a poem are so closely connected that the 'how' of the utterance contributes to 'what' is said, and the 'what' closely involved with the 'how' of the saying, then there will regularly be the sort of intimacy between conviction and machinery that Yeats describes. A conviction, or something vaguer but perhaps more gripping, a sense of orderliness,

[50] 'The Philosophy of Shelley's Poetry', repr. in *Essays and Introductions* (1961), 74.

readily lodges in the literary systems of ordering which are the machineries of poetry. A poet might in writing about, say, a moment's impulse to break free from the constraints of marital obligation arrange for the words to spill over the limits set by the predetermined metrical form, 'spill over' in that the sense-units of speech would be put out of joint with the line-units of the poem. Browning does so in the passage from 'The Flight of the Duchess' beginning 'Well, early in autumn . . .' which I quoted before. The cadence produced by the amassing fluidity of speech within the lines is itself melting and eloquent, and gives us rhythmic reason to believe that the poet endorses the momentum of this impulsive speech as against the prosodic constraints, the Duchess's liberty against the Duke's requirements. A different judgement could be made rhythmically implicit by making that relaxation of metrical regularity sound reckless or disarrayed. In either case, the constraint and release of speech within the poetic form writes in precise detail the poet's concern for proper value, a concern written large in the poem's story and shape.

Consider, for example, the contrasting effects produced by Elizabeth Barrett Browning and George Meredith when, both writing in abab rhyming pentameters, they finish a sentence before the end of the line. EBB:

> The face of all the world is changed, I think,
> Since first I heard the footsteps of thy soul
> Move still, oh, still, beside me, as they stole
> Betwixt me and the dreadful outer brink
> Of obvious death, where I, who thought to sink,
> Was caught up into love, and taught the whole
> Of life in a new rhythm.

Meredith:

> By this he knew she wept with waking eyes:
> That, at his hand's light quiver by her head,
> The strange low sobs that shook their common bed,
> Were called into her with a sharp surprise,
> And strangled mute, like little gaping snakes,
> Dreadfully venomous to him.[51]

---

[51] *Sonnets from the Portuguese* (1847–50), VII; I quote EBB's poems from the edition of F. G. Kenyon (1897) [hereafter referred to as Kenyon], 313. *Modern Love . . .* (1862); I quote Meredith from the edition of P. B. Bartlett (2 vols., New Haven, 1978) [hereafter referred to as Bartlett], i. 116.

In each poem, a rhythm and a pattern of rhyme are established, more firmly in the Barrett Browning because this is the seventh sonnet of a sequence and because she remains more faithful to the iambic pulse of the verse than Meredith, whose 'Modern Love' opens with these lines, lines which sound clearly iambic to my ears at only two points (the first and fifth lines quoted). The half-lines which complete the sentence in each poem deviate from the iambic norm (EBB; Of life in a new rhythm; Meredith: Dreadfully venomous to him) but in the first instance, the 'new rhythm'—named as well as exemplified in the verse—should be heard as a welcome modulation of the rhythm of a life as well as the rhythm of the line, a newness which sounds to us as more of felicity than could have been expected as does the 'free gift' of the internal rhyme in the previous line ('caught . . . taught'). In Meredith, on the other hand, the vehemence of the dactyls in his half-line signals an estrangement within the verse, a voice resistant to the regular metrical voice of the poem, as the wife's will stands hard and apart from her husband's in the represented situation.

So far these are plain examples of the analogical enactment of sense of the words by shape of the words. Such enactments would not of themselves demonstrate the collaboration of 'machineries' and 'conviction' which Yeats described. That collaboration begins to appear when we observe the kinds of enactment a poet manages within a verse form, and particularly whether the enactments are managed truly within the form rather than combatively against it. Pope, for example, can make his couplets various and mobile while preserving their integrity as couplets, but the couplets of *Endymion* or *Julian and Maddalo*, though often lively, gain their vitality at the cost of a reader's ability to hear the verse in the proper units of the couplet. Such considerations rhythmically bear the writer's sense of a verse form, his attitude to whether its constraints enable or hinder, even suffocate, the voice which speaks for a person through the form. The room for manœuvre the form affords serves as a means to discover the freedoms, the 'give', there may be in extra-literary forms. And it is in such discovery that poetic machinery carries a poet's convictions.

Barrett Browning writes elaborately conventional sonnets within which she attempts to produce irregular cadences so she may show simultaneously the binding conformity of a responsible love and the liberties given in its bonds. Meredith writes sixteen-line poems, which call up reminiscentially the standard form of a love-sonnet but exceed the conventional bounds, emphasizing formally the implicit struggle between nostalgia and sarcasm which lies within the title of his sequence. As 'Modern *Love*', it implies 'this is all that love comes to nowadays, and its current shabbiness shows up what was all along behind those fine old phrases'; as '*Modern* Love', it implies 'those old phrases *were* fine, and the modern world lets us down by betraying the aspirations they expressed'. The speech Meredith wishes to endorse as authentic sounds out against the demands of metrical regularity and against the conventions of the sonnet, Barrett Browning's makes itself heard in compliance with these forms.

What is true of the poets' working with the precedent rhythmic conventions of poetry is also true of their bearing on established formalities of diction. Take EBB and Meredith with the little phrase 'in truth'. 'Modern Love' describes a dinner-party at which husband and wife pretend all is well with their rocky marriage:

> It is in truth a most contagious game:
> HIDING THE SKELETON, shall be its name.

The thirty-third sonnet of *Sonnets from the Portuguese* makes a pledge:

> So let thy mouth
> Be heir to those who are now exanimate.
> Gather the north flowers to complete the south,
> And catch the early love up in the late.
> Yes, call me by that name,—and I, in truth,
> With the same heart, will answer and not wait.[52]

Meredith depicts a compounding of pretences: society's demand that all marriages at least appear to be happy, the social masquerade demanded by that claim on this particular occasion, the poetical dolling-up of the world's mendacities with lavendered phrases like 'in truth' which mean only 'indeed' and serve to introduce falsity into the polite world. The deceitful affectation of this 'in truth' comes from the fact that, in the speech community for which the poem is written, nobody says 'in truth', unless they are being pompous or poetical. ('In truth' is marked as archaic in the *OED*, sense IV. 13) The phrase's artificiality can be made sardonically to betray the complicity between poetry's verbal romancing and the empty upkeep of social appearances. Barrett Browning's 'in truth' carries her passionate conviction of the survival of old forms of address, old ways of speaking. This sonnet asks

---

[52] Bartlett, i. 126; Kenyon, 319.

the lover to call the beloved by the pet-name which had been hers as a child; making his mouth 'heir' to her dead relations, the beloved maintains a continuity of affection, catching 'the early love up in the late', a love whose reality is not cast into doubt but illuminated by being thus old-fashioned. The poet's own mouth inherits from the exanimate, inherits more than the equipment of deceit which Meredith's poem finds in the poetical tradition. The sonnet form of earlier loves is to breathe anew for EBB, and the phrase 'in truth' is both formal and heartfelt as a vow should be. Though Meredith's word had generally more weight with Thomas Hardy than EBB's, it is her lines here, her 'flowers', her 'early love' caught up in the 'late', her 'same heart', which point the way to that masterpiece of inward debate on the sweetness of love's old songs, 'After a Journey', and to its affirmation that the permanence of affection is not 'merely' poetical much as poetry assists Hardy, and others, to constancy: 'I am just the same as when | Our days were a joy, and our paths through flowers.'[53]

A more particularly Victorian link between convictions and machinery can be seen in the frequency with which these poets in their comments on prosody describe the relation of speech to metre as analogous to that of impulse to law, of individual passion to social institution. They were themselves conscious of the weight of attitude practically borne by the machineries of verse. For most poets of any period, terms of art exist in a dimension of conduct, of responsibilities and allegiance; what is notable in this group of poets is the extent to which their ethical imagination is centred on courtship and marriage. Coventry Patmore and Thomas Hardy in their explicit pronouncements on marriage stand as far apart from each other as Meredith and Barrett Browning—for the former, it is a sacrament, for the latter, a contract on questionable terms—but they share a conjugal understanding of prosody. Patmore declares that 'The quality of all emotion which is not ignoble is to boast of its allegiance to law. The limits and decencies of ordinary speech will by no means declare high and strong feelings with efficiency;'[54] and his words, especially that oddly fulsome 'boast', provide a good comment on his own poetic practice in its contrivance of a rendezvous between sociable locution and set form. He moves, in

[53] 'After a Journey', from *Satires of Circumstance* (1914); all quotations from Hardy's poems are from the edition of Samuel Hynes (3 vols., Oxford, 1982–5) [hereafter referred to as Hynes], ii. 60.
[54] 'Essay on English Metrical Law' (1857), repr. from the revised text of 1894 in *Coventry Patmore's 'Essay on English Metrical Law': A Critical Edition with a Commentary*, ed. Sister M. A. Roth (Washington, 1961), 7.

his essay on metric, again and again, from terms of art to the language of contact between selves. Thus: 'In the finest specimens of versification, there seems to be a perpetual conflict between the law of the verse and the freedom of the language, and each is incessantly, though insignificantly, violated for the purpose of giving effect to the other.'[55] This has a keen sense of the arduous and wearying negotiations of a practitioner of verse between the demands of speech and of metre, and it is at the same time an experienced account of human intimacies, in which each partner is 'incessantly, though insignificantly violated' by the mutual abrasion of selves with divergent needs. Or again: 'The language should always seem to *feel*, though not to *suffer from* the bonds of verse. The very deformities produced, really or apparently, in the phraseology of a great poet, by the confinement of metre, are beautiful . . . Metre never attains its noblest effects when it is altogether unproductive of those beautiful exorbitancies on the side of law.'[56] Patmore's concern for happy subjugation ('beautiful exorbitancies on the side of law') concerns more than the domestication of the erotic life within a marriage, but marriage is for him the best emblem and actually the best exemplar of his wider concerns, so that when, as a prosodist, he calls his own epoch 'an age of unnatural divorce of sound and sense',[57] he focuses his sharpest sense that marriage is the technical ideal of his writing as well as the subject of its celebration. He also recognizes the adversarial character of his work with respect to some social developments of his time, adversarial in its texture as well as in what it states. Basil Champneys noted of him that 'About the time when the English Legislature was, by the divorce laws, relaxing the marriage tie, Patmore was devoting his genius to idealizing it and asserting its permanence in time and eternity . . .'.[58]

Hardy would probably have countered Patmore's complaint about 'unnatural divorce' by remarking that marriage itself was an unnatural state, especially when it bound two people who did not love each other. He wrote in 1894 that it was questionable 'whether civilisation can escape the humiliating indictment that, while it has been able to cover itself with glory in the arts, in literatures, in religions, and in the sciences, it has never succeeded in creating that homely thing, a satisfactory scheme for the conjunction of the sexes.'[59] Very like Hardy, to put, in 'that homely thing', a tone of 'It's not *much* to ask after all . . .' so close to a utopian abstraction, 'a satisfactory scheme for

---

[55] Ibid. 9.   [56] Ibid. 8.   [57] Ibid. 18.   [58] Champneys, i. 35.
[59] *New Review* (June 1894), quoted in Millgate, *Thomas Hardy: A Biography*, 357.

the conjunction of the sexes' ('conjunction' is a bit prim in this context). What is so characteristic of him is the way he combines here an unfretful protest that he doesn't have high hopes with an eagerness for a horizon of engineered satisfactions, at once gleaming and neat. Elsewhere, he had a keener insight into the lack of satisfactory schemes for the human heart, the great insight he expressed so stringently in 1881:

> General Principles. Law has produced in man a child who cannot but constantly reproach its parent for doing much and yet not all, and constantly say to such parent that it would have been better never to have begun doing than to have *over*done so indecisively; that is, than to have created so far beyond all apparent first intention (on the emotional side), without mending matters by a second intent and execution, to eliminate the evils of the blunder of overdoing. The emotions have no place in a world of defect, and it is a cruel injustice that they should have developed in it.
>
> If Law itself had consciousness, how the aspect of its creatures would terrify it, fill it with remorse![60]

This is a long way from Patmore's 'beautiful exorbitancies on the side of law' but can equally be expressed in the poet's treatment of speech within metrical forms. Hardy's particular gamut of timbres in that matter arises from his remarkable achievement of a simultaneous sense that the metrical form is extrinsic to the speech, imposes on it, blinkers its view, and yet that this speech without such form is unimaginable because the vocal accent entirely relies for its production on the conforming institution of the verse.

Hardy's general practice as a prosodist could be described by saying that he tries to give metrical law a consciousness which will be terrified and filled with remorse by the aspect of its creatures. Doing so, he further dramatizes the lyric voice, further exploits a gap between metre and passion for, reading some of his more successful poems, we read expressly the drama of formality, as if the words were one speaker and their pattern another. He worked out rhythmically, then, the irritation he felt in life at 'the . . . necessity of conforming to rules which in themselves have no virtue',[61] rules of verse which like the laws of marriage he might want, outside the poems, to consider as 'not . . . essential laws, but . . . laws framed merely as social expedients by humanity, without a basis in the heart of things . . .',[62] but which the poetry discovers more soberly to be the substance of many things in the

---

[60] Hardy, *Life*, 153.   [61] Ibid. 114.
[62] 'Candour in English Fiction' (1890), Orel, 127.

heart. 'Poetry is emotion put into measure. The emotion must come by nature, but the measure can be acquired by art,'[63] he wrote; 'put into' here, by virtue of a grave pun, means not only 'put into' as in 'He put the manuscript into a box' but also 'invested' as in 'You would scarcely believe how much Hardy put into writing that poem'. The emotions he felt in and about measure are quite different from Patmore's but Hardy's 'emotion'/'measure' polarity remains in harmony with Patmore's imagination of verse as the relation of 'emotion' to 'law'.

An even more individual example of the relations between writing and conjugality comes in the extraordinary correspondence between EBB and Browning. The intensity with which they imagine absent interlocutors, their probing of what the other might reply, the developing sense of how the letters continued the talks they had in their meetings but continued that talk in a different and complicated medium—all this makes their correspondence the most inquiring study in the phenomenology of the dramatic monologue that we possess. (I am not suggesting that this was their main intention in writing to each other.) So, for example, the dramatic monologue hinges on a difference between the sorts of weight we can rightly give to something spoken and to the same words when written down. Writing down what has been said constitutes a judgement of speech, as the police caution that what one says may be taken down and used in evidence reminds us. In writing, speech is brought to book. Elizabeth Barrett Barrett feared Browning's memorizing and scrutiny of her remarks: 'do not avenge yourself on my unwary sentences by remembering them against me for evil'.[64] The request came too late, and she need not have been so apprehensive, for Browning dwelt on her words in an ecstasy of interpretative fondness, determined to make the best of them:

When I come back from seeing you, and think over it all, there never is a least word of yours I could not occupy myself with, and wish to return to you with some . . not to say, all . . the thoughts & fancies it is sure to call out of me: there is nothing in you that does not draw out all of me . . .[65]

She meant the world to him, and the world, for Browning, 'is not to be learned and thrown aside, but reverted to and relearned'.[66] The

[63] Note of 1899, in Hardy, *Life*, 322.
[64] Letter of 27 Aug. 1845, *RB/EBB* i. 174.
[65] Letter of 9 Nov. 1845, ibid. i. 261.
[66] 'Introductory Essay' to *Letters of Percy Bysshe Shelley* (1852), P i. 1003.

attentive return to her utterance which Browning gave, she returned to him, for, she wrote, '. . . I felt your letters to be *you* from the very first, & I began, from the beginning, to read every one several times over'.[67] Browning had begun to write dramatic monologues before he entered into correspondence with EBB, but I think he learned, among other things more important to him but not to us, some of the profundity of the form as they developed their intimacy through the interchange of meetings and letters, in the accumulation of ardent scanning and re-audition, glossing each other's words, eliciting the voice of a personality from fragments of conversation and comment.

The dramatic monologue constantly trades back and forth between ear and eye, between our imagination of the text as spoken and our reading of the text as written. This most distinctive of Victorian poetic forms works principally with the difficulties and opportunities set before the poet in the intonational silence of the written word, and the correspondence between Browning and EBB often reads like a schooling in the rich significances of the printed voice, as he called it, or in 'this talking upon paper', as she called it.[68] Before they met and heard each other speak, they were hearing voices in imagination. 'I have learnt to know your voice,' she told him in January, 1845, 'not merely from the poetry but from the kindness in it'.[69] After they had met, a distinction arose between 'you in your special capacity of being *written*-to' and the person 'spoken-to'.[70]

The distinction of the person in writing and the person in speech tasked their imaginations in these private circumstances as a dramatic monologue tasks the imagination of its reader:

You never guessed perhaps . . what I look back to at this moment in the physiology of our intercourse . . the curious double feeling I had about you . . you personally, & you as the writer of these letters, . . & the crisis of the feeling, when I was positively vexed & jealous of myself for not succeeding better in making a unity of the two.[71]

To meet that task of unification, Browning in particular deployed his vitality on the page so as to give it the air of the stage, to make the words maximally seem to pass *between* two people present to each

---

[67] Letter of 19 Feb. 1846, *RB/EBB*, i. 477.
[68] Letter of 3 Feb. 1845, ibid. i. 12–13. 'Talking on paper' is an ancient phrase for letter-writing; it dates at least from Seneca.
[69] Letter of 11 Jan. 1845, ibid. i. 5.
[70] RB to EBB, 3 May 1845, ibid. i. 53.
[71] EBB to RB, 4 Jan. 1846, ibid. i. 359–60.

other, though he could see that there was only one person there. It was
a version in actuality of that evoking of the second, silent person which
he practised in the monologues: 'I feel at home, this blue early
morning, now that I sit down to write (or, *speak*, as I try & fancy) to
you . . .'.[72] He pretends that he can see her about to interrupt him with
a protest against his ways of putting things, a move he will make again
at length in 'Bishop Blougram's Apology':

My life is bound up with yours—my own, first and last love. What wonder if I
feared to tire you—I who,—knowing you as I do, admiring what is so
admirable (let me speak), loving what must needs be loved, fain to learn what
you only can teach; proud of so much, happy in so much of you,—I, who, for
all this, neither come to admire, nor feel proud, nor be taught,—but only, only
to live with you and be by you—*that* is love—for I *know* the rest, as I say. I
know those qualities that are in you . . but at them I could get in so many ways
. . I have your books, here are my letters you give me,—you would answer my
questions were *I* in Pisa—well, and it all would amount to nothing, infinitely
much as I know it is; to nothing if I could not sit by you and see you . .'.[73]

That '(let me speak)' magnanimously invents the character he
cherishes—her ever-watchful modesty with its hints of petulance and
self-will even in its refusal to admit anything admirable in herself. The
whole passage, in its exuberant insistences ('only, only'), its markings
of the phases of its own progression ('as I say'), its gestural sense of the
writer pointing to objects in a space which is common to him and his
interlocutor ('here are my letters'), creatively breathes the represented
speech of Browning, as he wrote that her letters spoke to him in 'the
voice of you, which a letter seems'.[74]

The rest of us outside their developed intimacy can find these
wonderful, affectionately palimpsest letters, sprawling with vows and
cajolery, interjections, cross-references and private jokes, difficult to
read just because they so closely approach the condition of speech. R. T.
Lakoff identifies, from a linguist's point of view

. . . the problem that students of conversational strategy keep being bedeviled
with: that conversation, as taken off tapes and represented in manuscripts, is
fiendishly hard to understand and very hard to keep paying attention to . . .
One source of our difficulty lies in the reader's interpretation of what lies on
the page. . . . once we attempt to translate oral communication to the written
page, we find ourselves having to translate meaning, as much as form. The

---

[72] Letter of 21 Aug. 1845, ibid. i. 164–5.
[73] Letter of 12 Oct. 1845, ibid. i. 229.
[74] Letter of 17 July 1846, ibid. ii. 883.

characteristics that work in one medium are not necessarily ideal for the other; direct translation tends not to preserve sense, or effect.[75]

The subtlety and verve of their letters, though, like that of the dramatic monologue, consist in their not pretending to be transcripts of speech; they ask for the imagination and not the illusion of speech. In the letters, Browning's imaginative vitality is saved from fantasy by his acknowledgement that 'all would amount to nothing . . if I could not sit by you and see you'. All this creativity depends on something it cannot create, the existence of an other person, preciously and barely factual, and of such dependence Browning is right to say that it 'is love', in the imagination as in fact. The sign of this existence is that the person concerned may be absent from you, and writing, as Freud said, was originally the voice of an absent person.

Such absence instigates the verve of the letters between EBB and Browning, as it prompts the poetic imagination in the dramatic monologues. In each case, the imagination of the writer reaches out both to present a self and to elicit a hearer for that self. The aspect of this general task of address in writing which most particularly marks the monologue and the correspondence between Browning and EBB is the exceptional mutuality of relation between speaking and attending selves. The Duke of 'My Last Duchess' is heard only through the envoy's silence, Andrea del Sarto only through Lucrezia and what she does not say: Lucrezia and the envoy provide the acoustic space of the monologues' speech, but we guess at the character of that acoustic only through hearing such speech within it. The dramatic monologue stages with minute vitality the risks and slips of communication between its characters, but it does so with a superb reliance on its own ability to run, to manage, such risks and to convey their significance. The worlds of misapprehension or deceit or anxiety which are so often the scenes of Browning's monologues are themselves contained within a world where skill in speech is often both possible and benign. The very intentness of the creative imagination on the figure to be evoked challenges us to a recognition of that figure's distinct identity, and in recognizing a separate life makes a realization that such a life could be missed in communicative aim or even more entirely lost. The correspondence between the two poets undergoes in fact this rhythm of the imagination in which the very swell of trust in their own powers

[75] 'Some of my Favourite Writers are Literate . . .', in Tannen, *Spoken and Written Language: Exploring Orality and Literacy* (Norwood, NJ, 1982), 245.

draws on a faltering, a self-suspicion. This is the source of that correspondence's nervous life, a life which sometimes prompts the writers to an excess of zest.

Browning had his own explanation of this:

All this missing of instant understanding—(for it does not amount to *mis*understanding)—comes of letters, and our being divided. In my anxiety about a point, I go too much on the other side from mere earnestness,—as if the written words had need to make up in force what they want in sound and promptness . . .[76]

His words apply pointedly to their letters. If we re-apply them to the dramatic monologue, they describe, in their reference to 'letters, and our being divided', the plight and the opportunity which faces a poet like Browning in the printed voice. In that context, the 'force' which writing most needs to make up for its loss of 'sound and promptness' will be the illocutionary force of intention as well as the 'force' of vivid emphasis Browning meant (I take it that his 'promptness' means 'promptness of response in a known situation', the evident relation to a context of utterance which spoken words have and which written words may not have). 'We are the fools of language,' Patmore writes more despondently to a friend, 'I think that I am speaking my mind, when in fact I am doing no such thing. Words, even with the aid of the voice's emphasis, the body's gestures, and the face's expression, are poor weak things. What are they then without such aids?'[77] The psychological flexibility and inquisitiveness of the best dramatic monologues, though, thrive on the uncertainty created for us as to whether we are in fact hearing someone speak his mind, or whether we in part impute our minds to the speaker as we lend his text our voice as well as our ears. Particularly when a reader faces and tries to voice an intonationally ambiguous line, he is asked to reflect on the pull one reading rather than another exerts on him, and to ask why it does so. He comes to know himself in the act of becoming convinced that he knows the fictional speaker.

There are also times when the abundant supply of significances which a face-to-face confrontation gives may prove an embarrassing richness, so that the comparative paucity of indicators in writing may not be just a deprivation—a thought which Wordsworth gives particular weight to in his poems of encounter, and which, for

[76] Letter of 31 July 1846, *RB/EBB* ii. 919–20.
[77] Letter of 16 Mar. 1847, Champneys, ii. 147.

Browning in *The Ring and the Book*, that poem massively without encounter between its 'speakers', provides the conclusive justification of his art in the work:

> So, British Public, who may like me yet,
> (Marry and amen!) learn one lesson hence
> Of many which whatever lives should teach:
> This lesson, that our human speech is naught,
> Our human testimony false, our fame
> And human estimation words and wind.
> Why take the artistic way to prove so much?
> Because, it is the glory and good of Art,
> That Art remains the one way possible
> Of speaking truth, to mouths like mine, at least.
> How look a brother in the face and say
> 'Thy right is wrong, eyes hast thou yet art blind,
> Thine ears are stuffed and stopped, despite their length,
> And, oh, the foolishness thou countest faith!'
> Say this as silverly as tongue can troll—
> The anger of the man may be endured,
> The shrug, the disappointed eyes of him
> Are not so bad to bear—but here's the plague
> That all this trouble comes of telling truth,
> Which truth, by when it reaches him, looks false,
> Seems to be just the thing it would supplant,
> Nor recognizable by whom it left—
> While falsehood would have done the work of truth.
> But Art,—wherein man nowise speaks to men,
> Only to mankind,—Art may tell a truth
> Obliquely, do the thing shall breed the thought,
> Nor wrong the thought, missing the mediate word.[78]

It isn't an assured piece of writing, but the strain the lines are under shows the burden of the difficulty Browning feels and tries to describe here. He puts things both baldly and fussily. He repudiates Wordsworth's description of the poet at 'wherein man nowise speaks to men'—and 'nowise' is indeed not a very idiomatic term: the *OED*'s instances are from sermons, philosophical treatises and law reports. (Tweedledum it was who said 'Nohow' and his brother 'Contrariwise',[79] and Browning seems to pre-echo their controversial bumptiousness here.) The

---

[78] XII. 831 ff., Altick, 627–8.
[79] *Through the Looking-Glass* . . . (1872) in *The Penguin Complete Lewis Carroll* (Harmondsworth, 1982), 165–6.

repudiation would not matter if he didn't replace Wordsworth's hopeful man speaking to men with the altogether less believable man speaking to mankind, a definition of the poet which leaves us to wonder whether someone who speaks to mankind mustn't, now and again, also speak to men. He has too much to say about the fact that 'our human speech is naught'; the triple periphrasis he treats himself to there seems perfectly happy to go on and on talking in the teeth of what it itself says. And the representation of what actual speech is like is hopelessly stagey ('Thine ears are stuffed and stopped'), and so does not bring into the verse the true lineaments of the difficulty Browning tries to locate. These linguistic insecurities, this gaucherie with the elements of conversation, indicate how close the passage comes to stating explicitly Browning's understanding of his own predicament amidst 'the misapprehensiveness of his age'.[80] The same troubles of talking which he describes inhibit him from saying explicitly what they are and show his need for the dramatic self-consciousness of the voice which he finds in his great poems.

Writing to Elizabeth Barrett he had lamented his lack of personal words, 'if these words were but my own . . .!', but writing after her death, and to the public, it is something else he needs, 'the mediate word', a word which will mediate disagreements and do so by being not the immediate utterance of a self but something comparatively unselved, coming between people with a creative lack of the accents of any partial individual, 'mediate' then in order to mediate. The printed voice is a principal medium of the mediate word which 'our being divided' calls for; Browning's skill was to find such words frequently and also to know the cost of his finds when he regards his art as withdrawn from the prompt sounds of conversation.

One condition of conversation which follows from the facts of acoustics is evanescence; the spoken word perishes as it comes into communicative being just to make way for the next word. Written words stay on the page after we have passed over them, and we may have recourse to them at will. Among Browning's many notable explorations of the region between speech and writing which poetry inhabits is his discovery that this region is one in which the equal needs both to preserve words and to let them pass may be mediated. I have already quoted EBB's fear that he would weight and preserve her words in a medium of attention beyond what their lightness in

---

[80] 'Introductory Essay', P i. 1006.

utterance could bear. On the other hand, the wife of 'Any Wife to Any Husband' thinks what it would be like to survive her husband, and thinks that the grief of being left alone would be relieved by the opportunity solitude would give for contemplation of all those incidents which, at the speed of life as they happened, passed them both by in that inattention which is the atmosphere of a shared happiness:

> Why, time was what I wanted, to turn o'er
> Within my mind each look, get more and more
>   By heart each word, too much to learn at first . . .[81]

Print offers such time to revolve and absorb words which were too much for us in the prime of their occurrence, and the offer is made specially by the stilled talking on paper of a dramatic monologue. For the monologue mediates between speech and writing in that the writing asks us to imagine a particular act of speech which it pretends to represent. That act of speech is maintained for our consideration and re-consideration as an actual speech before the invention of phonography could not be. This fiction of the original speech and the written record corresponds to a truth which is imperative on the reader—that the words must be re-voiced if they are to be understood. I discussed before how Tennyson worked with the double time implicit in his poems, the time of a vocal rendition and of the abiding written text.[82] The potentiality of this double time was further exploited by Browning as the stylistic ground for an inward debate between cherishing and possessiveness, between the impulse to dwell on, to cling to, everything about the beloved, and the need to relax and let the beloved breathe.

Getting words more and more by heart, as the wife wishes to do in 'Any Wife to Any Husband', is an actor's task with regard to his script. The delicate relation of the text of a dramatic monologue to a play script permits this Victorian form to enquire into the relations to valued living of rehearsing and memorizing. A perfectly rehearsed life would be something one merely performed rather than lived; a life totally committed to memory, as a script can be, is a logical impossibility. A poem by Browning can evidently be committed to memory and spoken aloud, but the poem still remains distinct from a fragment of theatrical script, or from the recitation-pieces popular in the Victorian period. Distinct in that, for example, the imagination of

[81] (1855), P i. 565.   [82] See above, p. 124.

costume or gesture for most of these poems would be otiose or intrusive. But more seriously distinct in that the drama of a poem like 'My Last Duchess' rests in the interaction between a consciousness imagining in English (the writer's or a reader's) and a consciousness imagined as non-English; the dramatic quality of the poem comes through as it confronts these consciousnesses with each other, and makes them dwell within a single subject. It would not be foolish for an actor considering the role of Shylock to ponder decisions about the use of crape hair or some vocal form of disguise, an accent or speech-tic. And, because of Shylock's place in a pattern of theatrical alternatives, in a group of figures with competing claims to be represented, decisions must eventually be taken about such matters if the role is to be performed at all. Someone who proposed to read 'My Last Duchess' aloud would mistake the poem if he thought it might be a good idea to deliver it in an Italian accent. The reader of a dramatic monologue is not asked to be an intermediary for its fictional speaker but to be the medium in which its fictional speech meets poetic form. The poems are profoundly untheatrical just because they make of the reader himself a theatre, and so dispense with the externalized elements of theatrical performance. Certainly, there are points of contact between the experience of reading a play script to oneself and reading a dramatic monologue. The closest of such contacts is in the demand of both kinds of writing for imagined voice. Yet a fine, private reader of a script will block in moves for the characters, specify responses for those who are on-stage but not speaking, and so on. Just these things are what the reader of a dramatic monologue often must not do. Looking at scripts of the nineteenth century, we see an abundance of instructions from the author—set-design, expressive gesture, tone of voice, and so on—but writers of dramatic monologues are far from such theatrical developments. The drama of imagining here is in an abstention from the fullness of theatrical interpretation.

EBB had some right on her side when she thought that Browning might crush the things she said with his passion of interpretation. He protested:

Dearest, do you think all this earnestness foolish and uncalled for?—that I might know you spoke yesterday in mere jest,—as yourself said, 'only to hear what I would say'? Ah but consider, my own Ba, the way of our life, as it is, and is to be: a word, a simple word from you, is not as a word is counted in the world: the word between us is different . . I am guided by your will, which a word shall signify to me: consider that just such a word, so spoken, even with that lightness, would make me lay my life at your feet at any minute: should we gain anything by my trying, if I could, to deaden the sense of hearing, dull the medium of communication between us . . .[?]

and she replied:

As to a light word . . why now, dear, judge me in justice! If I had written it, there might have been more wrong in it—but I spoke it lightly to show it was light, & in the next breath I told you that it was a jest—Will you not forgive me a word so spoken, Robert?[83]

The 'Robert' is unanswerably tender and vocal. Robert's confidence that nothing could be gained by deadening his hearing is a convention of romance. He thinks there is no relation between words that pass between them and the words of the world, which, if true, would make it difficult for them to live in the world, and his preparedness to lay his life at her feet 'at any minute' would seem to make it difficult for them to live at all.

Any husband, though, might feel what the wife of 'Any Wife to Any Husband' feels, and Thomas Carlyle felt so when he read over the brilliant letters of his dead wife. He too wanted time after she had died: ' "five minutes *more* of your dear company in this world; oh that I had you yet for but five minutes, to tell you *all!*" this is often my thought since April 21.'[84] It might have been more considerate of Carlyle to have wanted five minutes more to hear what his partner had had to say (as the wife does in Browning's poem). Five minutes listening to her, at least, he owed her after receiving a lifetime of her attention and attentions (though it is not a debt which Jane Welsh Carlyle in her generosity of heart and intellect would have exacted from him). Certainly, he needed time to turn over what had been too much for him to learn at first:

The whole of yesterday I spent in reading and arranging the *letters* of 1857; such a day's *reading* as I perhaps never had in my life before. What a piercing radiancy of meaning to me in those dear records, hastily thrown off, full of misery, yet of bright eternal love; all as if on wings of lightning, tingling through one's very heart of hearts! Oh, I was blind not to see how *brittle* was that thread of noble celestial (almost more than terrestrial) life; how much it was all in all to me, and how impossible it should long be left with me. Her

    [83] Letter of 30 Aug. 1846, *RB/EBB* ii. 1023–4 and with her reply of 31 Aug. 1846, ibid. ii. 1028.
    [84] *Reminiscences*, 2 vols. (1881), ed. C. E. Norton (1887), repr. in one volume with an introduction by I. Campbell (1972), 155.

sufferings seem little short of those in an hospital fever-ward, as she painfully drags herself about; and yet constantly there is such an electric shower of all-illuminating brilliancy, penetration, recognition, wise discernment, just enthusiasm, humour, grace, patience, courage, love,—and in fine of spontaneous nobleness of mind and intellect,—as I know not where to parallel![85]

He cannot find words to negotiate between 'piercing radiancy of meaning' and 'hastily thrown off', to tie the *brittle* in with the 'celestial'. And so the pile of abstract nouns, 'brilliancy, penetration . . .', with which he tries to recall her, only disassemble her particularly enchanting character into the elements of an approbation which could be given to someone you did not know very well. The cluster of terms fails to become such a constellation of qualities as forms a loved individual. It is a heart-broken passage, and it is heart-breaking to read. Literary appraisal, narrowly conceived as the record and description of rhetorical mastery or imaginative triumph, might judge Carlyle's writing here as severely as he judged his own conduct to Jane Welsh Carlyle—the writing is splashy, out of focus, and poorly composed. But there is no good reason why literary response should on principle impose narrow views upon itself. Indeed, the price of strictness about what may or may not legitimately be considered in literary judgement (so that, for instance, the biographical context of writing, and the qualities of character shown by a writer in such a context, are excluded) is that a writer's agency becomes vaporous or incomprehensible, and his writing ends up as no more than a bag of more or less successful tricks. When Edward FitzGerald read these passages of Carlyle and the letters and memorials of Jane Welsh Carlyle which her husband had prepared for publication, he brought the writings fully to life, to their life as well as his. He found in Carlyle's reminiscences 'a sort of penitential glorification' of Jane Welsh Carlyle, a phrase which exactly weighs the authentic excesses of Carlyle's tribute by recognizing their motive. His judgement was more than narrowly literary; he made a humane response:

I am just finishing Mrs. C. Letters—enough to break one's heart, both for her sake, and his. In one respect it even raises him in my Esteem, that he should insist on publishing what he must have [known] the World could turn against himself; but whether such domestic Tragedy as *her* lifelong sufferings, and *his* 'post mortem' remorse should be laid bare to the Public is another question. What do you say?[86]

[85] Ibid. 138.
[86] Letters of Mar. 1881 and May 1883 in *The Letters of Edward FitzGerald* . . . , iv, respectively 418, 582.

Carlyle's writing (even as it fails to master his feelings), Carlyle's devotion to editing and publishing his wife's writing, are a proper tribute to that great woman whom he felt he insufficiently appreciated while writing his great works. FitzGerald's esteem is only just.

When Browning remembers the occasion on which EBB gave him the manuscript of *Sonnets from the Portuguese* his ability to recall her in an actual world, a world that was hers, does not impair a sense of EBB's reality but fosters it:

... there was the little Book I have here—with the last Sonnet dated two days before our marriage. How I see the gesture, and hear the tones,—and, for the matter of that, see the window at which I was standing, with the tall mimosa in front, and little church-court to the right.[87]

Gesture and tone are remembered but not heard again through the writing; 'how' starts an exclamation and not an account of the manner of memory. No attempt is made to reproduce a voice which cannot be replaced (perhaps EBB said nothing, just handed over the book into which she had put her tones) but the little indications of topography, the mimosa and the church-court, and the physical placing of the 'little Book'—it was 'there' and is now 'here' (how vital that distance is)— sketch her circumstances vividly because of the casual way in which they are mentioned, and give us an inkling of her 'as if she were alive'.[88] There is a continuity between Browning's act of revivifying imagination here and similar acts in the poems which also call on arrangements of peripheral, seemingly random detail to release a sense of contingency into the finished verse. That continuity matters because it reveals the strength with which his imagining linked him to an actual world.

Browning's own temperament assisted him to such small miracles of a memory both pious and fresh. The aptitude of the dramatic monologue to find words which are mediate between speech at its instant of occurrence and the constitutively recurrent silence of the written word, to which we may have constant recourse, helped him too, as we see and hear at the end of that beautiful poem of conjugal reminiscence, 'By the Fire-side', with its achievement of lyricism from 'the heart of things' in contrast with Carlyle's collapse into a lyricism divorced from the 'fact of things':[89]

---

[87] Letter to Julia Wedgwood, [?] 11 Nov. 1864, in Curle, 114.
[88] 'My Last Duchess', P i. 349.
[89] Jane Welsh Carlyle's journal for 21 Oct. 1855, in Bliss, op. cit. 249.

So, earth has gained by one man the more,
    And the gain of earth must be heaven's gain too;
And the whole is well worth thinking o'er
    When autumn comes: which I mean to do
One day, as I said before.[90]

This is idiomatically dextrous, a lithe inconsequence in the manner of saying it—from the marvellously informal 'So,' to the deep lightness of 'well worth'—makes the speech 'just happen' to fit the stanza, makes the lines seem spoken in the moment we read them. Yet any reading of the poem which properly remembers what has been said before will remember that this last stanza reverts with composed neatness to the opening lines of the poem, that its form repeats the form of previous stanzas with a symmetry impossible in ordinary talk, and that such reversions nourish our understanding of the poem as a created whole 'well worth thinking o'er', as lives lived for years without particular planning, but with affection, may equally come to compose a whole that bears and requires thinking through. The speech of these lines is at once alive and muted; it gives us something worth dwelling on poetically and preserves in that dwelling what Browning called the 'moment's flashing, amplified, | Impalpability reduced to speech . . .'.[91]

This is actually the ideal of the dramatic monologue as he conceived and practised it:

Love, you saw me gather men and women,
Live or dead or fashioned by my fancy,
Enter each and all, and use their service,
Speak from every mouth,—the speech, a poem.[92]

Combining the fluidities of speech with set forms is a constant task for poets, not just a discovery of Browning's, but what is particular to him is the importance of the way in which 'the whole seems to fall into a shape',[93] the questionable providence in such appearances of arrangement in one's life, the delusions that may arise from making, or spotting, a pattern. Browning gathers men and women as if they were

[90] (1855), P i. 561.
[91] *Red Cotton Night-Cap Country, or Turf and Towers* (1873), ll. 4236–7, P ii. 184.
[92] 'One Word More' (1855), P i. 741. This instance supports Barbara Everett's felicitous phrase about 'My Last Duchess'—'what the reader . . . sees and hears is a voice', in her 'Browning Versions', *London Review of Books*, 4–17 Aug. 1983, repr. in her *Poets in their Time* (1986), 162. I am throughout this chapter much in debt to Miss Everett's essay.
[93] 'Andrea del Sarto' (1855), P i. 644.

so many flowers, but he also gathers them with a less lordly ease of acquisition—he gathers them through slow and arduous inference of imagination, through a guessing of their drift. He enters them; he comes into possession of them but only to the extent that he is then inside them, contained by them, in their possession. The paradoxes of dependence and control within such processes of creation work into the texture of his verse the concern for the role of an individual will, in all its vagaries from the under-achieving to the over-weening, which runs throughout Browning's career. The drama of form here is akin to the drama of the lyric in Tennyson, and in each case, it is the fictional separation of the writer from the speaker of the poem which conducts the thinking in the poem.

When Marvell or Pope arranges a match between speech and poetic form, for example, the result of the match demonstrates the poet's adeptness in his language and world, his agility or his debonair grasp. In the dramatic monologue, the source of the pattern is distinct from the origin of the speech. Browning draws such a distinction variously: there are some monologues in which the fictional speaker is also fictionally writing the poem; there are also the cases where the dramatic character of the poem is abstract or attenuated, since the reader is given no specific reason sharply to distinguish the speaker of the poem from Browning himself. The distinction in all its varieties makes the monologues also particularly apt for representing how differently things may fall into patterns for two people who are in some sense close to each, married to each other, for instance. For that class of dramatic monologues—the supreme instances in my view—where the silence of the monologue's interlocutor is also composed by the poet, for poems such as 'My Last Duchess' or 'Andrea del Sarto', the distinction between source of pattern and origin of speech permits us to see in the text of the poem that things may look differently to the other person in the poem (this is not easy to voice in reading the poems aloud). Browning wrote 'you saw me gather men and women' where we might, taking the vocal rendition of the poem as primary, expect 'heard me gather'. It is of the essence of the dramatic monologue that it urges eye and ear to collaborate in imagining a person from print, while insisting on the demarcation of their activities, acknowledging how hard the labour is to create the conditions in which 'the speech' should be a poem, the poem, speech.

Such dwelling on what actually passes and is passed over is a great, though ambiguous, gift of art to the human affections. Art offers us memorable objects, ready for our returning attention, objects which less clearly present themselves in our lives, and especially not in our conversations. In the conning and mulling over of works of art, in that habit of re-frequenting the work which so characterizes our dealings with imaginative creations, we find an exercise for the tenacity of our imaginations and for the compulsion to repeat which Freud identified as characteristic of instinctual life, though the compulsion is rendered conscious and perhaps less driving in our relation to works of art. Such tenacity of impulse, when it is unleashed on day-to-day living, can become a rigid, distorting seizure of the real. Browning was as adept at spotting such petrifications in the manner of loving as he was at conveying the circumstantiality of fondness. 'Mesmerism':

> Till I seemed to have and hold,
>   In the vacancy
>   'Twixt the wall and me,
> From the hair-plait's chestnut gold
> To the foot in its muslin fold—
>
> Have and hold, then and there,
>   Her, from head to foot,
>   Breathing and mute,
> Passive and yet aware,
> In the grasp of my steady stare—
>
> Hold and have, there and then,
>   All her body and soul
>   That completes my whole,
> All that women add to men,
> In the clutch of my steady ken—[94]

The fantasy of possession which possesses the hypnotist in this poem travesties the security of emotional tenure which marriage ceremonially conveys, and performs that travesty by taking the words 'to have and to hold' which are said by priest, groom, and bride in the marriage service and turning the liturgical repetition into an obsessive fixity in the speaker's mouth. While he may believe himself in power through these words, through his appropriation of a public locution to private, incantatory ends, the stanza form into which his speech falls holds his words to its other purposes—a second kind of ceremony contains and judges his deviance from the first ceremony. Thus, the syntactic suspension which runs from 'I seemed to have and hold' over five lines before it reaches the object 'her' conveys, certainly, the speaker's

[94] 'Mesmerism' (1855), P i. 574.

emphatic sense of having attained the woman but it does so by
simultaneously rendering his speech as unnatural, compliant to the
demands of a stanzaic pattern beyond his willing. It is a drama of
jealous possessiveness such as Browning often staged: the speaker
clutches on to what he cares for, his speech expresses his own
conviction of the firmness of his grip; the patterning of his speech by
the poet shows him rather to be in the clutches of his own passion. The
effect is one of the speaker's desire to capture but also of his formulaic
entrapment, and the effect is secured by setting speech athwart what
Isobel Armstrong has exactly described as 'a sparing, tense stanzaic
form which Browning is so good at making both lyric and self-
mocking'.[95] That form belongs to the printed voice, and is drawn from
Wordsworth.

Remaining Faithful: Hardy

Hardy inherited this dramatic skill with the stanza from Browning,
though he developed its potential in more calculatedly grim directions
than is usual in Browning. From the early 'At a Bridal', his lines are
marriage lines, and particularly lines of marriages askew; the
adjustment of speech to lines in his work bears mostly his sense of
being 'grieved that lives so matched should miscompose'.[96] He wrote
of Browning: 'Imagine you have to walk [a] chalk line drawn across an
open down. Browning walked it, knowing no more. But a yard to the
left of the same line the down is cut by a vertical cliff five hundred feet
deep. I know it is there, but walk the line just the same.'[97] Hardy had
almost certainly not read the letter Browning sent to Julia Wedgwood
about his determination to remain faithful to Elizabeth after her death,
but Browning's words there coincide with Hardy's criticism of what he
took to be his predecessor's blinkered forthrightness: '. . . I see a plain
line to the end of my life on which I shall walk, unless an accident stop
all walking—I shall not diverge, at least.'[98] What Hardy says about
Browning is wilfully uncomprehending of the relation between such a
'line' and the 'dangerous edge of things'[99] in Browning's writing, and
attention to Browning's conduct of speech in poetic lines would have
shown him that Browning knew a good deal more than a straight 'chalk

---

[95] Armstrong, op. cit. 288.   [96] 'At a Bridal', dated by Hardy as 1866, Hynes, i. 11.
[97] Unpub. note, quoted in Millgate, *Thomas Hardy: A Biography*, 409.
[98] Letter to Julia Wedgwood, 27 June 1864, Curle, 35.
[99] 'Bishop Blougram's Apology' (1855), P i. 627.

line', though he may not have diverged from it. It is the sort of wilful incomprehension you sometimes find between members of a family, and it matters more as signalling the familial relatedness between them than as revealing that Hardy's sympathies were, unsurprisingly, partial.

In this note which he never published, Hardy credits himself with wisdom and bravura; he is both a sage and a tightrope-walker. Yet the melodrama of the distinction he draws between consciousness of the abyss and respectable conduct is neither balanced nor judicious. Even respectability can sometimes be a risk that needs taking. In his poems, he usually avoids the attitudinizing of this note; he is helped to do so by the way lineation in poetry can make at once a 'cut' into speech, put speech in view of its own precarious position, and at the same time hold a firm 'line' of regular and decent behaviour. By this simultaneity of 'line' and 'cut', the poems may then not merely skirt but constantly keep in view the risks in their order:

> I said to Love,
> 'It is not now as in old days
> When men adored thee and thy ways
>   All else above;
> Named thee the Boy, the Bright, the One
> Who spread a heaven beneath the sun,'
>   I said to Love.
>
> I said to him,
> 'We now know more of thee than then;
> We were but weak in judgment when,
>   With hearts abrim,
> We clamoured thee that thou would'st please
> Inflict on us thine agonies,'
>   I said to him.
>
>   . . . .
>
> 'Depart then, Love! . . .
> —Man's race shall perish, threatenest thou,
> Without thy kindling coupling-vow?
> The age to come the man of now
>   Know nothing of?—
> We fear not such a threat from thee;
> We are too old in apathy!
> *Mankind shall cease.*—So let it be,'
>   I said to Love.[100]

---

[100] 'I Said to Love' (1901), Hynes, i. 147–8; the ellipsis in the third stanza quoted is Hardy's.

Hardy's fondness for fusty diction makes the archaisms here more searching than they would be in the simpler sarcastic instances I gave earlier from Meredith. Hardy has to speak to 'Love' on and in Love's own pretty conventional terms, even if the point of the conversation is to effect a snub. An ambiguity comes into the words of the poem's speech, and provokes thought about those words. 'We clamoured thee that thou would'st please | Inflict on us thine agonies' may report a request employing an old-fashioned construction ('O Love, be pleased to inflict . . .') or a more current locution ('O Love, inflict agonies on us, please')—an ambiguity not so much of meaning as of how dated the speech is. Since datedness, of locution and of loving, is the subject of the poem, this ambiguity is substantial. Especially so when the confident 'We now know . . .' is retorted upon in the forecast 'The age to come the man of now | Know nothing of'. Again, there is an uncertainty of linguistic date about the lines: they may be paraphrased 'modern man will know nothing of the future' or 'the future will remember nothing about modern man' according to how extreme a poetic inversion is heard in the lines. These local instances of created insecurity about the poetical tenor of the lines affect the poem as a whole in that we must ask whether the repeated 'I said to Love' is nearer to a conventional refrain or a colloquial repetition ('I said to him, I said'). Asking that, we begin to ponder whether the stanzaic form has here a lyric elegance or whether its elegance should be heard as a threadbare formality through which break unlyrical accents and sentiments.

The poem may be thought to repudiate a Browningesque 'Lyric Love',[101] but its implication within a version of the style of such lyricism puts up a struggle between the new world and the old in the words of the poem themselves, as they struggle between two worlds of style. The poem formally shakes the confident direction of view it announces. That last 'I said to Love' does not rhyme as clinchingly with 'Know nothing of' as 'Love' has previously rhymed, at the end of the stanzas, with 'above' and, in the stanza I have omitted, 'dove'. This is partly because Hardy breaks the pattern of the stanza, putting greater distance between the 'a' rhymes (stanzas 1–3 rhyme abbabba, stanza 4, abbbabbba), but whether this should strike us as an intimation of a new scheme for loving as well as a new rhyme-scheme, or whether the new scheme comes across as just the old one writ

[101] *The Ring and the Book*, XII. 868, Altick, 628.

larger, remains open to question. If 'of' and 'Love' rhyme well in the last stanza, the poem comes to rest on a formal completeness in the moment it points to the incipience of an entirely new world—an old satisfaction of the imagination, the full rhyme, comforts us in face of the unimaginably new. If we have rather a half-rhyme, that might be felt as a faltering of the voice in the face of such blank and large prospects as are explicitly conjured. Any vocalization of the poem must opt for one or the other of these alternatives, but the printed text allows us to see the 'cut . . . five hundred feet deep' in our imagination of love as we continue to 'walk the line' of love-poetry 'just the same', as Hardy did. 'I know it is there, but walk the line just the same': 'I am just the same as when | Our days were a joy, and our paths through flowers'.

In Browning the poeticality of the poem (its correspondence to set forms, its patterns of image, its world of sound) can usually be seen as coming from the poet, and its speech (its communicative intent in the fictional situation, its emphases of meaning, speech stress as against metrical stress) can usually be taken as coming from the fictional speaker. Hardy's dramatic lyrics, on the other hand, at their best derive both poem and speech from the poet but from diverse attitudes within and around him, for he is both attached and averse to the forms through which he creates, so that, as Hardy wrote, 'much is dramatic or impersonative even where not explicitly so'.[102] His second collection of poems was called *Poems of the Past and the Present*.[103] The title meant in 1901 that the celebrated novelist was publishing poems he had written before he began his career in prose as well as poems he had written since revising *The Well-Beloved*. It also means something more than a note of chronological provenance. His insistence on the dates of the poems continues long after he had ceased to need to point out that he had been a poet before he became a novelist. The best Hardy poems, in this and later collections, are poems of the past *and* present because they dramatize the date of their own poeticality in order to weigh the issue between his hopes and his loyalties, his views about love and marriage and his experience of those states. As it is a good general rule that the better dramatic monologues of Browning involve the composed, significant silence of the fictional interlocutor to whom the fictional speech is addressed, and that the poems are weaker in the

---

[102] Preface (1901) to *Poems of the Past and Present* (1901, but post-dated to 1902), Hynes, i. 113.
[103] Originally to have been called *Poems of Feeling, Dream, and Deed*.

degree to which that silence is merely token rather than teasing us into thought, so the best Hardy dramatic lyrics are those in which his idiosyncratic leanings to archaic diction turn out to be a form of the past's substance even in the impetus of repudiation which often drives the poems.

He brought to public notice through the impersonated offices of his second wife and biographer the 'following notes' on the subject of 'New Year's Eve':

> We enter church, and we have to say, 'We have erred and strayed from thy ways like lost sheep', when what we want to say is, 'Why are we made to err and stray like lost sheep?' . . .
>
> Still, being present, we say the established words full of the historic sentiment only, mentally adding, 'How happy our ancestors were in repeating in all sincerity these articles of faith!' But we perceive that none of the congregation recognizes that we repeat the words from an antiquarian interest in them, and in a historic sense . . .[104]

The prose is sentimental as the poems mostly aren't—sentimental in opposing 'historic sentiment only' to 'all sincerity', sly in making out that 'antiquarian interest' and 'historic sentiment' are much the same thing. The poems manage something better, historic sincerity. This is partly because the words of poetry are not 'established' in the way those of Anglican services are, and so permit after full assent has been withdrawn of a formal loyalty which is not merely formulaic, and partly because the printed voice of his poetry allows for the kind of contemplated wistfulness with regard to the implausibility of what he still desires which characterizes Hardy and which defeats his powers of expression when he tries to articulate it as a view. His resourcefulness as a poet found again and again shapes for his suspected aspirations, notably by exploiting the gap between the seen and the said, so that he can achieve the close of a poem as simultaneously askew and satisfactory.

Take 'In Tenebris II':

> Let him in whose ears the low-voiced Best is killed by the clash of the First,
> Who holds that if way to the Better there be, it exacts a full look at the Worst,
> Who feels that delight is a delicate growth cramped by crookedness, custom,
> and fear,
> Get him up and be gone as one shaped awry; he disturbs the order here.[105]

The sudden assertion of a duple rhythm in 'he disturbs the order here'

[104] Hardy, *Life*, 358.   [105] (1901), Hynes, i. 208.

(it may even be regular trochaics) against the prevailing triple time of the poem mimics a disturbance of order but only by being orderly in a more familiar manner than the body of the poem—a very Hardyesque achievement of a self-shocked respectability, or of a remorsefully conscious Law. The poem works with the rhythms of Anglican psalmody, draws strength to turn away from the tyrannous optimism of the voices congregated against Hardy, and draws it from the metrical tradition of those very voices. Being rhythmically out of step with what comes before, 'he disturbs the order here' only with difficulty rhymes with 'crookedness, custom, and fear'. That is, the rhyme 'fear'/'here' is full but it is not easy for the voice to give it the effect of fullness because of the counter-tug produced by the rhythmic ill-assortment of the two phrases. It can be seen to be perfectly in place, though. This effect, of a full rhyme mismatched with a rhythmic divergence of the rhyming elements, is characteristic of Hardy, particularly at the ends of poems, and conveys to the eye and ear the tussle in his feelings while also fashioning that disquiet into something more than affliction.

His fondness for church music generally was of the greatest importance to him, as a man and as a writer; the family tradition of playing church music brings together in his life domestic security, religion, and music so intimately that his devotion to all three could never later be quite disassembled, however his mind changed. Particularly, he found, when older, that he could sing things he could not say, or could take pleasure in hearing things sung he did not much like to hear said. And this ability of music to preserve for Hardy what he could no longer assert explains his dwelling on religious music and folk-ballads, both of which often breathe a world of values he has abandoned, as in the beautiful 'A Church Romance':

> She turned again; and in her pride's despite
> One strenuous viol's inspirer seemed to throw
> A message from his string to her below,
> Which said: 'I claim thee as my own forthright!'

> Thus their hearts' bond began, in due times signed.
> And long years thence, when Age had scared Romance,
> At some old attitude of his or glance
> That gallery-scene would break upon her mind,
> With him as minstrel, ardent, young, and trim,
> Bowing 'New Sabbath' or 'Mount Ephraim'.[106]

[106] *Saturday Review* (1906), Hynes, i. 306.

The close does not exactly fit, for 'Ephraim' probably requires stressing as a dactyl and so does not fully meet the iamb of 'and trim'. It is like a tune you can't quite remember. Throughout the piece, Hardy shows a sentiment for the history of words. The archaic 'in her pride's despite' means 'despite her being proud' but also 'in that despite of others which her pride created' (*OED*: 'Despite. 1. The feeling or mental attitude of looking down upon or despising anything; the display of this feeling; contempt, scorn, disdain. *Obs.* or *arch.*'). 'Forthright' probably works as an adverb modifying 'claim' but the pressure of the idiom 'to claim something as one's right' makes it sound like a noun, she is not only his right but his forthright (*OED*: 'Forthright. C. *sb.* A straight course or path; *lit.* and *fig.*'). The poetical 'heart's bond' is brought into non-metaphorical currency when at the end of the line it is 'signed' as a bond may actually be. The wonderful, quiet dexterity of the style in its adjustment of the past and present of the words enables him to be more tender and inquiring about what happens when 'Age' scares 'Romance' than when he's speaking out on the matter, because the words remember minstrelsy with pleasure though they do not indulge in a pastiche: they keep in view what has gone out of earshot.

For Hardy, the relation between the spoken and written versions of a poem is often the relation of the present to the past. He wrote in his journal for 27 January 1897: 'To-day has length, breadth, thickness, colour, smell, voice. As soon as it becomes *yesterday* it is a thin layer among many layers, without substance, colour, or articulate sound.'[107] This shrinkage of the voice on the page, while the voice yet remains there asking to be resuscitated, perfectly contains Hardy's idiosyncratic dedication both to nostalgia and meliorism—it makes the poem stand in time as well as withstand time. Many of the poems live on recall, especially of words spoken long ago; the verse contains 'midnight scents | That come forth lingeringly' as in 'Shut out that moon'

> And wake the same sweet sentiments
>   They breathed to you and me
> When living seemed a laugh, and love
>   All it was said to be.[108]

'Said' dwindles in print as this poem takes seriously what was lightly said, but whatever the derision time has brought him to feel about love's reputation, he has not lost his vulnerability to 'the same sweet

---

[107] Hardy, *Life*, 302.   [108] Dated 1904 by Hardy, Hynes, i. 266.

sentiments', they may always be revived—and that is the strength of his poetry, both in its unsentimental disillusion and in its enduring attachment on the page to the 'echo of an old song', to 'the heart whose sweet reverberances are all time leaves to me'.[109] These out-of-kilter rhythms in Hardy on rhymes which centre on 'me' sound the adjustments of his self to his forms.[110]

Yet he suspected his memory and his memories, noting 'how our imperfect memories insensibly formalize the fresh originality of living fact—from whose shape they slowly depart, as machine-made castings depart by degrees from the sharp hand-work of the mould'.[111] Hence, the perpetual alertness to the process of sensible formalization which is required by the sharp irregularities of his rhythmic practice. It would be possible to regard the inherited literary tradition of writing about love and marriage as a massive accumulation of insensible formalities which have over the years so departed from the mould that they no longer resemble it at all. Such a possibility must be guarded against within literature itself, in the texture of the verse. His double attitude here expresses itself variously, for example, in the range of senses which 'mechanic' and its cognates may take in his verse, from the tiresomeness of 'household life's mechanic gear' to the recognition that in poetry 'mechanic repetitions please'.[112]

Consider how he rewrites Coventry Patmore's 'Departure' throughout the 1912–13 poems. The Patmore is not well known, and it is both distinguished in itself and an important source for Hardy, so I give it in full:

> It was not like your great and gracious ways!
> Do you, that have nought other to lament,
> Never, my Love, repent
> Of how, that July afternoon,
> You went,
> With sudden, unintelligible phrase,
> And frighten'd eye,
> Upon your journey of so many days,
> Without a single kiss, or a good-bye?

---

[109] Both quotations from *Moments of Vision* (1917), respectively, the subtitle of 'Sitting on the Bridge' and the last line of 'The Change', which is dated 1913 by Hardy, Hynes, ii. 192.

[110] Compare Wordsworth's cadence at the close of 'She dwelt among the untrodden ways'.

[111] Preface to the Wessex Edition (1912) of *Wessex Tales*, repr. in Orel, 22.

[112] 'The Dawn after the Dance', dated 1869 by Hardy, Hynes, i. 280 and 'Why do I?', in *Human Shows* (1925), Hynes, iii. 157.

I knew, indeed, that you were parting soon;
And so we sate, within the low sun's rays,
You whispering to me, for your voice was weak,
Your harrowing praise.
Well, it was well,
To hear you such things speak,
And I could tell
What made your eyes a growing gloom of love,
As a warm South-wind sombres a March grove.
And it was like your great and gracious ways
To turn your talk on daily things, my Dear,
Lifting the luminous, pathetic lash
To let the laughter flash,
Whilst I drew near,
Because you spoke so low that I could scarcely hear.
But all at once to leave me at the last,
More at the wonder than the loss aghast,
With huddled, unintelligible phrase,
And frighten'd eye,
And go your journey of all days
With not one kiss, or a good-bye,
And the only loveless look the look with which you pass'd:
'Twas all unlike your great and gracious ways.[113]

This is taken up by Hardy in 'The Going' which does not console itself
with Patmore's picture of the last hours as filled with loving converse
but bitterly recalls opportunities for speech not only missed but turned
down, and in 'Without Ceremony' which brilliantly changes Patmore's
idealized version of 'It wasn't like you . . .' into

It was your way, my dear,
To vanish without a word
When callers, friends, or kin
Had left, and I hastened in
To rejoin you, as I inferred.[114]

with its actuality of intonational ambiguity about whether such a 'way'
infuriated or endeared. The lines prickle with visible indignation (the
sarcasm of 'as I inferred', the self-pity of 'hastened') which the silence
of print permits us to imagine as now overcome, remembered with
integral fondness for the vanished life. Even the small phrase 'my dear'
is left by Hardy to tremble between the sardonic and the plangent,

[113] In *To The Unknown Eros, Etc.*, (1877), Patmore, *Poems* 362–3.
[114] (1914), Hynes, ii. 53.

where Patmore polishes it up, reverently, into 'my Dear'. The superiority of Hardy's poems over Patmore's comes from the richer and clearer bearing of posthumous idealization on lost realities in his work, and this clarity, this richness, arise from his creativity in the depths of the page.

If you put Hardy's 'The Going'[115] next to 'Departure', and look at the occasions when the poets move from the represented colloquialism which is the basis of the style in each case to a higher level of diction, you find that Patmore does so in a manner complicit with what he himself denounced as 'vulgar ideas of "greatness"'.[116] The poetical appears at 'nought other to lament' and 'growing gloom of love' and 'As a warm South-wind sombres a March grove' and 'luminous, pathetic lash'. These are elevations which make us conscious of the poet as an elevator, but they find a swift and replete consolation in that consciousness. They do not register loss in the voice. 'Growing gloom of love' *means* that her eyes are clouding over in death, but the phrase takes pleasure in shielding its eyes from that fact. The self-protective impulses in the verse are natural and, what is more, carefully handled, but the poem moves in a less searching way than Hardy's. Where Hardy raises the dictional level of the language, he does so with a thoughtful variety quite unlike Patmore's singleness of purpose. The most 'poetical' stanza, the fourth, where Emma is remembered as 'she who abode | By those red-veined rocks far West' and as 'the swan-necked one', is also the stanza most governed by severely insistent past tenses—'You were' . . . 'You were'—tenses which place these glamours in an actual past to which the literary pastness of the hyphenated adjectives corresponds. Or when Hardy goes out into 'the alley of bending boughs', his grove and its sombering require a language which utterly unfamiliarizes what he sees and what we read

> Till in darkening dankness
> The yawning blankness
> Of the perspective sickens me!

Patmore's simile neatly, indeed nattily, fits onto what it is a figure of, where Hardy's language sets skill the harder task of representing loss of orientation, revulsion from the arranged perspective in which vision has been at home (and does so very well in the rhyme of 'dankness' and 'blankness', of the minutely concrete sense of vegetable moisture in the

---

[115] (1914), Hynes, ii. 47–8.
[116] Champneys, i. 255.

one with the abstract expanse of the other, setting the mind at odds with itself by switching categories of attention).

Patmore relies on the poetical, but poeticality may stand in the Hardy poem also as something to be fervently disbelieved in. He thinks of Emma going

> Where I could not follow
> With wing of swallow
> To gain one glimpse of you ever anon!

'Follow'/'swallow' is a rhyme which comes readily to poetical hand, and so can seem reach-me-down. When Tennyson has it, as an internal chime, in *The Princess*, he marks it as a falsetto effect ('ape their treble') and plays to the full its twee potential:

> O Swallow, Swallow, if I could follow, and light
> Upon her lattice, I would pipe and trill,
> And cheep and twitter twenty million loves.[117]

Hardy may have remembered these lines. At least, his 'could *not* follow' (my emphasis) and his modest yearning—'one glimpse' as against 'twenty million loves'—gain point in the recollected context of Tennyson's chirpy lyric. Hardy gives the rhyme a bizarre double aspect, both tripping and dumpy, and particularly does so by mismatching the line-lengths (6 and 5 syllables) as he does only once elsewhere in 'The Going'. That double aspect is there to convey the bitterness with which Hardy in actuality turns on poetical machineries which loft the poet like a bird wherever he may want to go. When people die, they go somewhere not even poetic language can follow them, but the belief that they go anywhere is also poetical. Only the audacity of genius could have produced so intense a confrontation with the machineries of poetry and the convictions they nourish, a confrontation permanently fixed in the timbre of a rhyme, slightly foolish on the voice, visibly mourned for on the page.

Much the same could be said of the contrasting rhythmic practices of the two poems. Patmore's handling of variation in line-length (from two to twelve syllables) is often dextrous, as in the mimetic 'Whilst I drew near, | Because you spoke so low that I could scarcely hear' with the leaning-in of the alexandrine, but it seems always secure of its own pathos, of the charm of the lilts he produces. Hardy, on the other

[117] R ii. 235.

hand, manages to charge his shorter lines with a doubt of charm as well
as a longing for the unruffled world such charm suggests:

> We might have said,
> 'In this bright spring weather
> We'll visit together
> Those places that once we visited.'

'Weather' and 'together' sound happily mutual, but they are contained
within the Hardy effect of rhythm countering rhyme which prevents
'said' and 'visited' quite chiming together. The idyll of possible return
to the scenes of their first raptures is lodged within the wistful sobriety
of that failed meeting of the words, words which sit, as it were, in sight
of each other without being able to come out with all they wish to say,
like Thomas and Emma Hardy, in fact.

He noted to himself in 1868, 'Perhaps I can do a volume of poems
consisting of the *other side* of common emotions', and the *Life* adds,
'What this means is not quite clear.' It is indeed 'not quite clear' but
perhaps what he meant comes out in an other note from the journal of
about the same time 'How people will laugh in the midst of a misery!
Some would soon get to whistle in Hell.'[118] Hardy's poetic practice
with short lines and consequent lilts shows how whistling in Hell could
express the other side of common emotions by poetically shaping the
other side of common forms. That is, emotions such as deep grief are
subject to odd 'vicissitudes of passion',[119] gusts of indifference, absent-
heartedness, laughter in the midst of misery, and so on—this is their
'other side'. Equally, as Tennyson's *Maud* brilliantly showed, the
formalities of verse have an 'other side', viewed from which refrains,
say, may indicate not delighted recourse to a subject but obsession with
it, or lilting short lines may indicate derision, infantilism, calculated
unconcern rather than a buoyant lyricism—or better, they may
indicate such lyricism *as* these less pleasant things. Hardy brings the
two 'other sides' together and lets us see them in the simultaneity of a
page which we can read as vocally ambiguous but cannot, from the
nature of vocal ambiguity, voice at once as such:

> How she would have loved
> A party to-day!—
> Bright-hatted and gloved,
> With table and tray

[118] Hardy, *Life*, 59, 61.
[119] Dr Johnson, 'Preface' to *Shakespeare* (1765). I quote from the text of Arthur
Sherbo (New Haven, 1968) in the Yale edition of Johnson, vii. 67.

> And chairs on the lawn
> Her smiles would have shone
> With welcomings. . . . But
> She is shut, she is shut
>   From friendship's spell
>   In the jailing shell
>   Of her tiny cell.[120]

The poem starts by expressing a regret, though not so large a regret as might be expected from the title 'Lament'. The pressure of the inherited formality of a lament on the words in the poem, and their counter to that pressure, convey the incongruities with which grief is surrounded, in which it is steeped. The last three lines of the stanza tightly hold the bewildering variety of life which besets the singleness of a passion like grief—we may voice them very slowly as ritually intoned, or clip them out so that they sound in recoil from the loss they describe, or track them with the voice as one pursues a thought which keeps coming back to haunt you, or even crow them with a gloating 'little did she know'. That last possibility is nasty, but it is a possibility which these great poems regularly contemplate. The aggression imbued in love over years of living together is not shrunk from. So Hardy can invent marvellously the compound 'Bright-hatted', which at once presents her delight in the radiance the hat lends her face, his delight in her delight, his rage within the delight at the pettiness of its source, and the way pettiness is transfigured in death because it is little things we most lose. When someone dies, their virtues don't die with them, only their foibles. This is too much for the voice to learn at first, and more than it can ever say all at once, but the lines contain all this other side of the common emotion in their hellish whistling.

Carlyle shocked himself when writing about Jane Welsh Carlyle after her death:

My one solace and employment hitherto is that of sorting up, and settling as I judge *she* would have wished, all that pertained to her beautiful existence and her: *her* advice on it all, how *that* wish starts out on me strangely at many a turn; and the sharp twinge that reminds me, "No!" One's first *awakening* in the morning, the reality all stript so *bare* before one, and the puddle of confused dreams at once gone, is the ghastliest half-hour of the day;—as I have heard others remark. On the whole there is no use in writing here. There is even a lack of *sincerity* in what I write (strange but true). The thing I *would* say, I cannot. All words are idle.. . .[121]

---

[120] (1914), Hynes, ii. 53; the ellipsis is in the original.
[121] *Reminiscences*, 167; the ellipsis is in the original.

The situation grimly returns on the Carlyle who devoted his writing to 'the fact of things' when at last he faces a reality stripped bare by the loss of his wife. He finds the lack of sincerity in what he writes strange but true, but he cannot write that lack of sincerity, he can only be committed to his writing when he appears in it as whole-hearted. He has no 'half-absence' of tone to match his widowed state. Words are idle because they will not work for him as *his* words—they begin to show that they contain in themselves the 'other side of common emotions'. In this distress of the personal amidst the impersonalities of language, the discovery of what is common in what seems so particular to him ('the ghastliest half-hour of the day;—as I have heard others remark') is no consolation.

Hardy can meet and turn that discovery into a new achievement because the poems allow in their written silence for depiction of the simultaneous multiplicity of a moment's regret, say, with its combination of resentment that one should be grieved, shame at one's own resentment, remorse, self-exculpation, a longing to be comforted and a determination to cling to grief as to one's own identity. They allow for the possibility of a lack of sincerity which may cross the words one has to use, and not idly use, as at

> I drove not with you. . . . Yet had I sat
> At your side that eve I should not have seen
> That the countenance I was glancing at
> Had a last-time look in the flickering sheen,
> Nor have read the writing upon your face,
> 'I go hence soon to my resting-place;
>
> 'You may miss me then. But I shall not know
> How many times you visit me there,
> Or what your thoughts are, or if you go
> There never at all. And I shall not care.
> Should you censure me I shall take no heed
> And even your praises no more shall need.'
>
> True: never you'll know. And you will not mind.
> But shall I then slight you because of such?
> Dear ghost, in the past did you ever find
> The thought 'What profit,' move me much?
> Yet abides the fact, indeed, the same,—
> You are past love, praise, indifference, blame.[122]

---

[122] Dated by Hardy as 'December, 1912', Hynes, ii. 49; the ellipsis is in the original.

How should you say 'True'? It accepts what has been said only to begin a rebuttal. It may accept with gratitude a kindly forgiveness from the 'Dear ghost' or it may ruefully take on what is possibly an implicit complaint—'You neglected me while I was alive, and you may continue to do so when I'm dead, but at least now I shall not care, though *then* . . .'; it may converse or self-commune or both or neither exactly, though which it is doing is important for determining the nature of Hardy's belief in the 'Dear ghost', determining whether he thinks he's imagining things or receiving spirit-messages or whether he believes the imagination is a reception of such messages.[123] This searching multiplicity of attitude arises from writing—Emma did not speak, her words were 'the writing on your face'. Her feelings may have been written all over her face, but they remained unread, though now they are aptly preserved in what he has written of her.

Or take the 'voiceless ghost' created in the print of 'After a Journey':

> Yes: I have re-entered your olden haunts at last;
>   Through the years, through the dead scenes I have tracked you;
> What have you now found to say of our past—
>   Scanned across the dark space wherein I have lacked you?
> Summer gave us sweets, but autumn wrought division?
>   Things were not lastly as firstly well
>   With us twain, you tell?
> But all's closed now, despite Time's derision.
>
> I see what you are doing: you are leading me on
>   To the spots we knew when we haunted here together,
> The waterfall, above which the mist-bow shone
>   At the then fair hour in the then fair weather,
> And the cave just under, with a voice still so hollow
>   That it seems to call out to me from forty years ago,
>   When you were all aglow,
> And not the thin ghost that I now frailly follow![124]

This poem with its 'Trust me, I mind not' from Hardy to the ghost replies to 'Your Last Drive' where the ghost 'will not mind'. Here too, the probity of the lyricism essentially involves the recognized inclusion of lacks of feeling, or feelings counter to the declarative surface of the

[123] The classic discussion of this topic is in Donald Davie's superb essay on 'Hardy's Virgilian Purples', *Agenda*, x (Spring–Summer 1972), repr. in *The Poet in the Imaginary Museum*, ed. B. Alpert (Manchester, 1977).
[124] (1914), Hynes, ii. 60.

poem, feelings contained in unvoiced possibilities which lie within the written words. 'What have you now found to say of our past' contains a vexed 'What is it *now*?' as well as a more evident openness to listen to what she has to say. That vexed tone runs against the insensibly formalized notions of how we ought to speak of and to the dead, but it does so to discover all the more vividly Hardy's attachment to her, because, actually, this is a tone we adopt to people who are pressing on us the fact of their independent existence. Any 'now' is paradoxical between a ghost and a human being, Hardy's 'now', drawing her into the imagined ambit of a social world in which people nag you for things and you turn on them with a 'what is it now?', is especially so, and therefore the more alive to her death. The lack of sincerity possibly implicit in being so short with the ghost demonstrates only the sincerity of his lack. Similarly, at 'I see what you are doing: you are leading me on | To the spots we knew', there is a chance to voice suspicions of the ghost, knowing what it is up to ('*I* see what you are doing') as well as coming to an unsuspicious realization ('I see what you are doing'). The visible simultaneity of the guarded and the welcoming tones corresponds to what, I imagine, we should feel about a ghost doing this sort of thing, even the ghost of someone we have reason to trust. At the line-end, 'you are leading me on' briefly worries at the deceit there may be in her guidance, before agreeing to be taken on, and perhaps in. These vocal ambiguities in the lines reduce his voice as hers is reduced; he too assumes the condition of disembodiment which print offers to language as it strips it of the realities of timbre, pace, accent, contour—all the glow of physical existence. She has shrunk from what she was when all aglow and is now thin, but his imagination creates for himself a condition of frailty to answer hers, so that they move in the verse together in an imaginative equity.

Hardy finds a use in measured language to meet Carlyle's shock at the 'lack of sincerity' in what one writes on the death of the person one has loved. The inherent universality of words makes them seem unconcerned with the particularity of experience, especially when that experience is of another person in an extreme of their individuality, that is, in their death. The self-accusing conscience then takes this linguistic unconcern as a token of its own insufficiency of attachment, senses a gap between what it '*would* say' and what the words said. Carlyle's remorse at his lack of sincerity really confesses a lack of skill to create personal writing out of the fact that 'we seem forced to use "common forms" for the most individual and unique part of our

experience.'[125] This is not Carlyle's fault alone, since the insensible
formalization of our terms for love is such that, as it seemed to EBB,
'in nothing, men . . & women too! . . were so apt to mistake their own
feelings, as in this one thing. Putting *falseness* quite on one side . . quite
out of sight & consideration, an honest mistaking of feeling appears
wonderfully common—& no mistake has such frightful results—none
can.'[126]

It may also be that the conduct of literature in the matter of love's
words fosters just such 'honest mistaking of feeling' by creating a
world of emotional facility in which, as EBB says in the same letter, we
grow accustomed to hearing 'that word which rhymes to glove & comes
as easily off and on (on some hands!) . . .'. But when Hardy rhymes
'loved' to 'gloved' in the stanza of 'Lament' quoted above, the rhyme
comes off not with fluent automatism but delicately. Only in a creation
of language which.consciously lives in the region between speech and
writing, between the actualities and ideals of utterance, can the
unconcern of language's generality, as it complies in the social
moulding of what we feel, be considered without adopting the fantasy
of a descriptive language ethically neutral between forms of life.

In particular, the way literary conventions answer to social
conventionalities assists writing to find the pleasure and necessity of
'common forms' in something better than conformity. EBB was sure
that her marriage would be special: 'We could not ~~learn~~ lead the
abominable lives of 'married people' all around—you *know* we could
not—*I* at least know that *I* could not, & just because I love you so
entirely.'[127] You might be forgiven for wondering if one of the usual
abominations of married life was not beginning even as she wrote—the
uncertain slither from 'we' to 'you' to '*I*' with what that shows of how
difficult it is for passionate attachment to an other person to stop itself
from becoming an absorption of them. It is oddly normal for
selfishness to grow out of effusions such as EBB's here. Rather than
loving someone 'so entirely', it might be more loving to be a bit half-
hearted, and literature's reticence of voice finds the key for that saving
halving of the heart which is the real condition of shared lives.

We can observe the dangerous allure of 'so entirely' in the
inflections of the word 'all', of which Empson remarked that it is 'as

[125] Julia Wedgwood to Robert Browning, Good Friday, 1869, Curle, 197.
[126] Letter of 21 Dec. 1845, *RB/EBB* i. 340.
[127] Letter of 2 Nov. 1846, *RB/EBB* ii. 836.

suited to absolute love and self-sacrifice as to insane self-assertion'.[128]
Much depends, then on how you say 'all' when declaring love entirely.
Carlyle:

Oh, I was blind not to see how *brittle* was that thread of noble celestial (almost
more than terrestrial) life; how much it was all in all to me, and how impossible
it should long be left with me.[129]

Browning:

> I would that you were all to me,
>   You that are just so much, no more.
> Nor yours nor mine, nor slave nor free!
>   Where does the fault lie? What the core
> O'the wound, since wound must be?
>
> I would I could adopt your will,
>   See with your eyes, and set my heart
> Beating by yours, and drink my fill
>   At your soul's springs,—your part my part
> In life, for good and ill.
>
> No. I yearn upward, touch you close,
>   Then stand away. I kiss your cheek,
> Catch your soul's warmth,—I pluck the rose
>   And love it more than tongue can speak—
> Then the good minute goes.[130]

Hardy:

> Woman much missed, how you call to me, call to me,
> Saying that now you are not as you were
> When you had changed from the one who was all to me,
> But as at first, when our day was fair.[131]

Putting Carlyle's 'all in all to me' alongside Browning's and Hardy's
'all to me' makes the Carlyle seem marked by a failed insistence which
results only in cliché. Yet the awkwardness here is not merely literary.
The trouble with the passage is that it gravitates always to 'me' ('*I* was
blind', 'all in all to *me*', 'left with *me*—*my* emphases). It wants to be a
particular tribute to an other person, but actually exemplifies self-
concern.

   Browning's 'No', on the other hand, (an ancestor of Hardy's 'True'

   [128] *The Structure of Complex Words*, 101.   [129] *Reminiscences*, 138.
   [130] 'Two in the Campagna' (1855), P i. 729.
   [131] 'The Voice', dated Dec. 1912 by Hardy, Hynes, ii. 56.

in 'Your Last Drive') curbs the wish of his 'all to me' without dismissing it, and brings a firm energy to discriminating between ideals of mutuality and illusions of identity. The verse creates recognitions of the unentirety of love even or perhaps specially at times like this. For example, if you want to bring out fully the rhyme of 'all to me' with 'nor slave nor free', so that the words may seem full of themselves as they express the desire for full communion, then you must disproportion the emphasis of 'all to me' and bring the phrase dangerously close to 'all to ME', which could sound greedy—the 'Is that all for ME?' of a child eyeing a cake. The wish the words try to voice is that the speaker should be entirely devoted to the interlocutor, but on the voice, within the demands of the rhyme with 'free', this can sound like the desire that the interlocutor be entirely devoted to him. The page enables us to watch, and watch out for, that process where devotion turns into demand.

There is an exquisite domestic miniature contained in 'set my heart Beating by yours . . .'; the suggestion is that hearts may be set, as watches, by each other. The metaphor beautifully gives a hint of the actual, social surrounds of such conjugal lyricism; it keeps household maintenance in view. The 'value and significance of flesh'[132] is tested in the third stanza quoted by the quiet ordering in the rhyme of 'kiss your cheek' to 'tongue can speak'—what lips can do in a kiss, 'Catch your soul's warmth' (not for nothing that Browning was fond of Donne, as that quick compacting of body heat with spiritual ardour shows), the tongue cannot carry out in utterance: the human mouth centres such passions and cannot cope with them. This sort of thing passes, 'the good minute goes', just as what Browning says also goes, and it is because of that evanescence that the stanza embraces the kiss with the half-rhyme of 'No' and 'goes'. But then this stanza preserves evanescence in the returning shape of the verse, can thus record the minute goodness of what occurred. As the first utterance of the wish, 'I would that you were all to me', weighed the words 'to me', the poem concludes by taking the weight of 'all':

> Just when I seemed about to learn!
>   Where is the thread now? Off again!
> The old trick! Only I discern—
>   Infinite passion, and the pain
> Of finite hearts that yearn.

[132] 'Fra Lippo Lippi' (1855), P i. 547.

'Infinite passion' brings out what would be involved in one person's being *all* to an other and honestly faces the paradox of how such passion feels in 'finite hearts'. Carlyle does not reach to that when he conjoins 'noble celestial' and 'almost more than terrestrial', which tends to make 'celestial' collapse back into meaning no more than 'very noble indeed' and flinches from the full implications of the word by the paltering 'almost'. In Browning's lines, there remains a vocal ambiguity which exactly poises the wish and the despair of absolute union of selves. The voice can turn in several directions along 'Only I discern'. Allowing for a pause between 'Only' and 'I', the sense would be something like 'The only thing is this: I discern' (compare 'Only I'd just like to say . . .'); Browning might have suggested such a pause with a comma. If the half-line hasn't this kind of pause, and is read with a trochaic pulse, stress falls on 'I'. The implication may then become 'I'm the only one who realizes . . .'. Such solitary perception may be arrogant or bereaved, or any of the other possible attitudes to the fact that your wife, or companion, doesn't understand you. Speech rhythm will equally allow 'Only I discern', without a pause and so meaning 'I realize just this one thing' with no special implication that the discerning is lonely. The double inflection intimates the moment within love when one wonders what the other is thinking, whether the other too is now feeling *this*—and that is the question from which the poem began: 'I wonder do you feel to-day | As I have felt . . .'.

Hardy's 'The Voice' is much in Browning's debt, as Hardy's voice, and the absence of it, generally is. He removes the ambiguity of stress in 'all to me' by setting it in so firmly dactylic a rhythm that the momentarily threatening voracity of the phrase in Browning is dissolved. This is only apt for not many believe that they can entirely possess a ghost; it is the ghost who comes to possess. The slightly alien quality of dactylics in English verse is chosen to insinuate through Hardy's words the timbre of a fascinatingly dead voice, that of the 'woman much missed', she who was 'missed' in life (not centrally attended to) and is now the more 'missed' (felt lacking, needed). The pressure of her dead accents comes through in the way that the second verse of the poem cannot get away from rhymes and half-rhymes on 'you':

> Can it be you that I hear? Let me view you, then,
> Standing as when I drew near to the town
> Where you would wait for me: yes, as I knew you then,
> Even to the original air-blue gown!

It may or may not be you that Hardy hears, but the reader hears 'you' over and again: 'you', 'view you', 'drew', 'to', 'you', 'knew you', 'air-blue'.

When the next stanzas go on to be unable to believe his ears, the sound of her voice, as of the 'you', withdraws:

> Or is it only the breeze, in its listlessness
> Travelling across the wet mead to me here,
> You being ever dissolved to wan wistlessness,
> Heard no more again far or near?
>   Thus I; faltering forward,
>   Leaves around me falling,
> Wind oozing thin through the thorn from norward
>   And the woman calling.

In place of 'you' there comes the awful self-entrapment in which he would be caught if his was the only voice here, the self echoing back to itself, a mere 'me': 'breeze', 'mead', 'me', 'here', 'being', 'near'. 'You' appears again only in the grimly material 'oozing through'; this is perhaps what her voice comes down to. 'The Voice' rewrites 'Two in the Campagna', I think, turning Browning's 'champaign' haunted by 'Rome's ghost' into a wet English mead haunted by a more familiar ghost, taking the 'everlasting wash of air' from the Italian skies and giving it to Emma's 'air-blue gown', changing the 'light winds', which weave through Browning's thoughts and eventually wash them away, to the 'breeze' plaguing Hardy's hopes of a rediscovered conversation. He re-thinks and re-sounds the earlier poem's questions about the entirety of human satisfaction as he disperses the vocables 'me' and 'you' across the page, the dark, scannable space wherein he lacks her, his 'all'. Nowhere more poignantly than in the poem's quiet shift from the second person of 'Woman much missed' to the third person of 'the woman calling', as he resigns the fiction of address. He gives up the voice but he does not give up on it, for this imagined conversation (with whom Hardy does not quite know) may be 'only' a poem but the poem continues to speak to people he could not have known.

Poetry and 'the sphere of mere contract'

Elizabeth Barrett Barrett told Browning an anecdote about an unhappiness of marriage:

What, after all, is a good temper but generosity in trifles—& what, without it, is the happiness of life?—We have only to look round us. I saw a woman once, burst into tears, because her husband cut the bread & butter too thick. I saw *that* with my own eyes. Was it *sensibility*, I wonder!—They were at least real tears & ran down her cheeks. 'You ALWAYS do it'! she said.[133]

This brings Tithonus's predicament down to earth. Morn by morn he saw Aurora renew her beauty, and meal after meal this woman had seen the bread and butter cut too thick. One of the many reasons why it is difficult to understand other people's marriages is that marriage invests the deepest feeling in what is diurnal, and so may militate against generosity in trifles, as the minute attention of poets to the dust in the corners of language may equally militate against the generosity, or what is taken to be such, of those who speak more freely. In marriage, from the small action there may well up remembered sources of tenderness or vexation; outsiders to the marriage see the small action and may not be able to tap in imagination the sources it reaches down to. This is also true of long-standing friendships, though there the vexation and the tenderness get less under the skin because the relationship between the two friends is not amorous. Again, it is true of long-term cohabitation, though there the partners have not vowed to face such frictions or warmth permanently together. Each relation has in the facts of imagination its particular griefs and pleasures, as do other kinds I have not mentioned (between parents and children, for example). Though the Victorian period was one of grand friendship between its great writers, and though it produced much fine imagining of parents and children, it was in poetry less productive of works instinct with the rhythms of those relations. *In Memoriam* celebrates a friendship very beautifully, but it dwells on the loss, not on the daily continuance, of that friendship. And it is to the novels and nonsense-verse of the period that we turn to find writing alive with the reciprocal demands of adults and children. The poetic focus of so much of the most distinctive poetry of the period is sharply conjugal, focused on the way the actualities of marriage create private idealizations of the other—the One Who Always Cuts the Bread Too Thick, for example—within the public ideals which the marriage officially declares.

Poets and spouses need to know how to weigh their moments, and the move from page to voice and back again provides poetry with one such measure of weight. (They may also watch and weigh themselves

[133] Letter of 18 Dec. 1845, *RB/EBB* i. 326.

otherwise—by acts of patient comparison such as poets make when adapting their utterances to customary genres, or such as couples make when visiting their married friends.) Patmore thought heaven would be one long instant of illuminated delight, a profuse speech made permanent:

Only for a few hours, perhaps, of the million which is about the sum of the longest lifetime, has [any individual] easily and unaccountably found himself to be living indeed. Some accident, some passing occasion which has called upon him to be more than himself, some glimpse of grace in nature or in woman, some lucky disaster even, or some mere wayward tide of existence, has caused the black walls of his prison-house to vanish; and he has breathed in a realm of vision, generosity, and gracious peace, 'too transient for delight and too divine'. These prophetic moments—one in a million—pass; but, unless he has despised and denied them, they leave him capable, more or less, of understanding prophecy; and he knows that in him also there is a potentiality, realisable perhaps under other than present conditions, of becoming one in that great society in which such states of life appear to be not momentary crises but habits.[134]

By the same token, I suppose, hell is always having the bread cut too thick. In his writings about poetry, Patmore enthusiastically imagines poetry as a foretaste of heaven's transformation of momentary ecstasies into eternal habits; the poet's task was to confer on the unknown ideal 'a *sensible* credibility' and on the actually known 'a truly sacramental dignity'.[135] Neither of these tasks, of grounding or transfiguration, can be understood unless some account is taken of Patmore's Catholicism, however perfumed and unconvincing that sometimes seems. Conferring on the unknown a '*sensible* credibility' was for him part of the task of fostering the Incarnation, itself not a single event but the process of the Church, though, according to Patmore, a process still incomplete: 'The Incarnation, in fact, is still only a dogma. It has not got beyond mere thoughts. Perhaps it will take thousands of years to work itself into the feelings, as it must do before religion can become matter of poetry.'[136] On the other hand, the sacramental dignifying of the sensible is possibly only through that Incarnation which has already been achieved in the body of the Church because it is the Incarnation which makes a sacrament of the Church and so is the source of the Church's sacraments.

Patmore's imagination centres on marriage, and there is no hope of

[134] 'Possibilities and Performance', in *Principle* . . . , 300.
[135] 'Imagination', ibid. 306.   [136] Champneys, i. 260.

hearing his poetry unless it is heard as a marriage in the language, and a marriage regarded as sacramental:

Nuptial love bears the clearest marks of being nothing other than the rehearsal of a communion of a higher nature. 'Its felicity consists in a perpetual conversion of phase from desire to sacrifice, and from sacrifice to desire, accompanied by unchangeable complaisance in the delight shining in the beauty of the beloved; and it is agitated in all its changes by fear, without which love cannot long exist as emotion.' Such a state, in proportion to its fervour, delicacy, and perfection, is ridiculous unless it is regarded as a 'great sacrament'.[137]

The identity of his terms for marriage and for poetry may not augur well for the poetry itself, but, in fact, the poetry manages more adeptly than the prose to show 'complaisance' along with 'fear' by its sensitivity to the balance of what Patmore called the 'corporeal' and the 'spiritual' in verse, the balance of the metrical and the semantic.[138] Particularly vital here is his personal variant on Wordsworth's drama of the lyric. Patmore complained of the errors of earlier prosodists:

The most common and injurious of such errors is that of identifying metrical pauses with grammatical stops. Some of the early English poets were at great pains to try the experiment of making these two very different things coincide. Now, one of the most fertile sources of the 'ravishing division' in fine versification is the opposition of these elements—that is to say, the breaking up of a grammatical clause by a caesural pause, whether at the end or in the middle of a verse.[139]

That practical creating of such 'ravishing division' in verse tempers Patmore's fervent espousal of the analogy of the soul to a bride, and works a sociable astringency and clear-sightedness into what might otherwise have been wide-eyed and plush.

Take the scene in *The Angel in the House* where, shortly after their marriage, Felix buys Honoria a pair of sand-shoes:

> I, while the shop-girl fitted on
>   The sand-shoes, look'd where, down the bay,
> The sea glow'd with a shrouded sun.
>   'I'm ready, Felix; will you pay?'

[137] 'Love and Poetry', *Principle* . . . , 338.
[138] Roth, op. cit. 7.
[139] Ibid. 23–4.

> That was my first expense for this
> Sweet Stranger, now my three days' Wife.
> How light the touches are that kiss
> The music from the chords of life![140]

Something like this actually happened in Patmore's married life, for he wrote to his wife in 1850, 'I could kiss the steps of the shop where I paid for your sand-shoes and felt so like a husband for the first time.'[141] The verse comes smiling through fears and ridicule which, in his intimate correspondence with her, he did not consider. The abab rhyme-scheme naturally divides *The Angel in the House* up into quatrains but here, as often elsewhere in the poem, Patmore divides his lines up more variously than the fixed pattern would suggest (in this case 3+1+2+2). The effect is to set her voice apart from his versifying, as her practical concerns stand off from the lyricism of 'The sea glow'd with a shrouded sun'. There is an alert social comedy in the movement of the verse, conveying as it does the different aims of their attentions—that suspended brooding of 'I . . . | . . . look'd' against the brisk 'I'm ready'. The bearing of the lines on speech is equally mobile. At 'the shop-girl fitted on | The sand-shoes', there is a definite opposition of metrical and grammatical pause which relaxes the integrity of the line for the sake of the sentence, whereas, at 'How light the touches are that kiss | The music from the chords of life', the milder opposition of metre and grammar requires that speech rhythm cede to the stanza's needs, so that 'kiss' may take that light, extra touch which it speaks of and seeks. In the verse, the demands of diverse selves and their preoccupations make themselves felt, as Patmore said they should, while 'each is incessantly, though insignificantly, violated for the purpose of giving effect to the other'. The tact in Patmore's lightness of touch shows when he abstains from dwelling on the rhyme 'this'/'kiss'; grammar insists that 'this' be passed over by the voice so that it shall form part of 'this | Sweet stranger'. That tact makes up the probity of his verse, its ability to be sensible as well as sacramental.

Though scarcely of the quality of Hardy's practice, Patmore's versification in its amused mobility, its happy chances of speech in the stanza form, carries his attitude to the individual lives within society's forms as fully as Hardy does. Some readers now find his attitude, at best, dated or unsympathetic, but this is not a reason for failing to

---

[140] *The Angel in the House*, II. xii, Patmore, *Poems*, 203.
[141] Champneys, i. 138.

credit the significance of his skill. *The Angel in the House* is not the most significant of Victorian poems on love and marriage, but it is often admirable in delineating the significance of the unserious, especially in finding verse for love's 'momentary transfiguration of life',[142] verse which weighs the momentary against the transfigurative, as when Felix and Honoria, courting, discuss the events of the social calendar:

> Across the Hall
> She took me; and we laugh'd and talk'd
> About the Flower-show and the Ball.
> Their pinks had won a spade for prize;
> But this was gallantly withdrawn
> For 'Jones on Wiltshire Butterflies:'
> Allusive! So we paced the lawn,
> Close-cut, and, with geranium-plots,
> A rival glow of green and red;
> Then counted sixty apricots
> On one small tree; the gold-fish fed;
> And watch'd where, black with scarlet rings,
> Proud Psyche stood and flash'd like flame,
> Showing and shutting splendid wings;
> And in the prize we found its name.[143]

The move from the book on butterflies through the joke (not a very good one) about the girls' themselves being 'Wiltshire Butterflies' to the coming-across a butterfly which is at once and conventionally mythologized into the Psyche whom Eros loved and then metaphorically turned into a society *belle*, ostentatiously coquettish with 'splendid wings' (the butterfly acquired 'splendid fans' in later editions), and back to the work on butterflies which gives a name to the specimen—this is marvellously agile, humorous and good-humoured at once. It has generosity in trifles. Patmore likes to coddle the real, as one might a child, but his indulgences are adult. The verse is up to matching what his social eye remarks—the hint of cloddishness which the regular iambic tetrameter of 'Their pinks had won a spade for prize' gives off in a context where Patmore has repeatedly deployed pyrrhics to keep the texture light, the extreme closeness of a speech-rhythm to the

---

[142] 'Love and Poetry', *Principle* . . . , 334.
[143] *The Angel in the House*, 'The Morning Call', IV. i. I have preferred here the text of the first and second editions; after 1858, Patmore made several changes which seem to me to upset the humour and balance of the passage.

surface of the verse at 'Then counted sixty apricots | On one small tree' which creates, as Hardy often more bitterly creates, a *style indirect libre* echo of a voice in the lines—all this very well catches in the style how captivating these unseriously significant details may be in the eyes of love.

Indeed, the verse deserves the tribute it pays to Honoria's mother:

> Within her face
> Humility and dignity
>   Were met in a most sweet embrace.
> She seem'd expressly sent below
>   To teach our erring minds to see
> The rhythmic change of time's swift flow
>   As part of still eternity.[144]

It is largely only at the level of 'rhythmic change' that Patmore can secure a balance between the 'swift flow' of speech, the social world, and 'still eternity'. In its diction, his verse lacks Hardy's courage to date and query the poetical; in its larger structure, it misses the deep invention of Browning's silent dramas (though *The Victories of Love* goes some way towards the massive representation of the world of chatter in which *The Ring and the Book* places married intimacy). Yet the verse does permit us to 'see' a simultaneity of rhythmic change and stillness in its evocation of accents in its forms, and in the complaisance with which it notes the sacrifice of speech to the desires of writing.

Patmore quoted Hooker to the effect that 'The love which is the best ground of marriage is that which is least able to render a reason for itself'.[145] He was prone to read Hooker as if the cleric were encouraging people to be as imprudent as possible, and assuring them that folly was the best guarantee of married happiness. Certainly, an exorbitant reading, though not perhaps a beautiful exorbitance, because it turns the supra-rational trust of a sacramental relation into the superstitious idolizing of a social tie. The closest Patmore came to imagining the Hardyesque 'other side' of such emotions (it is a rare moment) was when he wrote of a woman's doubts about marrying, her attempt to reason herself into calm acceptance:

---

144 Ibid. I. i, Patmore, *Poems*, 67.
145 'Love and Poetry', *Principle* . . . , 334.

The maiden so, from love's free sky
   In chaste and prudent counsels caged,
But longing to be loosen'd by
   Her suitor's faith declared and gaged,
When blest with that release desired,
   First doubts if she is truly free,
Then pauses, restlessly retired,
   Alarm'd at too much liberty;
But soon, remembering all her debt
   To plighted passion, gets by rote
Her duty; says, 'I love him!' yet
   The thought half chokes her in her throat;
And, like that fatal 'I am thine,'
   Comes with alternate gush and check
And joltings of the heart, as wine
   Pour'd from a flask of narrow neck.
Is he indeed her choice? She fears
   Her Yes was rashly said, and shame,
Remorse, and ineffectual tears
   Revolt from his conceded claim.
Oh, treason! So, with desperate nerve,
   She cries, 'I am in love, am his;'[146]

This 'fatal "I am thine"' is the beginning of the bride's pure blush 'when she says | "I will" unto she knows not what'.[147] (We sometimes hear that the Victorians did not believe women had sexual desires but the opening lines of this extract clearly ascribe sexual desire to a girl who has remained a virgin.) The lines convey, in their elegant fluster, how such words come with 'alternate gush and check | And joltings' by their imaginatively sympathetic countering of metrical and grammatical pauses at such managed spasms as '"I love him!" yet | The thought half chokes her in her throat' where the throttle of 'her in her' rarely for Patmore gives up full-throated ease so as dramatically to convey the trouble her utterance makes in her. The verse keeps its formal calm through all this, because that formal calm is the register of Patmore's conviction that reason cannot answer these questions of hers (neither should she be reasoned out of them). Hence, the need both for a bending of the verse towards her quandary and for a transformation of that bending into aesthetic flexibility. The skill of the passage comes into focus on the word 'nerve'. As the *OED* shows,

---

[146] *The Angel in the House*, II. i, Patmore, *Poems*, 146-7.
[147] See above, p. 176.

'nerve' was just ripening in the language at Patmore's time into a poise between 'anxiety', a 'fit of anxiety' (in the plural, 1815), 'courage' (1826) and 'audacity'—and in the subtle vocal ambiguity of the last line quoted, an ambiguity produced by the abrasion of speech-rhythms against metrical forms which is his speciality as a writer. We may read a regular iambic tetrameter:

$$\times \quad / \quad \times \quad / \quad \times \quad / \quad \quad \times \quad /$$
She cries, 'I am in love, am his; . . .'

or

$$\times \quad / \quad \quad / \quad \times \quad \times \quad / \quad \quad \times \quad /$$
She cries, 'I am in love, am his; . . .'

The more regular line is the one which contains the greater doubt—'I *am* in love'—protesting either too much or just enough.

The way resonance takes over from reasoning about love and marriage in Patmore can properly be suspected. I mentioned before[148] that the difference between Patmore and Hardy on the subject of marriage could be expressed by saying that Patmore thought the estate sacramental and Hardy thought it contractual. That difference extends to a wider difference in regard to marriage between Victorian attitudes and our own. It would be fair to say that the predominant view of marriage now is that it is a contract, and that, like any other contract, it is characterized by its involving responsibilities of the contracting parties to each other, and by the fact that it may be dissolved if these responsibilities are not met. Like other contracts, it will always be a live question whether it is rational to enter into it. Much of our difficulty in hearing Victorian poetry of love and marriage aright comes from the unwitting imposition of this view of marriage onto writing which rarely recognizes it, and which, even in the case of Hardy, never whole-heartedly adopts it.

Defending himself against criticism of *The Woodlanders*, he wrote:

In the present novel, as in one or two others of this series which involve the question of matrimonial divergence, the immortal puzzle—given the man and woman, how to find a basis for their sexual relation—is left where it stood; and it is tacitly assumed for the purposes of the story that no doubt of the depravity of the erratic heart who feels some second person to be better suited to his or her tastes than the one with whom he has contracted to live, enters the head of reader or writer for a moment. From the point of view of marriage as a distinct covenant or undertaking, decided on by two people fully cognizant of all its possible issues, and competent to carry them through, this assumption is, of

[148] See above, p. 198.

course, logical. Yet no thinking person supposes that, on the broader ground of how to afford the greatest happiness to the units of human society during their brief transit through this sorry world, there is no more to be said on this convenant . . .[149]

One of his many, not entirely self-consistent, points in this passage is that the contractual view of marriage does not, at its extreme, permit of a forgiving attitude to misdemeanours or changes of heart. The partners are supposed to have known entirely what they were doing when they contracted, and, equally entirely, what they do if they break the contract—they are then less forgivable. He recognizes that this assumption, though 'logical', is not one that much promotes human happiness (his language at this point is a mix of the utilitarian—'units of human society'—and the pious—'this sorry world'), but he is a little gentle about its shortcomings, it seems to me.

If we try to work out clearly the suppositions implicit in a contractual view of marriage, they turn out to be either fantastically severe or muddled. In one light, the partners to a marital contract must be regarded as entering into it under no undue influence; the terms of the contract must be precisely stipulated in advance, and the reasonable penalties to be incurred on its breach settled. This does not sound like the general practice of marriage, even in a society where the contract has become an ideal. So, we should abandon the legend that the Victorians were sacramentalist idealizers of marriage whereas we nowadays get on pragmatically with the job we have contracted for. In an other light, we might acknowledge the unrealism of a strictly contractual model, and regard the notion of a contract as no more than a very approximate metaphor which serves to indicate some important aspects of one conception of marriage—its reciprocity and its dissolubility. But metaphors exact responsibility, and a contract, if a metaphor, is a metaphor of obligation. Vaguely metaphorical obligations can in their way be as crushing and tormenting as the most precisian demands.

Hardy's ironies in this Preface to *The Woodlanders* waver with the instability of his (and our) power conceptually to articulate the institutions through which we live. The plight of somebody who falls in love with a person other than his marriage partner is severely described as that of an 'erratic heart who feels some second person to be better suited to his or her tastes than the one with whom he has contracted to

---

[149] Preface (1912) to the Wessex Edition of *The Woodlanders*, Orel, 19–20.

live'. 'Erratic heart' makes an insinuation as 'cowboy plumber' would make an allegation, and 'tastes' underplays what might be at issue in such a case, underplays in the tones of someone icily saying 'You should have known better'. Hardy is parodying a full-blown contractualism here, and shows how punitive its would-be reasonableness might be.

On the other hand, when Hardy was nettled by criticism of *Jude the Obscure*, he aimed his retort at the sacramentalist view: 'As for the matrimonial scenes, in spite of their "touching the spot", and the screaming of a poor lady in *Blackwood* that there was an unholy anti-marriage league afoot, the famous contract—sacrament I mean—is doing fairly well still, and people marry and give in what may or may not be true marriage as light-heartedly as ever.'[150] Natural enough for him to want to pretend that he couldn't care less—'what may or may not be true marriage'—after a reception such as *Jude's*, but his nonchalance fails to convince. Hardy was always a studied writer, and especially in a preface to a volume of a collected edition, the slip of the tongue he pretends to make at 'famous contract—sacrament I mean' looks and sounds childish, like schoolyard games of kicking the shins. He has been so unfairly treated, he feels, that he need no longer be fair in his own prose, and so he treats himself to jokes such as the casual play of 'unholy' against 'sacrament', to skimming the issue at '*fairly* well' (my emphasis) and avoiding it at 'light-heartedly'. The two comments together show in detail with what difficulties Hardy brings himself to speak in favour of either a contractual or a sacramental understanding of marriage.

⟨There are, after all, problems with both views.⟩Contractual marriage faces the same problems which attach to a Lockean account of civil society as founded on an original contract. The risks involved are hard to calculate sensibly in advance, the commitments undertaken are necessarily somewhat inexplicit. The desire to enter into the contract doesn't lend itself to the sort of scrutiny which the terms of a contract demand. The practical carrying-out of the contract demands such instant responses that there may not be time to consult and interpret the document of alliance—'Wretched would be the pair above all names of wretchedness,' as Dr Johnson wrote, 'who should be doomed to adjust by reason every morning all the minute detail of a domestick day.'[151] In the light of these considerations, we can see more than

---

[150] Preface (1912) to the Wessex Edition of *Jude the Obscure*, Orel, 35.
[151] *The History of Rasselas, Prince of Abissinia*, 2 vols., (1759), ed. G. Tillotson and B. Jenkins (1971), 77.

delusive lyricism in the Victorian view of marriage as a sacrament.

Not all Victorians were charmed by such a view, and even those who were charmed by it did not always wish to see it steadily through the whole of its implications. The Anglican Church does not in its Articles recognize marriage as a sacrament but many of the Anglican clergy were vociferous in opposition to the reform of the divorce law, though divorce had long been permitted under the English constitution of which the Anglican Church is also a part. This is just one example, and it exemplifies more than the richly curious positions of the Anglican Church; it shows something more fundamental because it shows the strains imposed on a human relation when that relation is held to sustain a religious mystery. A sacrament, in so far as it is a religious mystery, surrounds itself with problems for thought, but it can scarcely claim credit for notifying us in advance that it will be only with difficulty intelligible. A sacramental view of a human relation cannot be assessed directly in terms of whether its consequences are humane or not, because a sacrament is, by definition, a means and standard of what is the good for human persons. This, however, makes such a view quite as difficult to defend as to assail. The view remains for those who hold it a matter of faith. Holding a particular faith does not itself guarantee that the faith will inform the conduct of the faithful for good rather than ill (judging good and ill from any standpoint but that of the faithful).

But it is not necessary to embrace, say, [Patmore's Catholic orthodoxy] to see a point in Victorian ideals of marriage. It would be enough to recognize that the contrast between the nineteenth and twentieth centuries on this matter is not a contrast between the dewy- and the clear-eyed, but between different optics. It would go some way to arrange a true marriage of the minds of the two periods if we could enter imaginatively into the serious Victorian paradoxes of F. H. Bradley's definition:

Marriage is a contract, a contract to pass out of the sphere of contract; and this is possible only because the contracting parties are already beyond and above the sphere of mere contract.[152]

Or, if F. H. Bradley gives too abstract a witness to these practical truths, we might take Browning's reply to EBB's attempt to leave him freedom of choice as concerned her:

---

[152] *Ethical Studies*, 174 n.

My own Ba, if I have not already decided alas for me and the solemn words that are to help! (Tho' in another point of view there would be some luxurious feeling, beyond the ordinary, in knowing one was kept safe to one's heart's good by yet another wall than the hitherto recognized ones,—is there any parallel in the notion I once *heard* a man deliver himself of in the street—a labourer talking with his friends about '*wishes*'—and this one wished, if he might get his wish, 'to have a nine gallon cask of strong ale set running that minute and his own mouth to be *tied* under it'—the exquisiteness of the delight was to be in the security upon security,—the being 'tied'.) Now, Ba says I shall not be 'chained' if she can help![153]

It takes some confidence in an other person to compare, even obliquely and as a joke, your attachment to her to being '*tied*' under a 'nine gallon cask of strong ale'. The confidence was in this case justified. The humour of his reply is robust, though not bumptious—witness the sobered self-awareness in recognizing that it might be to one's good to be 'kept safe to one's heart's good' by a formal safeguard, such as a marriage vow. The possibility of this humour in these circumstances speaks volumes for both of them and for the culture they spoke for as they found in writing that a version of 'the exquisiteness of the delight was to be in the security upon security—the being "tied" ', as the refinements of their imaginations eventually met, not without many misapprehensions, the world they worked in, surprised as each of them was by that world—EBB's 'I *saw* a woman once, burst into tears' corresponds to Browning's 'the notion I once *heard* a man deliver himself of'—before they married.

[People constantly fail to understand 'What he sees in her' or 'how she puts up with him'; and the attempt to understand people's amorous feelings and their reasons for those feelings (if reasons is what they have) is often thwarted from the early years of wondering about one's parents (one of the hardest things to imagine is the moment of one's own conception). Browning wrote to Carlyle, explaining, as best he could, why he had not asked for advice he would have respected:

When I was about to leave England I should have been glad to talk over my intentions with you, respecting my marriage, and all the strange and involved circumstances that led to it. I did not do so, however, not from any fear of your waiving the responsibility of giving counsel, but because, in this affair which so intimately concerned me, I had been forced to ascertain and see a hundred determining points, as nobody else could see them, in the nature of things.[154]

[153] Letter of 3 Mar. 1846, *RB/EBB* i. 510.
[154] Letter of 14 May 1845, in *Letters of Robert Browning: Collected by T. J. Wise*, ed. T. L. Hood (Yale, 1933, repr., Dallas, 1973), 17–18.

Something other than self-will, something really 'in the nature of things', gets in the way of advice at points like these, and what hinders advice by the same token gives reason pause, for it is in the giving of reasons in favour or against an action that advice distinguishes itself from persuasion or sheer influence. Browning neither gives himself airs nor takes untrammelled liberty of judgement because he has been '*forced*' (my emphasis) to determine on a 'hundred determining points' which it is not in the nature of any helpfully reasoning outsider to know how to encompass or orient into the significance these points hold for those whom they intimately concern. To take outside advice is to step outside the love on which one seeks advice. This does not mean that a lover, such as Browning, is condemned by the fact of loving to a mere gamble, to behaving in a way that even he could not decide was prudent or imprudent, but it does mean that the sphere of love cannot be co-terminous with the sphere of contract in which it must be true that judgements about the contract—whether it is wise to undertake it, whether it has been broken—are made at least as well by those who have not contracted. Contracts can be made only when law already exists, and to say of two people that they are a law unto themselves is to say that they are beyond or without the law.

Just the difference between love and admiration demonstrates the uncontractual character of love, its irrationality but not its unreason-ableness. In the case of rational admiration, it will always be in principle possible to say on what it rests and in what circumstances a person might forfeit the admiration one gave him. This is not so with love. Equally, it would be irrational to admire a person for possessing a particular set of qualities and then refuse to admire an other person possessed of the same qualities (other things also being equal). This is not so with love for it is the particular life of qualities dwelling in a specific individual which is love's object. Such a life cannot be reduplicated, so comparisons cannot be made. Not all marriages are based on such love, and its character may in fact not be such as to make good grounds for a marriage, but some marriages are, and the ideal Victorian marriage was. Outsiders to this love may look on it kindly or askance, but our familiarity with what other people's views on our own intimacies are like does not encourage the hope that such views are always the true form of rational or reasonable or decent purchase.

Michael Millgate records the story of the American novelist, Gertrude Atherton, who saw Hardy at a social gathering in the 1890s in the company of 'an excessively plain, dowdy, high-stomached woman with her hair drawn back in a tight little knot, and a severe cast of countenance.' 'Mrs Hardy', her companion, T. P. O'Connor, informed her, adding 'Now you may understand the pessimistic nature of the poor devil's work.'[155] A joke, and like better jokes, about a failure, and, like the best failures, a failure in love. The joke's confidence that it knows what it was for Hardy to be married to Emma is a comic fantasy; it is only charitable to suppose that Gertrude Atherton and T. P. O'Connor knew that they were joining in a fantasy—if they did not know that, they were vulgar and cruel. The conscious fantasy could have been painful, had it been overheard, but it remains funny.

It is less funny to read Hardy's second wife noting in the ancillary typescripts of the *Life*:

November 27th. 1927.
Anniversary of the death of Emma Lavinia Hardy, T. H.'s first wife. Thursday was the anniversary of the death of Mary, his elder sister. For two or three days—& this morning—he has worn a very shabby little black felt hat that he must have had for twenty years—as a token of mourning. It is very pathetic— all the more so when one remembers what their married life was like.[156]

Florence Hardy had better grounds than anybody else alive when she was writing to think she knew what it was like to be married to Hardy, and perhaps therefore also what it was like for Hardy to be married, but her grim, lip-pursing 'when one remembers' takes too much on itself (forgivably, perhaps, if she felt that he gave too little to her because of his absorption in the regret of Emma). It is right to feel, looking at these examples, that the detached observer should recognize the need to be reserved in speech about other people's loves, that the claims of reason are here best met by an acknowledgement that the sphere is not one of neutral adjudication, and that a serviceable language for our imagining in this matter, one which knows how to abate the presumptions of judgement, is the mediate word of poetic speech.

Browning builds some of his long poems out of the comforted sociability of third persons' opinions about the lives of others. In *Pippa Passes*, the gap between Pippa's expectations and the actualities she passes works benignly, but in *The Ring and the Book* and *The Inn Album*

---

[155] Millgate, *Thomas Hardy: A Biography*, 313.
[156] Reprinted as an appendix to *The Personal Notebooks of Thomas Hardy*, ed. R. H. Taylor (1978), 293–4.

the profusion of views about the protagonists adds up to a thick
atmosphere of mistaking, of idle cynicism and idler dreams, which
threatens to swallow their lives, a massy concretion of 'the gaping
impotence of sympathy'[157] in such outsiders' comments. In *The Ring
and the Book*, for example, the speaker of each one of the poem's books
usually has a notion of what is the key-moment in his narration of the
events. In Book III, it is the mutual recognition in need of Pompilia
and Caponsacchi that provokes the 'critical flash | From the zenith'; in
Book IV, when Guido wakes to find Pompilia has deserted him, there
occurs the 'critical minute' at which he 'has the first flash of the fact';
for Caponsacchi himself, his high point ('the ecstatic minute') comes
as he waits for Pompilia to arrive later on the night of their escape,
while for his spiritual father, Innocent XII, what sums up the affair is
an instant of illumination on a night no other participant or
commentator has known at all:

> I stood at Naples once, a night so dark
> I could have scarce conjectured there was earth
> Anywhere, sky or sea or world at all:
> But the night's black was burst through by a blaze—
> Thunder struck blow on blow, earth groaned and bore,
> Through her whole length of mountain visible:
> There lay the city thick and plain with spires,
> And, like a ghost disshrouded, white the sea.
> So may the truth be flashed out by one blow,
> And Guido see, one instant, and be saved.[158]

The Pope thinks a life may be saved by a moment's confrontation with
the truth, and perhaps Browning thinks so here too—certainly the
verse is worked up to give sensuous credence to the power of a quick
sight—but the filaments which tie this 'critical flash | From the zenith'
back through the poem to all the other diversely portentous moments
mentioned make us think otherwise. Each speaker, lacking a knowledge
of what any other speaker in the poem has said, lacks also the mediate
word to negotiate between his and others' patents on the truth. This is
also true of the Pope, a Protestant irony from Browning in 1869, just
before the declaration of Papal infallibility.

He works similarly with the speakers' attempts to answer the
question asked earlier in 'Two in the Campagna' about the failures in

---

[157] *The Ring and the Book*, IX. 1001, Altick, 458.
[158] Respectively, III. 1045–6; IV. 1183–4; VI. 1138; X. 2118–27, Altick, 139, 191,
296, 534.

the heart of love: 'What the core | O'the wound, since wound must be?'
In Book II, it is Pompilia's father's financial need of an heir which
centrally provokes events:

> Moreover,—and here's the worm i'the core, the germ
> O'the rottenness and ruin which arrived,—
> He owned some usufruct, had moneys' use
> Lifelong, but to determine with his life
> In heirs' default: so, Pietro craved an heir . . .

Whereas for Guido himself, it seems to be the existence of his wife
which is the root of his disease in life:

> I did
> God's bidding and man's duty, so, breathe free;
> Look you to the rest! I heard Himself prescribe,
> That great Physician, and dared lance the core
> Of the bad ulcer; and the rage abates,
> I am myself and whole now: I prove cured
> By the eyes that see, the ears that hear again . . .

—lines in which Browning once again uses his superior skill in speech
to score over the wicked aristocrat's words and so score off him, as
Guido's self-righteous vehemences produce kinked internal rhymes in
the blank verse ('did | God's bidding') and a ghastly half-rhyme of
'core' against 'cured'. By Book IX, Pompilia's child, which Book II
thought of as Pietro's heir, has become principally Guido's heir, and
her presentation of the child to Guido is a pure gift without rottenness,
such as the more confirms Guido's inner decay:

> Bestows upon her parsimonious lord
> An infant for the apple of his eye,
> Core of his heart, and crown completing life,
> The *summum bonum* of the earthly lot![159]

Browning may have been led to play with these divergently confident
identifications of the 'Core of [the] heart' because 'core' in Italian
poetry is another name for the heart (and Italian uses the word for
'heart', 'cuore', to mean the core of a fruit, as Browning hints in the
audacious placing of 'apple' and 'core' in neighbouring lines.)

The ingenuity writ large in the long narrative poems as people
comment on 'matrimony the profound mistake'[160] made by others is

---

[159] Respectively, II. 209–13; V. 1702–8; IX. 1315–18, Altick, 70, 252, 466.
[160] *The Ring and the Book*, III. 1036, Altick, 139.

tirelessly resourceful but this structuring brilliance is a lesser thing
than what Browning achieves in the dramatized lyricism of the great
monologues of marriage. There he does more than surround the
protagonist with third-personal voices which beset his speech; he
makes that speech itself sound third-personal to itself, a character in
slippage. This is a better thing because more truly imagined. For our
selves live on the air they breathe, and one current in that air is others'
opinion of our love, and of those we love—a fact Iago played on when
he conjured the air around Othello so skilfully that Othello no longer
knew himself and suffocated Desdemona. This turning of a self into its
own third person is done by dramatizing the silence of the interlocutor
through which we hear the speaker and through which he hears, if he
can't help it, himself.

Nowhere more richly than in Browning's 'Andrea del Sarto':

> But do not let us quarrel any more,
> No, my Lucrezia; bear with me for once:
> Sit down and all shall happen as you wish.
> You turn your face, but does it bring your heart?
> I'll work then for your friend's friend, never fear,
> Treat his own subject after his own way,
> Fix his own time, accept too his own price,
> And shut the money into this small hand
> When next it takes mine. Will it? tenderly?
> Oh, I'll content him,—but tomorrow, Love!
> I often am much wearier than you think,
> This evening more than usual, and it seems
> As if—forgive now—should you let me sit
> Here by the window with your hand in mine
> And look a half-hour forth on Fiesole,
> Both of one mind, as married people use,
> Quietly, quietly the evening through,
> I might get up tomorrow to my work
> Cheerful and fresh as ever. Let us try.[161]

The poem asks us to imagine what her silence means, as she sits there
with his world falling about her perfect ears, to imagine her silence
more dispassionately and less fitfully than he does, and yet to live with
the incompleteness of our imaginings on this point, to content
ourselves with virtual satisfactions of the desire to know what two other
people are like when they are alone together. How many times has she

[161] (1855), P i. 643–4.

heard this speech before? Perhaps she thinks, along with EBB's tearful wife, 'You ALWAYS say this', and perhaps he does. 'Tithonus' sets the recurrences of art, and our pleasure in them, against the dramatic plight of the speaker in the toils of a lived version of such permanence. 'Andrea del Sarto' also revolves for consideration the terrible appetite for stillness in art: what may be the same old speech for Lucrezia and Andrea must become repeatedly the same new speech for us. The possibility of such a double existence in time which the poem so deeply mines is opened only by its status as at once speech and text—the supplicatory instant of his speech, as perhaps he represents it to himself ('for once' she must bear with him, 'for once' he will ask that), the wheedling habit of saying this, as perhaps she hears him, weakly sunk evening after evening in self-pity and that adhesive despotism of the feeble-hearted which does not 'ask much' but asks it so often that to give way once would be to surrender everything.

These diverse auditions come out of the beautiful impartiality Browning creates on the page, an impartiality that is the perfection of conduct in listening to the woes of the married. The first line is abrupt as an opening for a poem but does not determine what sort of abruptness the speech of the poem has in its fictional situation: he may be interrupting himself in mid-tirade or interrupting her in a spate of obstinate demands. 'No, my Lucrezia' looks as if it might cut her off as she is about to object, but such a cut-off can sound in many ways—peremptorily, as a request for patience, with concessive force. To observe this of the opening lines is to say no more than that they are magnificently dramatic; they open themselves to a variety of interpretations in performance, and so show a many-angled realization of what they record. But the strength of these lines is not quite that Shakespearian strength which lends itself to many voicings on the stage, though it certainly derives from the practice of Shakespeare as a dramatist more directly and with greater creative novelty than Browning managed as a playwright. There is a difference. The text of a play may sustain various vocalizations in performance, as the text of this poem does, but because a script is designed for performance, it has in view that some set of the possible vocal alternatives must eventually be chosen, and that such a set will be coherently suggestive. The dramatic monologue is not designed for performance in this way (it has nothing in common with the 'recitation' pieces which were popular in the period) and there is therefore no need to call a halt at any particular time to the vocal ambiguities of the text, nor are they so arranged that some set of them may be chosen and presented. On the contrary, the essence of the form is that the poem most fully exists as the imagination of a set of mutually incompatible voicings between which we are not only not asked to choose but asked not to choose, and this not because of an aestheticized penchant for semantic abundance at any price but because the words have been made so completely to exist *between* the fictional speaker and interlocutor that opting for one interpretation against another is a form of taking sides.

All about Andrea lies a passionate uncertainty, to which he is also accustomed, and it seeps into his words. How does 'my Lucrezia' sound against the reiterated 'your face . . . your heart . . . your friend's friend' and 'his own subject . . . his own way . . . his own time . . . his own price'? Nothing seems quite to belong to Andrea in this context except, the next occurrence of the first personal possessive, 'my work', work in which (to speak bluntly as it is the poem's achievement not to do) he prostitutes himself to pay the price of her adulteries. Letting 'frank' speech like that in on these lines is like endorsing T. P. O'Connor's 'Mrs Hardy . . . Now you may understand the pessimistic nature of the poor devil's work'. The poem does not speak so plainly about him, or for him. An impersonation of Andrea del Sarto, using the poem's words as its text, would either expose him or espouse his cause, but we are not asked to take on the role, indeed we are told again and again by the impossibility of voicing the text's ambiguities that we cannot act his part. 'I'll work then for your friend's friend, never fear, | Treat his own subject . . .' may be paraphrased out as 'Do not worry, I'll work for your friend's friend as he wishes' or as 'I will work for your friend's friend, I will not be afraid (of doing so, or of the reasons why you ask me to do so), I will paint as he and you ask . . .'. There is nothing in the text to settle the semantic ambiguity of 'never fear', only the imagination of the vocal habits of the speaker which we bring to the text and find in it, vocal habits which are a lifetime in sound in between Andrea and Lucrezia. It is our judgement on what the relation between them would permit in the way of talk that decides whether we think it plausible that Andrea should use to her the colloquial 'never fear' as an imperative or the elliptically formal 'never fear' as part of a future indicative. But the poem is the evidence for that judgement as well as the substance of a verdict. What is so exceptionally tender and inquiring in this masterpiece is that this position of the reader, as counsel and judge at once, is Andrea's too. In every word and gap of words between them, he hears himself plead

his own case and judges his own eloquence as he is judged by it.
 Beckett:

I'm the clerk, I'm the scribe, at the hearings of what cause I know not. Why
want it to be mine, I don't want it. There it goes again, that's the first question
this evening. To be judge and party, witness and advocate, and he, attentive,
indifferent, who sits and notes. It's an image, in my helpless head, where all
sleeps, all is dead, not yet born, I don't know, or before my eyes, they see the
scene, the lids flicker and it's in. An instant and then they close again, to look
inside the head, to try and see inside, to look for me there, to look for someone
there, in the silence of quite a different justice, in the toils of that obscure
assize where to be is to be guilty.[162]

Beckett's prose has the ability, as does Browning's verse, to achieve
flurry and also to flurry achievement, an ability evident in the syntax of
this passage with its incessant turns of the self-controverting mind,
caught in its own qualifications, the visible drama of a voice shocked at
hearing itself. The dramatic monologue is such an 'obscure assize
where to be is to be guilty', generally, but also with sharp particularity
in the monologues of marriage, because it is when you are found
wanting by someone you are vowed to love that being is also being
guilty, it being not what you have done so much as who and what you
are which is judged. 'Andrea del Sarto' is the supreme instance of this
form. It is not only the observed who stands accused but also the
reader who stands in for him, taking on the charge of that other
personality in the fictional act of envisaging his voice. In the
interrogated self-consciousness of the process of voicing a monologue,
the reader witnesses his own complicity in setting up the character. He
comes back to the certainty of his reading, his sense of weight, accent
and contour, with a reasonable doubt: so deep is his own implication in
the way he sees, and then hears, the features of the case, that he has to
feel for sufficient and steady ground on which to settle his convictions
about the voiced character.

  The very incompleteness and ambiguity of the testimony in a
monologue (its constitutional one-sidedness) provoke us at once to a
passion for circumstantiality and a distrust of circumstantial evidence.
Our prejudices are abetted by the incompleteness of what lies before
us, we become avidly sure that, at *this* point the speaker really gives
himself away, or that *those* words are said with such-and-such a
betraying force. So we might note that Andrea del Sarto does not

[162] *Texts for Nothing*, v. trans. from *Textes pour rien* (Paris, 1954) by the author,
collected in *Collected Shorter Prose 1945–1980* (1986), 85.

speak for long without mentioning silver or gold or some money ✓
(twenty-three times in 267 lines, by my count),[163] and be inclined to
think him mercenary, but the patterns in the poem come from
Browning as much as from Andrea, and serve an aesthetic function as
well as a psychological one. Who is to say that he loves his art because
it brought him money or loves money because 'all is silver-grey' in his
art, or that he loves gold because it procures him Lucrezia's minimum
of fidelity or because—he is, we recall, a painter—it reminds him of
the colour of her hair: 'Let my hands frame your face in your hair's
gold'? The name 'Lucrezia' begins with the word 'lucre', but his love
for her may not begin and end with her name. We may, of course, say
the lines so as to give weight to one rather than the other of these
views. Doing that makes us outsiders to the voice this text so
considerately holds in abeyance, makes us ready to be assured in a way
that the writing neither rules out of court nor courts.

Much in the poem comes from Browning's own life. Lucrezia's
'small hand' which shuts on money reads like a parody of EBB's hand
which modelled for the 'spirit-small hand' of 'By the Fireside'.[164] And
the poet's letters to his beloved contain the beginnings of those great
lines

> Eh? The whole seems to fall into a shape
> As if I saw alike my work and self
> And all that I was born to be and do,
> A twilight-piece. Love, we are in God's hand.   ✗
> How strange now, looks the life he makes us lead;
> So free we seem, so fettered fast we are!
> I feel he laid the fetter: let it lie!

Browning had refused EBB's offer that he should not be ' "chained",[165]
and so had reason to know that contenting oneself with a fetter might
not be merely servile. He had also found it very important to sit with
the person one loves—Andrea: 'let me sit | Here by the window with
your hand in mine'; Browning: 'So I give my life, my soul into your
hand—the giving is a mere form too, it is yours, ever yours from the
first—but ever as I see you, sit with you, and come away to think over it
all, I find more that seems mine to give . . .'[166]—so that he need not
have thought it demeaningly mawkish of Andrea to ask for that favour
of all favours. 'God's hand' which stretches over these human hands

---

[163] Daniel Karlin first drew my attention to this.
[164] P i. 556.   [165] See above, p. 248.
[166] Letter of 15 Mar. 1846, RB/EBB i. 538.

given and withdrawn in marriage comes to the poem from the life, from his exultant 'and I do believe that we shall be happy; that is, that *you* will be happy: you see I dare confidently expect *the* end to it all . . so it has always been with me in my life of wonders,—absolute wonders, with God's hand over all',[167] where Browning thinks '*the* end to it all' is getting married. That is only the beginning of the poem. This hand stretches to the end of their marriage, in EBB's dying words, ' "My Robert—my heavens, my beloved"—kissing me (but I can't tell you) she said "Our lives are held by God." '[168] 'God's hand over all', 'Our lives are held by God', 'Love, we are in God's hand'—it is a sequence in which the hand gradually closes over the married pair, moving from a superintendent benediction through a judging disposition to the control which you have when you hold things in the palm of your hand (the sequence is not a chronological series). The letter of 1846 is sure that if she is happy, he will be happy too; the poem published in 1855 imagines an artist who has found that, greatly as he rests his happiness on his wife, her happiness does not entirely rest with him. We should draw from this no more certain conclusions about the Brownings' marriage than the poem allows us to draw about the del Sarto home life. Beautiful though it is that the poet's and the poem's life fall together into such a whole.

The poem contains a doubt of both the patterns it creates within itself and within its relation to Browning's own marriage. For it is a piece about the artist's consumption of the real as well as a study in the intertexture of love. They are terms of Andrea's art which confer on the life he lives a unity as a 'twilight-piece' or all 'silver-grey'. He is represented representing how he lives. Again, not because of an abstract pleasure in the recessively self-reflexive possibilities of art, but because the imagination is here engaged with and in and within the actual. When Andrea contents himself 'Because there's still Lucrezia, —as I choose', he both triumphs in the spirit and collapses into illusion. Nothing in the poem tells us that he does not know what he is doing. Browning, when younger, had made the word 'choose' at the line-end sound falsely confident, in 'My Last Duchess': 'and I choose | Never to stoop'.[169] Falsely confident, both because the narrative we guess at through the poem suggests that the Duke was ready to stoop to something very base indeed, and because, though his syntax

---

[167] Letter of 29 Oct. 1845, *RB/EBB* i. 253.
[168] Letter to Sarianna Browning, 30 June 1861, in Hood, op. cit. 62.
[169] P i. 350.

hearkens the word 'choose' forward to 'never', our acquaintance with
the rhyme-scheme doubles the word back to rhyme with, and grate
against, 'excuse'. Browning made the word rhyme beyond its speaker's
intentions, and so curbed the intended pride of choice. Andrea's
'choose' rhymes with nothing because the poem is in blank verse, and
also because Browning is less surely standing above his fictional
speaker here. He less now wishes, after a decade of marriage, to
demonstrate the personal will as if that were itself the Good.

Ottima (Italian: 'the best' (female)) asks in *Pippa Passes*

> The past, would you give up the past
> Such as it is, pleasure and crime together?[170]

Andrea cannot give it up, nor would he if he could, and he makes of his
incapacity a choice. Something like this must go in in any artist's
relation to the past, for the past is material to him, as the body of
someone loved always is, even when love is no more. Andrea's eventual
fabrication of his life as the product only of his own will can itself be
seen in diverse vocal lights: it may be conjuring of mastery out of the
medium of his helplessness, a confession that he can do no better and
can't break the habit of her, something like a personal spinning of a
Nietzschean myth of eternal recurrence which would give rise to a joy
indistinguishable from resignation. The absolute asceticism of the
page as regards his voice sets before us the drama in these lyrical
propensities of a life. His 'as I choose' anticipates and meets Hardy's 'I
am just the same'; they are both acts—deeds and performances—of
the imagination on the actuality of a life devoted to imaginings and so
brought up in the ideal, for better for worse.

In Browning's translation of Euripides, the chorus is made to say:

> Cry aloud, lament,
> Pheraian land, this best of women, bound—
> So is she withered by disease away—
> For realms below and their infernal king!
> Never will we affirm there's more of joy
> Than grief in marriage; making estimate
> Both from old sorrows anciently observed,
> And this misfortune of the king we see—[171]

Nor do they affirm that there is more of grief than joy. The verse

---

[170] (1841), P i. 309.
[171] *Balaustion's Adventure; including A Transcript from Euripides* (1871), ll. 654–61, P i.
886.

makes estimate but not from the standpoint of an enlightened neutrality which can compute the cost of ideals in terms of pains assessed on a scale indifferent to any contention about why pains are undergone, and whether some pains are worth putting up with. The 'old sorrows' are not seen from a novel vantage which supersedes the terms of the past, they are 'anciently observed'.

Browning has himself not always been translated so well into the modern world. A commentator writes of the poems in *Men and Women*: 'In one way or another, each of these poems on love and alienation within marriage implies serious criticism of the institution. Passion dies; men are frustrated or unfaithful; women are trapped or left alone . . .'[172] Any of these things might happen to someone who was not married, so it would be foolish to think their occurrence proved something about marriage rather than more generally about the things people are capable of doing to each other, whether they are married or not. That does appear to be the commentator's thought, though, and it is a representative thought of our day, which is very impressed by the fact of pain. Pain is a brutal fact, but it is not a brute fact which settles any issue, as if, citing evidence of miseries one proved at once the savagery of the ideals which exacted them. All cultures involve and produce pain; the degree to which they do so is one question about them. But it is not *the* question. There is also the question of what significance a culture can attach to the pain it inevitably produces, the question of whether it comprehends its own distress. These Victorian poems do that remarkably, and this is something which should interest not only literary historians. Browning had recourse to the Greeks again when anciently observing the sorrow of his own marriage, from which he did not shy or stray:

The general impression of the past is as if it had been pain. I would not live it over again, not one day of it. Yet all that seems my real *life*,—and before and after, nothing at all: I look back on all my life, when I look *there·* and life is painful. I always think of this when I read the Odyssey—Homer makes the surviving Greeks, whenever they refer to Troy, just say of it 'At Troy, where the Greeks suffered so.' Yet all their life was in that ten years at Troy.[173]

---

[172] W. S. Johnson, *Sex and Marriage in Victorian Poetry* (Ithaca, 1975), 208.

[173] Letter to Isabella Blagden, 22 May 1867, in *Dearest Isa: Robert Browning's Letters to Isabella Blagden*, ed. E. C. McAleer (Austin, 1951), 267. The editor of the volume notes of Browning's quotation from *Odyssey* that it is 'not found'.

4

HOPKINS: THE PERFECTION
OF HABIT

Early Hopkins

FROM an early age, Hopkins had ambitions of humility in his poems:

> Elected Silence, sing to me
> And beat upon my whorlèd ear,
> Pipe me to pastures still and be
> The music that I care to hear.
>
> Shape nothing, lips; be lovely-dumb:
> It is the shut, the curfew sent
> From there where all surrenders come
> Which only makes you eloquent.
>
> Be shellèd, eyes, with double dark
> And find the uncreated light:
> This ruck and reel which you remark
> Coils, keeps, and teases simple sight.
>
> Palate, the hutch of tasty lust,
> Desire not to be rinsed with wine:
> The can must be so sweet, the crust
> So fresh that come in fasts divine![1]

These lines may be no more, or no less, than a further instance of that poetic doctrine of reserve which appears in Tennyson as in Browning. Whatever Hopkins might have wished in his Anglo-Catholic zeal, 'The Habit of Perfection' sounds not only like a monastic habit but, more generally, like the literary habit of making eloquent repudiations of

---

[1] 'The Habit of Perfection', 1866, revised at a later, uncertain date. All quotations from Hopkins's poems are from the edition of Catherine Phillips (Oxford, 1986), hereafter referred to as Phillips, followed by a page number. Thus, for this reference: Phillips, 80–1.

eloquence. Yet Hopkins took on this youthful pose with entire seriousness—he burned his poems on entering the Society of Jesus, and more than twenty-five years elapsed between the composition of 'The Habit of Perfection' and the partial breaking of the poem's 'elected silence' on its publication.

Nobody, probably, heard him read his poems aloud; at least, no record of the experience survives, to my knowledge. This is curious, because he specifically asked for a hearing of his poems: 'The rhythm of this sonnet . . . is altogether for recital, not for perusal . . .' or again '. . . you must not slovenly read . . . with the eyes but with your ears, as if the paper were declaiming it at you'.[2] Coventry Patmore, who scarcely warmed to the poems, thought this might have been because he had not heard them: 'It struck me, however, at once, on reading your poems, that the key to them might be supplied by your own reading of them; and I trust some day to have the benefit of that assistance.'[3] Hopkins tried to supply some help to fetch a voice from out the page with his notation of accents, bar-lines, pauses, counter-pointings, and so on, but Patmore thought this did not help. The marks rather hurt his pride: 'I fancy I should always read the passages . . . as you intend them to be read, without any such aid; and people who would not do so would not be *practically* helped by the notation'.[4] It was inconsistent of Patmore on the one hand to insist that he did always read the passages as Hopkins intended and also to wish that he could hear Hopkins read them so that he might know how they were supposed to sound, but he was right to feel that the notation does not '*practically*' help a reader in search of the poems' voices. The notation does not solve problems of performance and reception, questions of tune and drift which the poems raise. Indeed, it often raises new difficulties, not only because of its idiosyncrasy on occasion, but because it becomes itself part of the text we try to render vocal, an abiding sign of the difficulty with which this writing gives itself to the air of speech. The notation makes explicit the constant neediness of script, the fact, amongst others, that there are as many ways of saying 'AND' as ways of saying 'and'.

    [2] Letters of 11 Oct. 1887 and 21 May 1878 to Robert Bridges, in *The Letters of Gerard Manley Hopkins to Robert Bridges* (Oxford, 1935, rev. 1955), ed. C. C. Abbott [hereafter referred to as *GMH/Bridges*], 263, 51–2.
    [3] Letter of 20 Mar. 1884, in *Further Letters of Gerard Manley Hopkins* . . . (Oxford, 1938, 2nd, rev. and enlarged ed., 1956), ed. C. C. Abbott [hereafter referred to as *Further Letters*], 354.
    [4] *Further Letters*, 353.

It would be odd for a poet who wished his readers to be all ears to determine to keep his own mouth shut on principle. Hopkins wrote 'The Habit of Perfection' in his early twenties when he knew little of what his vocation, either priestly or poetic, would ask of him, and this poem's lush symmetries sometimes read only as reveries of a fervour as yet unexperienced. Richly synaesthetic, the piece does not always embody the abstemiousness it praises. At times, that very richness is embarrassing, each rift overloaded with ore. The eyes are to be 'shellèd', that is, encased in double darkness; Hopkins probably chose the word to hark back to and pick up from 'whorlèd' in the first stanza. The verb, 'to shell', though, also means to discase, to extrude, as peas from a pod. Meaning that, it crosses the calm imagination of pious blindness in the lines with memories of Gloucester's terrible, unanaesthetized blinding, the comfortable 'double dark' shadowed by 'All darke and comfortlesse?'[5] Similarly, the semantic density of 'the hutch of tasty lust' is awkward, and, in its awkwardness, sounds greedy. The phrase means that the mouth is like a little house in which the pleasures of taste dwell, but 'lust' was obsolcte in the language as meaning (innocently) 'delight' by the time Hopkins wrote the poem; its senses of 'sinful desire' and 'sexual desire' had come to predominance, just as 'hutch' more readily denoted a rabbit hutch than a small house in nineteenth-century English. 'Can' was still a few years short of usages such as 'a can of beans' when Hopkins wrote the poem, but his use of the word to mean 'a drinking-vessel' already smacks of archaising, especially in conjunction with the ellipsis which makes 'sweet' refer to the vessel rather than to its contents. 'Crust' does have a sense with regard to wine, referring to the lees deposited in a bottle or cask; this makes for a nicely condensed hyperbole—'the delights of asceticism are so great that even the dregs of ascetic wine are refreshing'—but against this, there presses the stronger tie of 'crust' to bread which has not previously been mentioned in the stanza. Hopkins writes here with startling agility, cramming the verse with suggestion, and, because he does so, the poem remains only enthusiastic and not devout.

The stanza I quote on the sacrifice of voice, though, is more than that; it expresses patience and supplication. It differs from the stanzas around it by sensing precisely the aim of ascetic discipline, its aim at something beyond itself, its possible fertility. When lips *'shape* nothing'

[5] *King Lear*, III. vii. 84; Folio, l. 2160.

(my emphasis), they pronounce a zero, and forming a circle they begin to say 'O', the prime, calling word of poetry. Lips in such shape of nothing are the opposite of tight-lipped, just as to be 'lovely-dumb' is not to be plain dumb. Hopkins makes the turn towards an outgoing asceticism on the word 'come' in the third line of the stanza; it means 'come at' ('arrive') and not 'come from' ('derive'). A paraphrase should run: 'what sustains your eloquence is the curbing of speech imposed on you from that source to which all acts of surrender lead'.

Such a source for Hopkins was actually a person—Christ—who provided the pattern of silence as of other surrenders. The 'elected silence' which the poem chooses is meant not as a self-protective hoarding of the voice but as a gift to Christ; the lines look for a manner of replying to the demand of Christ's example. They envisage an imitation, often proposed in the tradition of the Church, and urged again by Newman in his sermons preached before the University of Oxford, a 'setting the pattern of the Son of God ever before us, and studying so to act as if He were sensibly present, by look, voice, and gesture . . .'.[6] The element of Christ's pattern which the poem looks towards is one omitted in Newman's list of 'look, voice, and gesture'— Christ's silence before his accusers. Donald MacKinnon brings the example of that silence to bear sharply on a more recent predicament of speech:

A year or two ago there was an outburst of feverish clerical protest against the way a London vicar was treated on one of the B.B.C.'s satirical television programmes, against the way in which this vicar's words were drowned by a clamour of loud anti-Christian protest on the part of the other participants in the show. Inevitably every effort was made to ensure that such ill-mannered silencing of the Christian view should not occur again; I say 'inevitably'. For it is a mark of the pathetic intellectual (and spiritual) naivety of the clerical mind not to see that it may be that in the present situation such enforced and publicly demonstrated aphasia is one of the only methods of effective communication open to Christians. I do not mean that in such circumstances a dignified silence has its own impressive quality; I mean more profoundly that we must recognize our inability significantly in speech to transcend the frontiers of intelligible discourse, and that in this present, a certain sort of silence may be a means of communication . . . For a Christian the memory of Christ's reported silence before his accusers and judges is and must be a

---

[6] Sermon of 13 Apr. 1830, in *Fifteen Sermons preached before the University of Oxford* (1843) [hereafter referred to as *University Sermons*], 36. All quotations from Newman's works, unless otherwise stated, are from the Uniform Edition (UE), 36 vols., (1868–81).

paradigm of the authentic *marturia tei aletheiai* [martyrdom for the sake of truth]. 'Answerest thou nothing'?[7]

There is a comic and pitiful gap between what happened to 'a London vicar on one of the B.B.C.'s satirical television programmes' and Professor MacKinnon's detection in that event of 'our inability . . . in speech to transcend the frontiers of intelligible discourse'. Something merely contingent, a little knockabout fun on a late-night show, opens up the full Kantian dilemma of pure reason, and it seems hard to credit that such a metaphysical conflagration could be kindled from a passage of social friction. But then, the cultural life of religious convictions and counter-convictions often issues in such stories where conversation, just in so far as it is casual, or thought from one side to be casual, reaches to unforeseen extremity, whether at Christ's trial or in those irritable exchanges, at once credal argument and sociable bickering, which surrounded the Victorian convert to the Church of Rome. Hopkins also asked MacKinnon's question of which 'methods of effective communication' are 'open to Christians' and equally felt that there must be a Christian conduct of speech which might involve 'a certain sort of silence'. It is necessary to say which sort.

'Silence' is not exactly the right word for it. As Professor MacKinnon does not mention at this point, Christ was not absolutely silent before his 'accusers and judges'. Though each of the synoptic gospels insists that he 'answered . . . never a word' to the questions put him, each of them also records that he replied to the question 'Art thou the King of the Jews?' with 'Thou sayest' or 'Thou sayest it'.[8] (John's gospel gives him more to say,[9] but in no case is he entirely silent.) The relevant 'sort of silence' in and under question appears in the fact that Jesus's words to Pilate, 'Thou sayest it', return his own speech onto Pilate; Jesus creates the conditions under which Pilate might hear his own voice. That is, the reply itself questions the terms and act of the questioner, in an attempt to show that what has been asked as if it were unequivocal was actually contentious. It is not always possible to give a 'straight answer' to a question (e.g., 'Have you stopped beating your

[7] Introductory essay to *Borderlands of Theology and Other Essays* (1968), ed. G. W. Roberts and D. E. Smucker, 32.

[8] The accounts are in Matthew 27; Mark 15; Luke 23.

[9] See John 18, esp. v. 37: 'For this was I born, and for this came I into the world, that I should give testimony to the truth. Every one that is of the truth heareth my voice.' (Douay Bible.) The Johannine narrative would have had a particular importance for Hopkins and Newman after their conversion, for it is John's gospel which is read in Catholic churches on Good Friday.

wife?' or 'Art thou the King of the Jews?') for what is at stake is the question of what, in such circumstances, it is to answer or to deal 'straight'. Christ's reply exemplifies conduct which Newman recognized as essential when he was an Anglican, and which became the essence of his style as a Catholic, most celebratedly in the exchanges with Kingsley. Speaking of the fact that the world triumphs over weak Christians because 'it assails their *imagination*', Newman sought as a necessary corrective a means whereby 'we may retort upon the imaginations of men'.[10] Both Hopkins and Newman in their different ways created sorts of silence which should work as retorts upon the imagination of the world, and Newman's own practice of 'holiness embodied in personal form . . . the silent conduct of a conscientious man' in the face of 'versatile and garrulous Reason'[11] was for Hopkins the decisive instance near to hand of Christ's pattern.

Newman comes very near to Hopkins's poetic quandary and creative skill in a letter to his diocesan, the Bishop of Oxford, which he wrote in March 1841. Some wanted Newman to 'state his position', others that he should keep decently silent, but he was not sure that even silence was still decent:

. . . if I write I have a choice of difficulties. It is easy for those who do not enter into those difficulties to say, 'He ought to say this and not say that,' but things are wonderfully linked together, and I cannot, or rather I would not be dishonest. When persons too interrogate me, I am obliged in many cases to give an opinion, or I seem to be underhand. Keeping silence looks like artifice.[12]

Characteristically, Newman puts a shade of nettled humour into his prose; 'things are wonderfully linked together' takes 'wonderfully' both in its biblical sense where it applies to the mysterious intricacy of God's providence and also in a more social sense of the inextricability of one human concern from another, a social sense at which Newman is mildly bemused while he is awed by the biblical sense. The capacity of 'wonderful' to appear in a hymn by Faber ('My God, how wonderful Thou art!', 1840) and in Blackmore's *Mary Anerley* ('he trimmed his whiskers and put on a wonderful waistcoat', 1880—both instances, *OED*) provides an instance of how the language compacts spiritual

---

[10] Sermon of 27 May 1832, *University Sermons*, UE 132, 134.
[11] Sermon of 22 Jan. 1832, ibid. 92.
[12] Reprinted by Newman in *Apologia pro vita sua* (1864–5, variously revised until about 1886). I quote from the edition of M. J. Svaglic (Oxford, 1967) [hereafter referred to as Svaglic], 156.

elevation and chatty familiarity in single words, and so suggests how complexly larger linguistic units may twine those human concerns which can so readily be thought separable. The way spirituality is domesticated in words, and the domestic spiritualized, gives one ground for that startling conjunction of metaphysical strife and petty hostilities which Professor MacKinnon makes in the case of the silenced vicar. For writers such as Newman and Hopkins, the reciprocal adjustments of sociable locution and liturgical forms make up a central part of learning to live as English Catholics rather than as Anglo-Catholics. Known ways of speaking, linguistic habits, had to be faced, and turned in a new direction, had to be perfected, as a convert might have said, but, equally, the unfamiliar language of Catholicism had to be seen and heard as truly an English speech, not as something unutterably alien, if the convert's voice was to be persuasive to the as yet unconverted—converted eloquence had also to be the perfection of habit. This was the 'choice of difficulties' Newman had to make, and Hopkins too grew skilled in such choice when he, following Newman, learned through conversion the weight of 'the bearing of the Christian towards the world, and . . . the character of the reaction of that bearing upon him',[13] a 'reaction' to which he had to summon a 'retort'.

The voice in Hopkins's early poems has not undergone very much, and so can sound quaint; this may happen, as in 'The Habit of Perfection', because of a cultivated remoteness of idiom in uses like 'lust', but it can equally occur when Hopkins's writing is uninquiringly accommodated within the convenient fluencies of an inherited speech, poetic or otherwise. Though Hopkins frequently attends from the beginning of his career to the subject which was to release his genius, '*afflictio*, affliction, properly *wrecking*',[14] he does so with a style as yet unwounded, reckless of cost. Consider 'The Escorial' in which he tries to think about Philip II's decision to build as a memorial to Saint Laurence, who was grilled to death, a palace on the plan of a gridiron:

> For that staunch saint still prais'd his Master's name
> While his crack'd flesh lay hissing on the grate;
> Then fail'd the tongue; the poor collapsing frame,
> Hung like a wreck that flames not billows beat—

[13] Newman, letter to Kingsley, 7 Jan. 1864, repr. in Svaglic, 343.
[14] Hopkins, retreat notes of Nov.–Dec. 1881, repr. in *The Sermons and Devotional Writings of Gerard Manley Hopkins*, ed. C. Devlin (Oxford, 1959) [hereafter referred to as *GMH Sermons*], 187.

> So, grown fantastic in his piety,
> Philip, supposing that the gift most meet,
> The sculptur'd image of such faith would be,
> Uprais'd an emblem of that fiery constancy.[15]

He tries to negotiate between what was done to Saint Laurence and what Philip II made of that. He sets out clearly enough the actuality of 'hissing on the grate' in contrast to the emblematic figure of 'fiery constancy' but his attitude to such devout transforming of pain into an object of celebration remains unclear. The phrase 'grown fantastic in his piety' only muddles through the grotesquerie of Philip's religious conceit, leaving the suspicion that the young Hopkins was as impressed by Philip's bizarre preciosity as by his piety. 'Crack'd flesh' admirably notices the detail of Saint Laurence's martyrdom (the skin splitting like a roast chestnut) but the observant calm of the phrase is merely juxtaposed with the commiseration in 'poor collapsing frame'. The two attitudes, of detached watching and of urgent pity, which Hopkins will recognize in his mature work as co-equals in any attempt to understand suffering as both humanly undergone and part of a divine plan, do not spark thoughtfully off each other here as they do in 'Felix Randal', a poem which confronts the unsearchably calm and neutral 'O is he déad then?' with the emphatically tender 'child, Felix, poor Felix Randal'.[16] The ability of Hopkins's mature style to encompass English from a wide dictional range, and so to dramatize a gamut of attitudes, intimately connects with its rhythmic suppleness. If one looks at these lines from 'The Escorial', the occasions when syntax is altered from the norms of speech seem to have in view only the securing of a regular iambic pulse: 'Then fail'd the tongue' instead of 'Then the tongue fail'd' or 'Then the tongue failed'. The rhythm preserves the poet's fluency unimpaired as it speaks of Saint Laurence's loss of speech, the style declines to bend to its subject. Hopkins claimed for Philip's construction that it was to be 'The pride of faith, and home of sternest piety' to 'remotest ages', but 'The Escorial' lacks a historical sense of the developments in devotional practice, of how remote in fact his own sensibility was from such baroque triumphalism, just to the degree that it is replete with a sub-Keatsian, dictional practice of archaising—'zeal-rampant', 'engemming', 'high casements', 'blazoned groins', 'rich blazonry', 'damasqu'd'—

[15] 'The Escorial', 1860, Phillips, 1.
[16] 28 Apr. 1880, ibid. 150.

diction which fills the poem with echoes of *The Eve of Saint Agnes* and makes its textures far from stern or martyred.

In his astonishing unfinished poem on the martyr Margaret Clitheroe (canonized only after Hopkins's death), he treats martyrdom with a radically dramatic lyricism, quite beyond the stilted verse of 'The Escorial'. He arranges a conflict of tones in the lines to carry the actual variety of response to her suffering, from the hagiographical gossip of 'She held her hands to, like in prayer; | They had them out and laid them wide | (Just like Jesus crucified)', where the parenthesis clucks and whispers between rumour and pious legend, to the punning indifference of her tormentors:

> Within her womb the child was quick.
> Small matter of that then! Let him smother
> And wreck in ruins of his mother[17]

The act of mind which conceives the possibility of calling the murder of an unborn child a 'small matter' makes imaginative entrance into the lives of men who are capable of treating that child as just a little bit of matter. The language of the poem takes on and suffers under the savage coarseness it recounts. A coarseness which Hopkins meets with daring linguistic honesty when he chooses to place 'ruins' near to 'mother' and risks the consequent chime of 'mother's ruin'. (It isn't certain when the phrase 'mother's ruin' came into the language—Partridge's *Dictionary of Slang* says 'late 19thC.', which is perhaps too late for this piece. Anyway, the fierce rhythmic jocularity of the lines remains clear, along with their excruciating assonance of 'quick' against 'wreck', their callous literalism about what it is to be a 'ruined woman'.)[18]

These grim slithers of the word around her crushed body do more than mime the surrounding jeers, though they do that brilliantly. They also resign the poet's voice to hostile idioms, a submission of his silenced personality under the pressure of actuality which attempts to meet, though not to match, Margaret Clitheroe's own authentic martyrdom for truth: 'On trial at York in 1586 for sheltering Catholic priests . . . [Margaret Clitheroe] refused to plead and was therefore pressed to death with heavy weights, the penalty for remaining mute.'[19]

[17] ? 1876–7, ibid. 127.

[18] I am indebted here to Geoffrey Hill's discussion of related issues with reference to the recusant martyr, Robert Southwell, in his 'The Absolute Reasonableness of Robert Southwell', in *The Lords of Limit*.

[19] Christopher Devlin's note, *GMH Sermons*, 279 n.

In these vagaries of idiom, composed by Hopkins as in conflict with each other, he responds more fully to the historical setting of this martyrdom than he had in 'The Escorial' and, by the same token, he creates a keener sense, through the form of the broadsheet ballad which his poem recollects, expands, and twists, for the relation between the religious sensibility of her age and of his. That is, as it were, the social aspect of the quarrel centred on Margaret Clitheroe; Hopkins's response to the metaphysical dimension of her suffering as she is silenced is again sharper than in his early verse. If the poet's voice undertakes on one side a representation of the executioners' farcical boisterousness, on the other it faces a celestial calm which has a different chill:

> And every saint of bloody hour
> And breath immortal thronged that show;
> Heaven turned its starlight eyes below
> To the murder of Margaret Clitheroe.

'Thronged that show' for a moment imagines the company of the blessed as a crowd gawping at a side-show; the clash of 'starlight eyes' against 'murder' violently contrasts the radiant security of those above and the vulnerability of creatures on earth. Yet these heavenly onlookers, for all their distance from the scene of Margaret Clitheroe's pain, are not uninvolved in what she suffers; her suffering is not only a spectacle for them, but also a memory, for it is 'every saint of *bloody* hour' (my emphasis) who looks on her, that is, every saint who suffered martyrdom. The lines live with the paradoxes of martyrdom: the very finitude and consequent fragility of the human creature which leave it helpless against torture and violent death ('bloody hour') also enable the creature to discover within itself, by rendering up its own helplessness as a sacrifice, something beyond the reach of violence, a 'breath immortal'.

Hopkins relies here with great sophistication on the simplicities of popular piety; he is not implying a complaint about the failure of God to '*do* something' nor is he unaware how oddly these narrative habits of the ballad sit with a developed theology. Story-telling tussles with theological comprehension; the timed experience of such pain as this strains against its timeless significance and eternal reward. These conflicts form the compositional matter for a poetic and religious imagination more intimately engaged as well as more troubled than we find in Hopkins's verse before his conversion. The poem dramatizes

the array of speech around Margaret Clitheroe's sacrifice, and does so
with a passionate commitment of poetic voice to the impartiality of the
page. The conflicting idioms are all just *there*, arranged in a style of
imaginative retort which enables them to hear each other. The more
moving, then, when Hopkins at the end of the fragment recovers his
own accents, turns to speak to the dead martyr, addressing the woman
who has been till the close the third-personal object of the poem:

> When she felt the kill-weights crush
> She told His name times-over three;
> *I suffer this* she said *for Thee.*
> After that in perfect hush
> For a quarter of an hour or so
> She was with the choke of woe.—
> It is over, Margaret Clitheroe.

The lovely intimacy of that last line—it is the only one in the poem as
he left it which contains an internal rhyme on her name—is won
through the language of the world, a language Hopkins has allowed to
glance across the poem its lights of derision and unconcern. And the
address of 'Margaret Clitheroe' responds to her only speech in the
poem, to the only 'Thee' in the poem, her words to the pattern and
goal of her silence. It is as if, in that last line, Christ as well as
Hopkins spoke to Margaret Clitheroe.

When he wrote 'Easter Communion' in 1865, he was within a year
of seeing 'clearly the impossibility of staying in the Church of England'[20]
but the poem is still the work of the Puseyite Hopkins and, as the Jesuit
Hopkins was to say of Puseyites, it is without 'common sense':[21]

> You vigil-keepers with low flames decreased,
> God shall o'er-brim the measures you have spent
> With oil of gladness, for sackcloth and frieze
> And the ever-fretting shirt of punishment
> Give myrrhy-threaded golden folds of ease.
> Your scarce-sheathed bones are weary of being bent:
> Lo, God shall strengthen all the feeble knees.[22]

One foolish thing here is the attempt, laudably devout no doubt, to
bring the sonnet to its conclusion with a rhyme on 'knees', but the

[20] Journal entry for 17 July 1866, in *The Journals and Papers of Gerard Manley Hopkins*,
ed. H. House and G. Storey (Oxford, 1959) [hereafter referred to as *GMH Journals*],
146.
[21] Letter of 6 Jan. 1877 to A. W. M. Baillie, *Further Letters*, 240.
[22] Mar.–June 1865, Phillips, 60.

deeper failing of the poem lies in its inability to distinguish between the incidental pleasures and the real purpose of religious observance. 'O'er-brim the measures you have spent' comes both from the Bible— 'Give, and it shall be given unto you; good measure, pressed down, and shaken together, and running over, shall men give into your bosom'[23]—and from Keats's 'To Autumn'—'Summer has o'er-brimmed their clammy cells'.[24] The Bible speaks of an abundantly just reward; Keats was writing about a natural excess which deludes the bees who receive it into thinking that 'warm days will never cease'. If Scripture governs our sense of the lines, they confirm an orthodox hope for heavenly reward; if Keats tells more on the poem, it raises an arch eyebrow at a pious belief of the faithful. It might well do both these things, but 'Easter Communion' does no more than one or the other. 'One or the other': the alternatives merely proliferate, they do not mark points in a range of responses between which the poem moves (or is torn). The poem arranges no meeting of these divergent senses, and so the wealth of Hopkins's linguistic inheritance issues in a dissoluteness of attitude in the verse.

Phrases which have a dense life to come in his writing remain inert in these lines. Compare 'weary of being bent' with 'I wéar— | Y of idle a being but by where wars are rife'.[25] The later line compacts the sense of 'being' as an abstract noun with 'being's function as a participle. The vocal collocation might be 'I weary of idle-a-being but by where wars are rife' or 'I weary of idle-a-being-but-by where wars are rife'; print does not decide the voice here. We can then paraphrase, taking the first collocation, as 'I am tired of my useless self and of the fact that, wherever I am, people are at war with each other' (he was writing in Ireland), or, taking the second collocation, a paraphrase would read, 'I am tired of standing about, a mere spectator, helpless to do anything, in the midst of all this fighting.' The two senses, taken together, create a nuance which understands the way in which a particular sense of impotence (the second paraphrase) may turn into a general conviction of one's own futility (the first paraphrase), and this perfectly delineates that petrifaction of the self in a state of depression, when it seems that what you happen to be now is, bleakly, what you in essence were, are, and always will be.

---

[23] Luke 6: 38 (King James).

[24] Written about Sept. 1819, text of *The Poems of John Keats*, ed. M. Allott (1970, 3rd, corrected imp., 1975), 651.

[25] 'To seem the stranger . . .'? 1885–6, Phillips, 166.

Again, 'the ever-fretting shirt of punishment' in 'Easter Communion' looks as if it revelled in the discipline. 'Fret' was a word that drew Hopkins early and late[26] but it was not until *'Justus quidem tu es, Domine'* that he arrived at the union he needed of active with passive fretting, of anxiety about one's own state with the duress which makes that state. When he got there, he could see fretting as affording grace:

> See, banks and brakes
> Now, leavèd how thick! lacèd they are again
> With fretty chervil, look, and fresh wind shakes
>
> Them . . .[27]

'Fretty' imagines the natural indentations of the chervil as something worried into being, altering the eagerness of 'the ever-fretting shirt of punishment' into a new asceticism integral to the lyric itself, a sense and a texture of existence under strain. The lines are created out of self-thwarting as at the violent achievement of the rhythm in 'fresh wind shakes | Them' which sets the grammatical impetus forward of 'shakes', trying to join its object, 'Them', against the word's desire to hark back to its rhyme, 'brakes'. Two claims are made on the voice, claims which it can with difficulty meet simultaneously; it must hark back from 'shakes' to 'brakes' so that the rhyme may sound out and it must press on from 'shakes' to 'them' so that the syntax may flow. This is something different from the intonational ambiguities which I discuss elsewhere. In those cases, where two or more divergent voicings cannot be said at once, the impossibility of utterance represents a permanent dilemma of attitude. In an instance of vocal strain such as Hopkins creates on 'shakes', the voice represents rather the process of developing towards a solution of a dilemma; it is excruciated but not without hope. The rhythm of the poem works a sense of the daunting as well as the exhilarating energy which blows in this wind which is both 'fresh' but also 'shakes'. Gardner and MacKenzie annotate 'fretty chervil' as 'cow-parsley . . . which has richly serrated, lacy leaves'[28] but this is not enough. The *OED* reveals

[26] For earlier fretting in Hopkins, see 'dainty-delicate fretted fringe of fingers' in 'A Vision of the Mermaids', about 1862, Phillips, p. 13, where the delicacy of the lacey fretwork predominates, and 'those wastes where the ice-blocks tilt and fret' in 'I must hunt down the prize . . .', July–Aug., 1864, ibid. 28, where the verse is given over to the grind in 'fret'.

[27] 17 Mar. 1889, ibid. 183.

[28] *The Poems of Gerard Manley Hopkins* (Oxford, 1918), 4th ed., rev. and enlarged, by W. H. Gardner and N. H. MacKenzie (Oxford, 1967, corrected, 1970), 296.

the richness of the poet's sense of what it is to be 'richly serrated'. 'Fretty' is both an heraldic term—'Covered with a number of narrow bars or sticks . . . interlacing with each other' (an example is given from Mrs Jameson's *Legends of the monastic orders as represented in the fine arts* (1850) which sounds like a book Hopkins might have known)—and also an adjective meaning 'Of persons: Fretful; irritable. . . . Of a sore: Inflamed, festering' (1844). This conjunction of the formalized and the pained is searchingly contained in Hopkins's mature use of the word. The sharp intelligence of 'fretty' in 'Thou art indeed just, Lord' rebukes the posturings of the Puseyite Hopkins, and shows again how the mature style discovers an asceticism which is widely attentive rather than in a swoon, and discovers it by inventing an acoustic for the social drama of the disparate voices within the English language of his day.

About the time he began to wonder if he had a future as an Anglican, he wrote 'My prayers must meet a brazen heaven'. The second verse reads:

> My heaven is brass and iron my earth:
> Yea, iron is mingled with my clay,
> So harden'd is it in this dearth
> Which praying fails to do away.
> Nor tears, nor tears this clay uncouth
> Could mould, if any tears there were.
> A warfare of my lips in truth,
> Battling with God, is now my prayer.[29]

It is the first poem he wrote which shows that its author lived in an industrialized country. Before this, the world of Hopkins's writing is agricultural, a scene of activities which have biblical precedent and, hence, a ready potential as emblems. As far as concerns the natural world, the early poems inhabit an idyll which he left behind in such poems as 'God's Grandeur'. An idyll, in fact, like that of Keble's *Tract Eighty-Nine*, as of Keble's poems. Stephen Prickett acutely comments:

In *Tract Eighty-Nine* Keble goes into great detail over the mystical significance of the visible world. The sky, he tells us, represents 'a canopy spread over the tents and dwellings of the saints'; birds are tokens of 'Powers in heaven above who watch our proceedings in this lower world'; and waters flowing into the sea are 'people gathered into the Church of Christ'. . . . Whatever Keble

---

[29] 7 Sept. 1865, Phillips, 74.

himself seemed to think, this is clearly a very un-Wordsworthian universe in its elaborate system of significations.[30]

It is the neat, emblematic fixity of the system of significations which is un-Wordsworthian rather than its elaborateness. In 'My prayers must meet . . .', features of Hopkins's developed style begin to appear, features which will eventually enable him to write Wordsworthian poems of dramatized and developing analogy such as 'The Windhover', 'As kingfishers catch fire' and 'The Blessed Virgin compared to the Air we Breathe'. Three small features of the poem show Hopkins at work in the language in ways which provide grounds for the metaphorical imagination of the later verse. There is, first, his repetition of 'Nor tears, nor tears', a repetition which comes somewhere between a stutter and a caress on the word, between finding the word an obstacle and finding it a place of rest. The page sets the required scene for such a 'coming between', impartial as it is between requiring the voice to enact an impeded or a fond speaking of the words. Hopkins characteristically and purposively creates his repetitions, as also the exclamations in his verse, to have such a double aspect, at once of baffled and of heightened fluency; repetition would not of itself suggest a range of possible bearings on vocalization but the short-term density of repetition in Hopkins and its frequent occurrence at points where it interrupts syntactic cohesion or strains liaison between words often lend it this dramatic aspect.

He also has a tendency to write in such a way that his lines at times give an effect of etymological or phonological analysis of their constituent words. At 'this clay uncouth | Could mould', we feel a chain of phonemes mutating into one another, out of which the words appear to grow, with what may be experienced either as hideous parasitism or vivid germination. This feature of the style certainly derives from Hopkins's philological interests, and it may embody some of his convictions about the history of language, but it also has the important literary effect of giving to the sequences of his words a simultaneous character of independent life and of willed contrivance, as they look both like compiled anagrams and evolutionary processes. The 'look' of the words matters as well as the sounded connections (as here in 'Could mould', a rhyme to the eye though not to the ear). Emphasizing this point runs counter to many of Hopkins's declarations

[30] *Romanticism and Religion: The Tradition of Coleridge and Wordsworth in the Victorian Church* (Cambridge, 1976), 106.

about the supremely vocal character of his style, but those declarations often ring with a worried exaggeration because he feels himself so misapprehended by his readers. I will argue later that his poetic reservation of voice is an essential invention of his genius as he faces in verse questions of religious authority—its source, its nature, its possession by any visible church, the response it demands of a believer—questions which were, in many ways, central to his career as a poet and as a Catholic. This double character of independent life and willed contrivance in the words is a basic constituent of the drama of analogy in the mature poems. It stages a sense both of the world's separateness from the figuring mind and of its susceptibility to that mind. So too, the tendency of Hopkins's word chains to read partly as sets of anagrams and partly as evolutionary series contributes within the phonetic texture of the verse to the interaction of the timeless with time. As we read, empirical noticing develops into religious apprehension.

The third nascent feature of the developed Hopkins style which appears in this stanza comes in his typically sharp focus on a phrase which is often colloquially familiar:

> A warfare of my lips in truth,
> Battling with God, is now my prayer.

I discussed in the third chapter Meredith's use of 'in truth' as a pompous way of saying 'really' or 'indeed' and contrasted that use with Elizabeth Barrett Barrett's exultant solemnity with the phrase.[31] Hopkins inclines more towards her usage, but his 'in truth' also dips into the sociable world of Meredith's idiom, because, as a poem about the approach to conversion, 'My prayers must meet . . .' revolves particularly a possible conflict between religious and social claims. Earlier in the stanza, he also nods towards the language of the world when he sets 'Nor tears, nor tears this clay uncouth | Could mould' with its deliberated, ungainly inversion against 'if any tears there were'. The effect can be brought out in paraphrase: the poem says that he is so far from God that his heart is hard, and praying does not help, nor would even a burst of tears help by moistening his 'clay' and so making it more malleable to the divine will. 'If any tears there were' verges on the laconic when it adds to that self-analysis, 'not that I seem much given to tears'. Moments of rhythmic and dictional abrasion like this begin a style in which Hopkins truly meets his subject, whether a 'brazen heaven' or 'clay uncouth', truly 'meets' in the relevant range of

[31] See above, p. 197.

senses—'to come face to face with, or into the company of a person', 'to encounter or oppose in battle', 'to cope or grapple with', 'to satisfy the requirements of a particular case'.

'In the Valley of the Elwy' has a comparable dwelling on the double tune implicit in a phrase which may be either lightly passed over or fastened on and stressed:

> I remember a house where all were good
> To me, God knows, deserving no such thing . . .[32]

If you follow the tune of 'all were good to me', ignoring the lineation and modelling it on an idiom like 'My parents have always been good to me', 'me' takes no particular stress. If you speak the words so as to bring out the Miltonic syntax of the qualifying phrase, 'me . . . deserving no such thing', 'me' must be given a stress. The colloquial voicing lies within the verse as written but can't be spoken out if the demands of the poetic syntax are also to be met. The word 'me' is pressed into a warfare of the lips as they seek to do justice both to the social ease Hopkins felt in this company and to the self-reproach which tells him not to take such kindness as no more than his due. Swivelling on the self, he realizes two dimensions of the incident, the domesticated amiability and the sense of a more severe judgement (God's) behind the surfaces of life, two dimensions, each with a corresponding language, of familiar habit or of calling to account.

It is characteristic of him to jam these two senses together in the reciprocal shock at 'me, God', as he does to open 'The Wreck of the Deutschland'—'Thou mastering me | God'—or in 'The Bugler's First Communion'—'the Lord of the Eucharist, I'.[33] Putting the first person pronoun so hard up against the Divine Name, he wants to sound the severance between a language oriented to the self in the world and a language attuned to God, and then to bind those two languages together again, showing their intimacy, in a poetic realization like that he makes doctrinally about the Resurrection at the end of 'That Nature is a Heraclitean Fire . . .': 'I am all at once what Christ is, since he was what I am . . .'.[34] After all, you can say 'God knows' in 'In the Valley of the Elwy' either as a shrug ('God knows when I'll finish this job!') or as a recognition of His powers of judgement ('God knows the secrets of the heart'). The possibilities of voicing here mark out a

[32] 23 May 1877, Phillips, 131.
[33] Respectively, 1875–6, ibid. 110 and July 27 (?), 1879, ibid. 148.
[34] 26 July 1888, ibid. 181.

range of coherent response to the fact of one's own unworthiness, a range from the habituated despondency in a shrug to the shock of a sudden recognition. Though you cannot say both at the same time, the double direction the words may be seen to be capable of taking can represent the mysterious and eventual acceptability of the created world, and a particular self in that world, to its Creator, their meeting each other.

'Eventual' is the important word, for it draws attention to a complex of differences between Anglo-Catholic writings such as those of Keble and the work of a Catholic convert like Hopkins. Differences in literary practice do not simply correspond to credal or ecclesial divergences any more than they correspond to differences of class or gender; Faber's hymns are more like Keble's than they are like Hopkins's poems, which are closer to Milton, while Hopkins's prose is more akin to that of the Anglican Newman than it is to *All for Jesus*. But as a writer acquires a practical self-consciousness of what it is for him to write, his literary conduct increasingly forms part of his sense of what it is for him to act in ways other than by writing. It may be that the writer deliberately tries to conform his writings to convictions he arrived at without thought of their possible literary consequences, or that he recognizes and convinces himself only in writing and in order to write. On the other hand, things may happen to him which so change his understanding of himself as an agent that his literary activity is changed too, without a plan, perhaps without his realizing that it has changed. There are other possibilities, not readily susceptible of categorization as 'voluntary' or 'involuntary', which make up the concrete intermittences and the dim continuity of self-conception in the act. Delicate though these considerations of an agent's self-understanding are, they are required of us in reading and interpretation, as I argued in my first chapter.[35]

So, in Hopkins's case, it may be suggested that the 'eventual-ness' of his mature style is part of his understanding as a poet, of the ecclesiological implications of Newman's *An Essay on the Development of Christian Doctrine* with its sense of the life of the Church as best evidenced in traditionary change—'here below to live is to change, and to be perfect is to have changed often'[36]—rather than in a supposed, primitive purity. For Newman, revelation continues to occur, and history retains an incarnational significance which it has not lost or left

---

[35] See above, p. 50.
[36] (1845), UE 40.

behind in the period of Christ's life or the lives of the Apostles or in some, variously determinable period before the Reformation; that persistent occurring also sounds in Hopkins's style. Keble's *Tract Eighty-Nine*, looked at in this light, is a manual of Protestant style in harmony with the established decorums of his hymnody in *The Christian Year*. The 1827 'Advertisement' to that collection breathes a confidence Newman ceased to share:

> Next to a sound rule of faith, there is nothing of so much consequence as a sober standard of feeling in matters of practical religion: and it is the peculiar happiness of the Church of England to possess, in her authorized formularies, an ample and secure provision for both.[37]

It is not that Newman came to dislike *The Christian Year*; he did not (and it is likely that, had he ever read Hopkins's verse, he would not have preferred it to Keble's). Nor is it only a matter that, as his *Tract Ninety* demonstrated, he found the Articles more ample and less secure than many Anglicans wished them to be. Rather, the meditation in his *Essay* on development in Church doctrine and practice broke for him the connection Keble maintained between sober standards of feeling and 'authorized formularies', not only the Anglican formularies but formularies *per se*:

> The stronger and more living is an idea, that is, the more powerful hold it exercises on the minds of men, the more able it is to dispense with safeguards, and trust to itself against the dangers of corruption. As strong frames exult in their agility, and healthy constitutions throw off ailments, so parties or schools that live can afford to be rash, and will sometimes be betrayed into extravagances, yet are brought right by their inherent vigour. On the other hand, unreal systems are commonly decent externally. Forms, subscriptions, or Articles of religion are indispensable when the principle of life is weakly.[38]

Thus, over two decades, two friends and collaborators, Newman and Keble, develop into an estrangement, and a philosophical divergence so sharp that they would even attach conflicting senses to the fact of that very 'development'. Such a process goes on within Hopkins himself during his career, a process at once cultural and cultic, so that he understands his own poetic practice in terms of that process and its wide ramifications. Keble's iconic correspondences can then seem to

---

[37] Keble, 'Advertisement' to *The Christian Year: Thoughts in Verse for the Sundays and Holidays throughout the Year*, 2 vols. (Oxford, 1827), vol. i, p. v.

[38] *Essay on Development*; I quote here from the text of the 1st ed., 76–7. The passage is omitted from UE.

fit the world as it is to a set of divine meanings with the merely social tranquillity of a functionary of the State Church, though it is more than a social judgement to take up such an attitude to the Anglican Establishment and those in its orders.

Or it might seem that *Tract Eighty-Nine* offered a dilute, allegorical pantheism not only because Keble was sedately contented with most of the political constitution in which he lived but because he fundamentally mistook what the created world had to tell of the creator. As Newman wrote: 'The truth is that the system of Nature is just as much connected with Religion, where minds are not religious, as a watch or a steam-carriage. The material world, indeed, is infinitely more wonderful than any human contrivance; but wonder is not religion, or we should be worshipping our railroads'.[39] Newman's opponent here was not, of course, Keble but Sir Robert Peel and the liberal theism which Newman detected in Peel's reliance on works of popular scientific edification to lead the lower classes by imperceptible degrees from marvelling at Nature to the higher reverence. Yet Keble and Peel share what was to strike Newman as a secular reliance, distinct though their varieties of that reliance were, a reliance which appears in their being prompt to identify the signs of Providence in empirical fact, whether the facts of natural or of church history. Years later, his joke about the adorability of the railways hardened into a reluctant sense of the extreme paucity of natural signs for the truths of revelation: 'What strikes the mind so forcibly and so painfully is, His absence (if I may so speak) from His own world. It is a silence that speaks.'[40] This is extremely beautifully put; it recognizes an identity of purpose in Christ's silence before his accusers and judges and God's abstention from demonstrative presence in His own creation. And this is the kernel of Hopkins's mature writing about the natural world, something different from what is still commented on when he is admired as Hopkins the nature-poet.[41]

[39] *The Tamworth Reading-Room* . . . (1841), UE 302.

[40] *An Essay in Aid of a Grammar of Assent* (1870). I quote from the edition of Ian Ker (Oxford, 1985) [hereafter referred to as *Grammar*], 255–6.

[41] Consider, for instance, M. McLuhan's remarkable claim that 'Hopkins looks at external nature as a Scripture exactly as Philo Judaeus, St. Paul and the Church Fathers had done', in 'The Analogical Mirrors', *Gerard Manley Hopkins: A Critical Symposium by the Kenyon Critics* (1945, enlarged ed., 1975), 19. On the other hand, N. H. MacKenzie rightly insists, with particular reference to 'God's Grandeur': 'The poet constantly emphasizes . . . that God's glory is hidden except to the enquiring eye or on special occasions', *A Reader's Guide to Gerard Manley Hopkins* (1981), 63.

In 1878, he wrote with regret to Dixon of 'this want of witness in brute nature',[42] and a year earlier he had written 'God's Grandeur':

> The world is charged with the grandeur of God.
> It will flame out, like shining from shook foil;
> It gathers to a greatness, like the ooze of oil
> Crushed. Why do men then now not reck his rod?
> Generations have trod, have trod, have trod;
> And all is seared with trade; bleared, smeared with toil;
> And wears man's smudge and shares man's smell: the soil
> Is bare now, nor can foot feel, being shod.
>
> And, for all this, nature is never spent;
> There lives the dearest freshness deep down things;
> And though the last lights off the black West went
> Oh, morning, at the brown brink eastwards, springs—
> Because the Holy Ghost over the bent
> World broods with warm breast and with ah! bright wings.[43]

The world is not only invigorated or fuelled ('charged') with God's grandeur, it is accused ('charged') with it because the world, and particularly man in the world, has the responsibility ('is charged with') to meet and manifest that grandeur but often fails to live up to its responsibilities. Hence, the 'want of witness' in creation.[44] This reflective wealth of sense in 'charged' comes out of the line gradually, though the line itself strikes out with an axiomatic suddenness and boldness. We might say that the first line has the relation to the body of the sonnet of a text to a sermon, but saying that with assurance would involve a decision about how to voice the line which the line does not itself support; it would ascribe an act of pronouncement to the opening which would then be followed by explication or brooding. Yet the sonnet broods from beginning to end.

---

[42] Letter of 13 June 1878, in *The Correspondence of Gerard Manley Hopkins and Richard Watson Dixon* (Oxford, 1935, 2nd rev. imp., 1955), ed. C. C. Abbott [hereafter referred to as *GMH/Dixon*], 7.

[43] Feb.–Mar. 1877, Phillips, 128.

[44] Compare Newman's criticism of the importance attached to the physical creation by some pious writers: '. . . how are we concerned with the sun, moon, and stars? or with the laws of the universe? how will they teach us our *duty*? how will they speak to *sinners*? They do not speak to sinners at all. They were created *before* Adam fell. They "declare the *glory* of God," but not His *will*. . . . We see nothing there of God's *wrath*, of which the conscience of a sinner loudly speaks.' I quote from the sermon, 'The Religion of the Day', preached 26 Aug. 1832, *Parochial and Plain Sermons*, 8 vols., UE i. 317–18. Newman's contrast between '*glory*' and '*wrath*' prefigures Hopkins's contrast between 'grandeur' and 'rod'.

'It' in the next lines is usually, and rightly, taken as referring to 'the grandeur of God' but though this is the prime reference, reading this way produces a wrinkle or two in the verse, and the wrinkles suggest that 'it' refers to something else as well. For instance, not much is said of any 'grandeur' by saying that it becomes ('gathers to') a 'greatness'; we might also wonder how *God's* grandeur resembles anything which can be 'crushed'. Though the rules of pronominal reference in English indicate 'grandeur' as the referent of 'it', verse-syntax could override these rules and offer instead 'the world'. What would the lines mean if 'it' referred to 'the world'? They would be apocalyptic lines about the final perfection of the world, crushed and flaming out, in its coming-to-an-end on the Day of Judgement; the Divine Majesty has been invested in creation but creation will not unfold all that grandeur to created beings until heaven and earth have passed away. Either reading of the lines provides a reason for the question which follows them: 'Why do men then now not reck his rod?' Men should fear God's power to chastise because creation can at times display his magnificence and/or because they know that the world will end. The point of joining these possible senses, writing them in the same breath, is to assist the reader as he contemplates the mystery of God's power, assist him not to lose the perception of created beauty in terror at the thought of that beauty's end, while—at the same time—assist him to check the desire to believe that the world could not possibly end because it is so beautiful ('God just wouldn't let it happen'). More than this: when we read lines 2–4 as referring to 'grandeur', we should note that the lines speak of human ability in manufacture as able to display or image God's creative power; when we read them as referring to the world, they require an imagination of all the world, including man's works, as just instances of material fashioned by a superior being, showing nothing special about God in themselves though He may show through them. The opening of the sonnet, then, at once sets a range of the aspects of human skill (including poetic skill) before us, skills variously considered as actual indices of providential grandeur or as incomplete signals whose sense will be legible only in a futurity when they themselves will have ceased to be.

The second quatrain explains why men ignore these potent but ambiguous evidences of what God has done for them and what He might do to them. They do so because manufacture humanizes the natural world and makes it seem to man entirely his own product. The world so testifies to man of himself that it speaks of nothing else, and man witnesses only himself in creation. Hopkins here makes no complaint specifically against industry or urbanization (as he does in 'The Sea and the Skylark') because manufacture in 'God's Grandeur' includes both metalwork and oliviculture; images of production by aid of fire ('flame out . . . seared') receive equal treatment with images of agricultural labour ('ooze of oil . . . trod'). Despite this ceaseless human exploitation of the earth, with its result so cruelly parodic of the Hegelian claim that man spiritualizes his world through labour, that non-human, resourceful world of nature is not exhausted through commerce ('never spent' taking up from 'seared with trade'). Even were the world worn out, this would still not be the Last Day.

The sestet does imagine an eco-catastrophe, for in 'though the last lights off the black West went' the verb is a subjunctive, and the line should be paraphrased as 'if the Sun were to be extinguished'. However, the end of man's world is not God's end of the world, as Hopkins makes out in the most astonishing moment of this great sonnet — the moment of the turn at 'went | Oh, morning'. What that moment says is that even were the earth as we know it to be destroyed, there would be a new dawn 'Because the Holy Ghost over the bent | World broods', because God's care for the world exceeds ours, because our destruction of the world could not impair his power to contemplate his creation and hatch ('broods') something more from it. The astonishing thing in the final tercet is Hopkins's audacity in using the natural rhythm of sunset and sunrise to stand for the recovery of creation by its creator even when all the creaturely rhythm that could be known to a man such as Hopkins had stopped, a recovery and perpetuation in the new significance of eternity according to Hopkins and which he yet pictures as the absolute type of temporality as he knows it, the turns-about of night and day. Such a power of re-invention in Hopkins is beyond Keble's formulaic correspondences in *Tract Eighty-Nine*, but the wonder of Hopkins's skill to make creation mean, given his poem's testing and dread of man's skills, does not tempt us to worship the railroads (or Hopkins). The sonnet's exultant agility, its affording to be rash, and its security of mind amidst melodic extravagance (to adapt terms I quoted before from the *Essay on Development*), attune it to ancient and orthodox traditions, for that entirely individual and syntactic move—'went | Oh, morning'—with its arch of human concept and experience from their setting in temporal existence across to the eternal and transcendent recovers the words of Revelation: 'And I saw a new heaven and a new earth' (21: 1; Douay Bible).

The three features of Hopkins's style which I pointed out in 'My prayers must meet . . .' come into their own here. 'And for all this' opens the sestet and so constitutes the fulcrum of the poem, but the phrase itself hinges between lightness and a pondered, judging weight. If thrown off on the voice (stressed, perhaps, And for all this or And for all this, it sets aside the laborious complaint of the previous four lines, as who should say 'That's nothing to fuss about'. If pressed on (with a stress such as And for all this or And for all this), it recognizes how terrible man's depredations have been but, in the moment of that recognition, how the divinely sustained resources of nature have managed to be awesomely greater. Between these poles of shrugging-off or gravely shouldering, the phrase contains a range of attitudes to human responsibility with regard to the created world and, because the written line contains this gamut but any voicing of the line sets a task of balance for the reader, reading the poem aloud provides a model in our response to what Hopkins has made in the poem of our more general conduct towards the world in which his making and our reading both take place.

Taking the weightier inflection of these words, we hear that 'all this' in line 9 harks back to 'all is seared' in line 6, and that should alert us to the modulations at the level of syllable which Hopkins, here as elsewhere, weaves into his lines so that the phonetic texture itself in its reminiscences and anticipations shall 'time' the process of religious illumination and poetic meaning ('time' in the sense in which an actor 'times' his lines or a general 'times' the moment of attack). In one respect, the sequence which courses through 'seared with trade; bleared, smeared with toil; | And wears man's smudge and shares man's smell' obviously mimics a slither into pollution, some travestying effluence of manufactured abundance which coats everything in reach, words as things. Yet he also creates it as such, a virtuoso sign not a regrettable but inevitable by-product of necessary semantic processes; the slither is also a glide, detritus—'man's last dust'—and yet fruitful—'man's first slime'.[45] The assay of human prowess which goes on conceptually in the structure of the poem equally affects the sonic material implied by the writing. The sestet talks about God's

[45] 'The Sea and the Skylark', May 1877, Phillips, 131.

ability to sustain his creation even when creation seems most derelict, but it does not only 'talk about' it, for the sustenance is there within the very language which the poem speaks, as the downslide of 'seared', 'bleared', 'smeared' provides the upswing of 'dearest', 'deep', 'eastward'. In both octave and sestet, as in 'My prayers must meet . . .', eye-rhymes matter as well as chimes which can be voiced: the shift from 'seared' and 'bleared' to 'shares' and 'bare' goes through 'wears' which eye-rhymes with the preceding words and ear-rhymes to what follows, just as the long haul of the poem eventually reposes in the 'breast' of the Holy Ghost as much because of an eye-rhyme with 'eastward' as because of the audible link to 'West'. Between them, eye and ear cover the world from East to West and find its unity in the over-arching care of the Holy Ghost. Hopkins may also set eye and ear at odds, as in those moments when he courts the stutter, writes a line which reads without trouble to the eye but which would trip up an incautious reader-aloud, as at 'Why do men then now not' with its shockingly impacted vocables (terminal and initial 'n' set against each other), the swerving internal rhyme of 'men then', the move of 'n' from a final position to become an alliterating key-note ('men then now not'), the illusion of semantic contrast between 'then' and 'now' with the momentary puzzle as to why, if 'then' (in the past) contrasts to 'now' (in the present), it governs a verb in the present, a puzzle dissolved when we realize that 'then' is here logical rather than chronological, dissolved but still undergone. That mouthful of quandary, brain-teaser and tongue-twister at once, eventually opens up into the spacious lyricism of the last lines with their chiastic alliteration, a melody of a mind no longer clenched but arrived at understanding. The sonnet closes in gratitude and makes itself grateful to the voice which reads it out:

> the Holy Ghost over the *b*ent
> *W*orld *b*roods *w*ith *w*arm *b*reast and *w*ith ah! *b*right *w*ings.

It takes time to write like this.

Eventually, the world will stand God's judgement, because it remains His world, whatever men may make of it. So this poem concludes. The poem's greatness does not lie in its conclusion but rather in its working to give weight to that 'eventually', to span the historical process of conforming the world to its Creator, a process of development within the history of the Church and in the lives of individuals within the Church. It would be partisan to identify

Hopkins's blend of theological depth and social delicacy in his handling of language with the character of the Church he entered and, conversely, to attribute an Erastian cosmology and class-bound gentility to the Anglican communion which he left. Elements of a particular comparison between Keble and Hopkins hardly provide the basis for a general law about the Roman and Anglican clergy in nineteenth-century England. None the less, in Hopkins's own case, the distinctiveness of his mature achievement does rise from such unformulable meshes of religious conviction and social experience. His move from Canterbury to Rome was also a move from Oxford to Liverpool, from the refined world he had known to a Church the majority of whose congregations consisted of Irish immigrant workers and their families.[46]

But Hopkins did not exactly have the common touch. Patmore was in a way right to be 'a little amused by your claiming for your style the extreme of popular character',[47] and when Hopkins tried directly for popular appeal, as in some of his sermons, he was not always judicious: Henry Marchant remembered that 'once he compared the Church to a milk cow and the tits to the seven sacraments'.[48] He himself noted ruefully that he had preached in 1880 on 'our Lord's fondness for praising and rewarding people': 'I thought people must be quite touched by this consideration and that I even saw some wiping their tears, but when the same thing happened next week I perceived that it was hot and that it was sweat they were wiping away.'[49] The entry shows how well he knew himself in his awkwardnesses. The undemonstrative sentences pass in review his enthusiasm as he composed the sermons on his own in the study ('I thought people must be quite touched'), his eagerness to be received, to 'get across' ('I even saw some wiping their tears'), and then come down to earth amidst the discomforts of people who had not shared his uplift ('I perceived that it was hot and that it was sweat') with a slight shiver of recognition and disappointment that where he had believed himself in touch he had only marked his distance (the slight, prissy elevation of diction at that 'I *perceived*'—my emphasis). The incident is both comic and very sad.

Even Newman, who so much more securely held the public ear throughout his career, viewed the prospect of being sent by his

[46] See Edward Norman, *The English Catholic Church in the Nineteenth Century* (Oxford, 1984) [hereafter referred to as Norman], 7.
[47] Letter of 5 Apr. 1884, *Further Letters*, 355.
[48] *GMH Journals*, 421.   [49] *GMH Sermons*, 81.

Catholic superiors to Birmingham with an apprehension edging on disdain:

... I ... feel more bound to London than Birmingham, if we had any work with the poor, or the many—And again, my line has hitherto lain with educated persons, I have always had a fancy that I might be of use to a set of persons like the lawyers—or again I might be of use to the upper classes; now London is the only place for doing this in. London is a centre—Oxford is a centre—Brummagem is no centre.[50]

'Brummagem' had been a pejorative form of 'Birmingham' since the seventeenth century when counterfeit groats were coined there but the word extended its sense in the Victorian period to cover all shoddy and showy ware. Newman's prose winces away from such a world of mass-produced vulgarisms rather as Arnold exclaimed in 'The Function of Criticism at the Present Time' against 'the natural growth amongst us of such hideous names,—Higginbottom, Stiggins, Bugg!'[51] For those such as Hopkins and Newman whose 'line' had 'hitherto lain with educated persons', the loss of 'centre' was severe, and they are hardly to be blamed for in some measure dreading it. (On the other hand, records of their activities as priests once they had gone to Birmingham and Liverpool show their unstinting dedication to the welfare of their parishioners.) Within the Church they joined, they found a variety of approaches to the relation of clergy to a laity many of whom were not only lay but also illiterate—men such as Wiseman adapted continental practices of popular devotion to English needs, and were followed by some converts, such as Father Faber, with Italianate fervour; others, such as Manning and Vaughan, concentrated the hierarchy's attention on social and educational reform. Hopkins and Newman, though not at all unconcerned with these efforts, remain largely apart from them, with other work to do.

Though the experience of conversion importantly faced Hopkins with harsh realizations of the social and linguistic disparities in England and in Ireland, faced him with conflicts of class and region, cult and political nationalism, the finding of apt styles for such disparity was a task in which being a Catholic offered no direct assistance. A socially responsive religious language remained as

<hr>

[50] Newman, Letter to J. D. Dalgairns, 9 Nov. 1845, in *The Letters and Diaries of John Henry Newman*, ed. C. S. Dessain, I. Ker, T. Gornall, and G. Tracey, 31 vols. (1961, in progress), xi. 30.

[51] (1865). I quote the text of R. H. Super, *The Complete Prose Works of Matthew Arnold*, 11 vols. (Ann Arbor, 1960–77), iii. 273.

difficult to attain for those outside as for those within the Established Church. Thus, for example, Wiseman and Newman were capable of a cruel, oratorical inattentiveness in their remarks about industrial England. Here is Wiseman, addressing Salford:

I believe that the truest and surest type of a great and prosperous nation will be the union of the two symbols which are here: on the one side those vast and darkened piles of building, which fill your city, with their full columns above which the banner of industry ever streams in the wind; and, on the other, the vast magnificent church of God, with its spire bearing the symbol of peace and salvation.[52]

This is a preacher's conceit with a vengeance. The eloquence disposes of the actualities it fluently transforms to emblems. What, we should ask, is 'the banner of industry' which 'ever streams in the wind' above the 'vast and darkened piles' of factory chimneys? It is not hard to guess (though Wiseman's prose does not help us to realize this) that the banner of industry is a plume of smoke, the sort of smoke Hopkins walked in: 'I was yesterday at St. Helen's, probably the most repulsive place in Lancashire or out of the Black Country. The stench of sulphuretted hydrogen rolls in the air and films of the same gas form on railing and pavement.'[53] Easy enough for Wiseman to imagine a 'union of . . . symbols' when his symbols were so remote from perception of fact, and also easy then to slump into the politician's cant of 'great and prosperous nation' (are these adjectives synonyms?), and the more easy, after having managed that, to refer to 'the symbol of peace and salvation' in such a way that it is almost as difficult to guess that this means the Cross as it is to guess that 'the banner of industry' is pollutant smoke.

Or take Newman in a discourse of 1849 to Birmingham:

You know there are persons who never see the light of day; they live in pits and mines, and there they work, there they take their pleasure, and there perhaps they die. Do you think they have any right idea, though they have eyes, of the sun's radiance, of the sun's warmth?[54]

He does not seem to know much about most mining communities if he imagines that they 'live' and 'take their pleasure' in the pits. There is nothing here as smoothly superficial as in Wiseman's sermon but still Newman is guilty of exploiting industrial exploitation for rhetorical

[52] Nicholas, Cardinal Wiseman, *The Social and Intellectual State of England* (1850), 19.
[53] Letter to Bridges, 8 Oct. 1879, *GMH/Bridges*, 90.
[54] *Discourses addressed to Mixed Congregations* (1849), UE 85.

purposes which depend for their success on throwing the actual roots of the figure into obscurity. His interest, as he speaks, is directed exclusively towards turning the industrial plight into a Christian version of Plato's Cave; that rhetorical interest then leaves not so much open as vacant the question of whether these conditions of work should be regarded as part of the universal condemnation to labour imposed on man as a result of the Fall (Genesis 3: 17–19) or whether individuals sin by treating their fellow creatures in such a way for the sake of profit. This is a notable blank in the decade when public debate about punitive conditions of work had been agitated around the Mines Act (1842) and the Ten Hours Act (1847).

Hopkins preached a sermon in Bedford during which he used miners as an image for quandaries in the spiritual life and, for once, his public rhetoric is more assured than Newman's because it less ignores the derivation of its metaphors:

. . . the Scripture in one place calls life night and in another calls it day. But these do not disagree. In respect of truth and the clearness we see it with / life is night and what comes after life is day; in respect of doing work in God's service and earning a reward hereafter life is day and what comes after is night. But you yourselves, brethren, are some of you well aware of this: to most men the daylight is the place to work in but those that work in the pit go where all is darker than night and work by candlelight and when they see the light of day again their work is over, as if day were night to them and night day, so then this life is dark, a pit, but we work in it; death will shew us daylight, but all our work will then be done.[55]

This is magnificent. It brings together the persuasive fluency of the pulpit—'death will shew us daylight, but all our work will then be done'—with an awareness of the divisions in working practice even within his working-class audience. Hopkins knows only 'some' will be aware what it is like to work in the night 'darker than night' of the pit. The metaphor is absolutely in touch with realities of work which are not shared by all, and his genius with the metaphor enables him to understand apparent contradictions in the scriptural texts by attending to the disparate rhythms in the lives to and for whom he speaks. The sermon meets the difficult requirements of Christian understanding. It comments on texts which seem at odds with each other but which have suspended their contradictory implications, as narratives do, over time. Theological interpretation of scriptural story has to try to harmonize

[55] Sermon of 30 Nov. 1879, *GMH Sermons*, 39.

these implications into the atemporal coherence of a quasi-philosophical argument, but, in doing so, it is in danger of falsifying the texts which it interprets, for they are not articulations of an intellectual system but records of events. Hopkins responds at an elevated intellectual level to such difficulties and his manner is equally adjusted to the actualities in which on this occasion he strives to make himself and his text heard. The sermon attains for once in his career as a preacher those necessary, mediate words which inform his mature verse. There is no record of what his audience thought of it.

Hopkins expertly recognized the divisions in English life and the corresponding varieties of English usage. That recognition gives some support to F. R. Leavis's influential praise of the poet's work and his defence of its occasional oddities on the grounds that 'the peculiarities of his technique appeal for sanction to the spirit of the language'.[56] Much in Hopkins does indeed respond self-consciously to the divergences within English, and this was a common response in the England of Murray and Sweet, of William Barnes and Thomas Hardy. If any imaginative writer creatively focuses all the endeavour of Anglo-Saxonists, philologists, students of dialect, lexicographers, phoneticians, comparative linguists, an endeavour which is so notable a feature of the intense linguistic self-consciousness of the Victorian period, then that writer is Hopkins, and, to that extent, Leavis's testimony is only just. Yet the Leavisian phrase, 'the spirit of the language', scarcely admits the strain under which 'English' and 'Englishness' came from these investigations or from Hopkins's creative activity. Both scholars of the language and imaginative writers discovered an abundance of diverse idiom and practice within English. In so far as they knew what to 'do with' the rich disparity of these materials, the discovery was a source of pleasure. The same discovery could give cause for an anxiety that the community of language-users might fall apart (or perhaps had fallen apart) into mutually uncomprehending groups, locked in ways of speaking which people did not always recognize were alien from each other. In the second section of this chapter, I give some examples of how Hopkins and Newman felt that at least one such source of incomprehension was the existence of two Englishes—a Protestant and a Catholic version. To which of these could Hopkins have appealed for sanction, which represented *the* 'spirit of the language'? And then, Hopkins sought his justification not only from 'the spirit of

[56] *New Bearings in English Poetry: A Study of the Contemporary Situation* (1932, repr., Harmondsworth, 1963), 125.

the language' but also from the language of the Spirit, and it is not clear that he thought those two were one, as Leavis seems sometimes to have done.

Leavis's many and eloquent descriptions of what he considered the Shakespearian use of English[57] constitute a passionate defence of the language-using community as the authentic source and aim of individual imaginative creation and an equally passionate insistence that the right conduct for the individual writer is one of responsive collaboration with that community and 'the living, the spoken language'[58] which it has made. As Leavis admired Hopkins, and as he considered Shakespeare the principal instance of all that is admirable in literary art, he thought Hopkins a Shakespearian writer. Yet what emerges from Hopkins's notebooks and letters is that the English poet most vital to him as a precursor was not Shakespeare but Milton; Hopkins's relation in his poetic practice to spoken English is quite as often Miltonic as it is Shakespearian (to accept Leavis's terms for the purposes of this argument), and this is because, like Milton, Hopkins was an artist consciously at odds with some parts of the community of English-speakers and so needed to resist as well as to collaborate. Leavis's account of Hopkins stands in need of correction, as does his more general notion of literary excellence. English itself should have told Leavis that even 'collaboration' can in some circumstances become a dirty word.

'This poem was to be an act of devotion, of religion', Hopkins wrote to Patmore of a projected Patmore work, 'perhaps a strain against nature in the beginning will be best prospered in the end'.[59] Part of the nature Hopkins himself strained against was that Leavisian 'spirit of the language', as the numerous preciosities and dare-devillings of his poems against the norms of English speech show. James Milroy makes a suggestion of great interest when he points out that Hopkins may have been influenced by Max Müller's *Lectures on the Science of Language* (1864) and their claim that linguistics was a physical science

---

[57] The most extended instance of such description is throughout *Revaluation* (1936, repr., Harmondsworth, 1964); see esp., 50–5, 210–13, 229, 245. Compare also Leavis's remark that Hopkins's influence was beneficial in promoting the feeling for 'what may be called the Shakespearian (as opposed to the Miltonic) potentialities of English', 'Metaphysical Isolation', in McLuhan, *Gerard Manley Hopkins: A Critical Symposium by the Kenyon Critics*, 134.

[58] 'Gerard Manley Hopkins', in *The Common Pursuit* (1952, repr., Harmondsworth, 1962), 46.

[59] Letter of 4 Apr. 1885, *Further Letters*, 359.

because natural languages were a part of the material creation.[60] Milroy's evidence, both from the etymological notes in Hopkins's journals and from his poetic practice, is convincing. We should then expect Hopkins to be wary of the language while also schooled in it and responsive to it. Otherwise he would be in danger of confusing a proper sense of the ends which mortal beauty can serve with the worship of railroads. He did not make a religion of the English language(s); in this respect, Leavis's praise of him, though alert and stimulating, has not the generosity to be accurate, to hear Hopkins as he wanted to be heard.

He wrote in a notebook while on retreat:

For, to speak generally, whatever can with truth be called a self . . . is not a mere centre or point of reference for consciousness or action attributed to it, everything else, all that it is conscious of or acts on being its object only and outside it. Part of this world of objects, this object-world, is also part of the very self in question, as in man's case his own body, which each man not only feels in and acts with but also feels and acts on. . . . A self then will consist of a centre *and* a surrounding area or circumference, of a point of reference *and* a belonging field . . .[61]

Language is such a 'part of this world of objects' which is 'also part of the very self in question', and specially so because a language essentially informs the mediacy between the self and other selves. The conduct of poetic style within a language, then, forms part of the broader activity of 'knowing one's place' in a culture. The moment at which a poem is consigned to another's voice is the moment at which the poet most clearly stands as both 'a point of reference *and* a belonging field', a point and a field which reciprocally locate each other. (There is, actually, no such 'moment' but rather a permanent possibility of coming to hear oneself as others hear one.) To stipulate for one kind of poetic activity, as Leavis appears to do in his esteem for the 'Shakespearian', whether he describes it rightly or not, turns out to be a requirement for uniformity of conduct irrespective of the circumstances in which action takes place. Such a requirement, given the celebrated ability of circumstances to alter, amounts to nothing else than, in Hegel's phrase, 'giving instruction as to what the world ought to be'.[62] That is always a seductive pastime, for a poet or for a critic,

[60] See his *The Language of Gerard Manley Hopkins* (1967), 50.
[61] Notebook entry for 20 Aug. 1880, *GMH Sermons*, 127.
[62] *Naturrecht und Staatswissenschaft* . . . (1821), trans. as *Hegel's Philosophy of Right* by T. M. Knox (Oxford, 1952), 12.

but it is rarely compatible with 'knowing one's place', and certainly not compatible with knowing oneself to be, partially yet essentially, displaced—'A self then will consist of a centre *and* a circumference'/ 'Brummagem is no centre'. If we want to approach Hopkins's 'very self in question', we need to know more about his experience of the language which was to him like his own body, a predicament and a means of action.

The Conversion of Eloquence

One aspect of Hopkins's mature writing can be described in some early lines he wrote about a nightingale:

> I thought the air must cut and strain
> The windpipe when he sucked his breath
> And when he turned it back again
> The music must be death.[63]

The poem is written as a ballad-monologue, giving the worries of a woman whose lover has gone to sea, and who finds the nightingale, commonly welcome to poets' ears, a bird of fear to her because she counts the cost to it of its song. It is characteristic of Hopkins to imagine bird-song in this excruciatedly physical way, as an attrition of the atmosphere on the bird's 'windpipe' and a retort on that attrition, a turning-back of the air, in the song. The crucial pun on 'strain' as meaning both a melody and a duress, which will be so important to him later, both in the poetry and in writing about his poetry, already starts work here in the imagination of music as a kind of death.

The poem ends with a narrator's comment:

> Thus Frances sighed at home, while Luke
> Made headway in the frothy deep.
> She listened how the sea-gust shook
> And then lay back to sleep.
> While he was washing from on deck
> She pillowing low her lily neck
> Timed her sad visions with his wreck.

'The Wreck of the Deutschland' is still nearly a decade away but this minor poem anticipates tremors to come; Frances's situation will be that of Hopkins at rest under a roof while 'lives at last were washing

[63] 'The Nightingale', 18, 19 Jan. 1866, Phillips, 80.

away', and the later poem will also time its 'sad visions' with the 'wreck' in the first of Hopkins's major workings-out of his central concern, the providential place of human anguish as that may be understood in the poetic composition of suffering. The reader has to be quick to see that 'Made headway in the frothy deep' vitally differs from 'Made headway on the frothy deep'; the phrase makes a minutely sardonic shift of the human purpose, when we realize that the head is making its way vertically downwards rather than horizontally onwards. These extremely light touches of tonal unconcern correspond, as they will do in 'The Wreck . . .', both to the detached comfort in which, with whatever anxiety, the disaster is contemplated and to the peculiar similarity of Christian resignation to assured callousness (from which Hopkins strains to distinguish it). The technique in these lines of allowing the page to contain but not voice intonations of a laconic or gloating indifference already looks forward to the 'Margaret Clitheroe' ballad, to the late sonnets, and to the visionary wreck of the *Deutschland*.

Hopkins was able to imagine music as a kind of death because he went through a death to arrive at his developed style, and the death he went through was his conversion to Catholicism. This sounds melodramatic but the words of his time on this matter substantially bear out the claim. Newman described what happened to converts in the eyes of 'the Prejudiced Man' (or the 'staunch Protestant' as he might less harshly be called):

They cease to have antecedents; they cease to have any character, any history to which they may appeal: they merge in the great fog, in which to his eyes everything Catholic is enveloped: they are dwellers in the land of romance and fable; and, if he dimly contemplates them plunging and floundering amid the gloom, it is as griffins, wiverns, salamanders, the spawn of Popery, such as are said to sport in the depths of the sea, or to range amid the central sands of Africa. He forgets he ever heard of them; he has no duties to their names, he is released from all anxiety about them; they die to him.[64]

This retorts on the prejudiced imagination, and it does so with some alacrity, comically transforming the bigot's sense of the alien and the outrageous into a source of linguistic exoticism and stirring fable. The second sentence quoted moves into something less sportive: 'he forgets . . .', 'he has no duties . . .', 'he is released . . .', all that is to the airy good of the Prejudiced Man; clearly, these things don't bother *him* but the consequence is suffered by others, 'they die . . .'.

[64] *Lectures on the Present Position of Catholics in England* (1851), UE 245.

This is a death in polemic, but Newman said without complaint, when he was preceded into the Church by some of his friends, 'when they went, it was as if I were losing my own bowels'.[65] Indeed, it is possible to find parallels in fact for most of Newman's fictional accounts of reactions to conversion; these parallels show he did not exaggerate. Thus, of the estrangement depicted in *Loss and Gain* when Charles Reding's sister realizes that he is beginning to leave the Established Church, Newman writes: 'At first it quite frightened and shocked her; it was as if Charles had lost his identity, and had turned out some one else. It was like a great breach of trust.'[66] This should be compared with the letter Newman's sister wrote to him in October 1845: '. . . for myself, who have always looked to you for comfort and support, to feel cut off by such a barrier!'[67]

When Newman mimics a Protestant father speaking to his convert son, it might be thought only a seizing of opportunities for parody in a manner not compatible with delicate feeling for the troubles of families who went through such an experience, but actual letters, and the conversations one guesses at behind them, suggest he was not engaged in parody at all. Here is the Protestant father:

'My dear John or James,' the father says, calling him by his Christian name, 'you know how tenderly I love you, and how indulgent I have ever been to you. I have given you the best of educations, and I have been proud of you. There is just one thing I cannot stand, and that is Popery; and this is the very thing you have gone and taken up. You have exercised your right of private judgment; I do not quarrel with you for this; you are old enough to judge for yourself; but I too have sacred duties, which are the unavoidable result of your conduct. I have duties to your brothers and sisters;—never see my face again; my door is closed to you. . . .'[68]

The little sarcasm about the 'Christian name' seems to me a bit of slyness which is otherwise absent from the passage. The claim that converts were exercising just that 'private judgement' in religion which they criticized in Protestantism is typical of the arguments of the period about conversion, and an instance of the extreme difficulty with which the converts made themselves understood to those who remained in the Established Church.

[65] Letter to H. Wilberforce, 7 Oct. 1845, *Letters and Diaries of J. H. Newman*, xi. 3.
[66] *Loss and Gain* (1848), UE 262.
[67] Letter from Mrs John Mozley, 6 Oct. 1845, *Letters and Diaries of J. H. Newman*, xi. 13 n.
[68] *Present Position*, UE 185.

The 'sacred duties' which require the banishment of the convert son from the family home were obeyed by many. Hopkins and his father both evidently thought it generous of Manley Hopkins not to forbid his son his house:

You are so kind as not to forbid me your house, to which I have no claim, on condition, if I understand, that I promise not to try to convert my brothers and sisters. Before I can promise this I must get permission, wh. I have no doubt will be given. Of course this promise will not apply after they come of age. Whether after my reception you will still speak as you do now I cannot tell.[69]

It is painful, it must have pained them, to see father and son forced into drawing up the precise terms on which they would meet each other, as if they were strangers. It was a kind of death. Hopkins's father writes to the Reverend Henry Liddon to beg that Liddon should save Hopkins from 'throwing a pure life and a somewhat unusual intellect away in the cold limbo which Rome assigns to her English converts'[70] and exclaims to his son in the accents of bereavement, 'O Gerard my darling boy are you indeed gone from me?'[71]

At his parents' insistence, Hopkins asked Pusey to see him and try to argue him out of going over to Rome. Dr Pusey was understandably reluctant to agree to such a meeting because he knew that Hopkins would come, as previous intending converts had come, 'with a fixed purpose not to be satisfied', and in the hope only of obtaining credentials of his unshakeable conviction, of being able 'to say to his relations "I have seen Dr P, and he has failed to satisfy me" '.[72] Pusey's refusal was a kindness to Hopkins and his parents, but it was not kind of Pusey in the same letter to refer to Hopkins as a 'pervert'. This witticism for 'convert' had considerable popular success in the period (the joke, according to the *OED*, dates from at least 1716). Undistinguished men such as Dr Cumming used it, 'Prodigious efforts are being made by the Ritualists to enlist converts, or rather I should say perverts'.[73] 'Pervert' did not acquire its sexual connotations until the late nineteenth century, meaning in the instances I quote only 'someone who has adopted a false doctrine; an apostate'. Still, the play on words did its work, saying something was so made it so: a convert

[69] Letter of 16 Oct. 1866, *Further Letters*, 94.
[70] Letter of 15 Oct. 1866, ibid. 435.
[71] Letter of 18 Oct. 1866, ibid. 97.
[72] Letter of 10 Oct. 1866, ibid. 400.
[73] *Ritualism, the Highway to Rome* (1867), quoted in E. R. Norman (ed.), *Anti-Catholicism in Victorian England* (1968) [hereafter referred to as *Anti-Catholicism*], 194.

might have claimed to be embracing the true faith but a pervert could make no such appeal.[74] Gladstone preferred an alternative playfulness about those who went over to Rome: 'The conquests have been chiefly, as might have been expected, among women; but the number of male converts, or captives (as I might prefer to call them), has not been inconsiderable.'[75] He went on in his next sentence to make a sympathetic noise: 'There is no doubt, that every one of these secessions is in the nature of a considerable moral and social severance'. This commiserating sagacity could not stop him making his joke about female converts as the victims of a seduction ('conquests') and male converts as enslaved ('captives'), a joke which probably did not amuse those who had undergone 'a considerable moral and social severance'.

Gladstone was not the only Prime Minister of the period to speak against the converts, indeed, he was restrained and thoughtful in comparison to Lord John Russell and Benjamin Disraeli. These men were only doing their job. They had a duty to their own careers which demanded that they express the opinion of the State Church. Newman had reason to know what that opinion was: 'Some years ago, when certain members of the Establishment were contemplating a submission to the Holy See, the Anglican prints suggested to them, that in that case their becoming course was, to quit the country for ever, and not to embarrass their friends with their presence.'[76] As Prime Ministers,

[74] A more refined occasion on which saying made it so appears in R. C. Trench's contrast of the Douay/Rheims and the King James translations of the Bible in his *English Past and Present* (1855; 4th, rev. ed., 1859), 32–4. He comments on the 'immense superiority' of the King James version, which he regards as a 'blessing' to the English nation, and he quotes 'a remarkable confession to this effect, to the wisdom, in fact, which guided [the King James translators] from above, to the providence that overruled their work . . . made by one . . . who has abandoned the communion of the English Church . . .'. He refers, in fact, to an article in the *Dublin Review*, 34 (June 1853), 441–82, which, while acknowledging the 'pure English idiom' of the 'Protestant version', describes it as 'miserably and notoriously unfair where doctrinal questions are at stake'. The article quotes F. W. Faber's remarks on the King James Bible; it is these Trench offers in an edited version as the 'remarkable confession' of divine wisdom and providence from 'one . . . who has abandoned the communion of the English Church'. Amongst the remarks which Trench omits from his citation are: 'it [the King James version] is worshipped with a positive idolatry, in extenuation of whose grotesque fanaticism its intrinsic beauty pleads availingly with the man of letters and the scholar' and, referring to the appeal of the version, 'All this is an unhallowed power!' Faber clearly repudiates the claim that the King James Bible is a 'blessing': '. . . and who would dream that beauty was better than a blessing?' The relevant passage appears on p. 466.

[75] *The Vatican Decrees in their bearing on Civil Allegiance: A Political Expostulation* (1874), quoted in *Anti-Catholicism*, 219.

[76] *Lectures on Certain Difficulties felt by Anglicans in Submitting to the Catholic Church*

they also had a duty to the 'Protestant Constitution'. When Hopkins joined the Society of Jesus in 1868, part of that constitution was 10 George IV cap. 7, *An Act for the Relief of His Majesty's Roman Catholic Subjects* (13 April 1829). Section XXXIV of the Act read:

And be it further enacted, That in case any Person shall, after the Commencement of this Act within any Part of the United Kingdom, be admitted or become a Jesuit, or Brother or Member of any other such Religious Order, Community, or Society as aforesaid, such Person shall be deemed and taken to be guilty of a Misdemeanor, and being thereof lawfully convicted shall be sentenced and ordered to be banished from the United Kingdom for the Term of his natural Life.[77]

Section XXXIV was never used to relieve any of His Majesty's Roman Catholic Jesuit Subjects of the right to live in England, but the power was formally there, and it gave Hopkins one ground for feeling no longer entirely at home in England, though he could not make anywhere else be home to him.

He made something of the feeling:

> To seem the stranger lies my lot, my life
> Among strangers. Father and mother dear,
> Brothers and sisters are in Christ not near
> And he my peace / my parting, sword and strife.[78]

It is poignant how 'Father and mother dear' may sound at first like the poeticized opening of a letter ('Dear father and mother'), as if it were going to have that 'strain of address, which writing should usually have'[79] according to Hopkins's ideal. Address has broken under strain, though; the 'Father and mother dear' turns out to be only the third-personal subject of an indicative sentence, the poem being not directly addressed to his parents or anyone else. The poignancy it has is like that of a letter that was never sent. The unspeakable gravity of what conversion meant for him can be seen in the oblique which cuts the line at 'And he my peace / my parting'. Had he put a comma there, it would have seemed as if Christ were simply and cumulatively these two distinct things (compare 'Pass me my coat, my umbrella, and my

---

(1850; repr. and rev. in 2 vols. as *Certain Difficulties felt by Anglicans in Catholic Teaching Considered*, 1879, 1876). I quote here from the text of the 1st edn., 139.

[77] Quoted in *Anti-Catholicism*, 139. Dr Norman discusses the background to this clause, Norman, 64.

[78] Phillips, 166.

[79] Letter of 20 Oct. 1887 to Patmore, *Further Letters*, 380.

gloves'); had he put a hyphen, it would have looked as if it was always Christ's nature to split families in this way (he often used a hyphen to express paradoxes which he considered permanent, especially the paradoxes of Christ's nature as 'God-made-flesh'[80]). But it was Hopkins's experience as a Catholic convert to live a life whose essence hung on a contingency, a blunt, historical fact which had assumed for most of his countrymen the status and potency of a spiritual truth: the schism from Rome. The impact of Christ on that life was then neither a revelation of the essential nature of Christ's bearing on His world, but nor was it possible to extricate the peace Hopkins found in Christ from 'moral and social severance'. The circumstances compounded the essential and the extrinsic in a manner wonderfully linked together so that it was hard to speak without becoming contentious and even keeping silence could not preserve neutrality. The '/' in 'my peace / my parting' replies to such a situation; it cannot be spoken and yet it is that in the line which most needs to be said.

He had not at first imagined that things had to be this way with his family. He had urged his father to call on Joseph, Mary, and Jesus in the confidence that 'the prayers of this Holy Family wd. in a·few days put an end to estrangements for ever'[81]—the conjunction of 'in a few days' and 'for ever' hopes for too much too quickly, as he was to learn. In the poem, he sustains an instant of hope through the ambiguity which arises from a poetic inversion: 'Father and mother dear, / Brothers and sisters are in Christ' says, under its breath, as it were, before the impetus of syntactic completion takes the hope away, that his relations *are* 'in Christ', though Hopkins knew that orthodox Catholic doctrine on the absence of salvation outside the Church had by some Catholics been interpreted as contradicting such hopes. The syntax of the hope approaches spoken English, the curb to the hope turns the poem's syntax away from speech: 'Brothers and sisters are in Christ not near' (rather than the more idiomatic 'Brothers and sisters are not near in Christ')—the line makes a tentative approach to a returned languge of domestic ease, but the words for a rapprochement are askew and escape from what he might personally like to say, turning into a poetic conventionality, a discipline beyond him, which breaks up the nearest and dearest by laying stress through the rhyme on the fact that the 'dear' are 'not near'. The poeticality of the line then flushes with just that impetus to plain speaking which it thwarts. 'Near

[80] 'Yes. Why do we all . . .', Aug. 1885, Phillips, 168.
[81] Letter of 16 Oct. 1866, *Further Letters*, 94.

and dear', 'nearest and dearest'—when he wrote this sonnet, these had long (for at least a century) been common reflexes of the language to refer to one's family. Tennyson has a trochaic lilt on the rhyme in his 1866 poem 'The Victim': 'Were it our nearest, | Were it our dearest'.[82] The rhyme has behind it an amassed force of habitual usage which presses on Hopkins's lines, which they long to accommodate, but which they turn from as the emphatic, alliterating 'not' snatches away the chime of 'near' back to 'dear'; the proximity in sound of the words makes only the more resonant the fact of separation. Such moments recur in the mature Hopkins, moments which gauge the distance of his poetry, his created self, from the habitual readiness of the English language to a native speaker's hand and mouth, moments at which across that elected distance he recovers the possibility of a distinct utterance.

The language stands for his family, and he can do in it what he could not do for them—bring it over to himself without fracturing its identity. His style here is not 'at home with' the English language, nor is it homely, but it 'comes home', though Hopkins's sense of where his home lay, in his art as in his calling, was not with his parents or with their ideas. It was where all surrenders come home: 'As we drove home the stars came out thick: I leant back to look at them and my heart opening more than usual praised our Lord to and in whom all that beauty comes home'.[83] His style lies, with minute consistency, along that oblique, unvoiceable line between 'peace' and 'parting', a line which marks the substance of his plight and the grounds of his achievement.

Newman too in his later writing went far from 'home' (in 1832, just before the Tractarian movement began, he had written a poem, entitled 'Home', about the Anglican Church[84]). In his prose, as in Hopkins's work, the word, especially in the phrase 'come home', has a sense of loss in it, for example when he writes about how time makes lines of verse which had seemed commonplace to the boy who first learned them touch the older man who reads them again and anew:

Passages, which to a boy are but rhetorical common-places, neither better nor worse than a hundred others which any clever writer might supply, which he gets by heart and thinks very fine, and imitates, as he thinks, successfully, in his own flowing versification, at length come home to him, when long years have

---

[82] R ii. 695.
[83] Journal entry for 17 Aug. 1874, *GMH Journals*, 254.
[84] See 'Home', dated 11 Nov. 1832, in *Verses on Various Occasions*, UE 62.

passed, and he has had experience of life, and pierce him, as if he had never before known them, with their sad earnestness and vivid exactness.[85]

In his own halting and accumulating prose, whose syntax is kin in its quiet way to Hopkins's vehement string and stumble of possessives, Newman conveys the time he talks about through his rhetorical suspension of 'Passages . . . at length come home to him . . . and pierce him'. It is one of his finest sentences, and its polish makes a form of retort of the imagination on troubles he had been through himself, as he moved from the schoolboy to the brilliant don, on a path of apparently indeflectible success in unworldliness, to the exile in 'Brummagem', convicted of libel, ignored or thwarted by many powerful co-religionists, gaining thereby what he mildly calls here 'experience of life'. The Church of England had been 'home' to him but he knew with bitter clarity in his mature work that 'Real assent, then, as the experience which it presupposes, is proper to the individual, and, as such, thwarts rather than promotes the intercourse of man with man. It shuts itself up, as it were, in its own home . . .'.[86]

It was not only a loss to live in such division from the language which had been, and still remained in a sense, one's home. There were divisions within that language, such as those which sprang from the different rhythms of work of its people, to which religious estrangement could lend a creative ear, and which men like Newman and Hopkins might otherwise not have much heeded. In *Loss and Gain*, the declared benefit of conversion is religious, but Newman also cuts a stylistic zest from the linguistic displacement of the Catholic, a style of 'retort' which is less gravely abstemious than Christ's 'Thou sayest it' but deals in its own spikey way with the imagination of the world.

Sheffield, the friend of the man who eventually converts, marvels at the earlier conversion of one of their acquaintance:

'. . . the idea of his swallowing, of his own free will, the heap of rubbish which every Catholic has to believe! in cold blood tying a collar round his neck, and politely putting the chain into the hands of a priest! . . . And then the Confessional! 'Tis marvellous!' and he began to break the coals with the poker. 'It's very well,' he continued, 'if a man is born a Catholic; I don't suppose they really believe what they are obliged to profess; but how an Englishman, a gentleman, a man here at Oxford, with all his advantages, can so eat dirt, scraping and picking up all the dead lies of the dark ages—it's a miracle!'[87]

Sheffield's metaphors run wild with indignation: 'swallowing' is a

[85] *Grammar*, 56–7.   [86] Ibid. 60.   [87] UE 117–18.

cliché for believing something absurd, but it does not fit well with a
'heap of rubbish'; 'tying a collar round his neck' sounds as if it is
something which an Oxford man ought always to do with *sang-froid*,
though not perhaps 'in cold blood', until Sheffield manages to make it
clear that he means a slave's collar. His sense of the slight to his own
dignity offered by this conversion reveals itself in the precise, upward
gradation of 'an Englishman, a gentleman, a man here at Oxford', for
Sheffield knows exactly when and especially where he's well off. The
comic insight peaks at 'it's a miracle'. The word has a different sense
for a Catholic who believes that miracles still happen and for a
Protestant who confines their occurrence to the primitive ages of
Christianity. Sheffield rests so familiarly in the language of his
upbringing that he cannot hear how his words would sound to the
Catholic whose conversion he wonders at—as the simple truth about
the miraculous operation of grace in adult conversions. Newman did
not make up this attitude to miracles, though he played it to the hilt.
Charles Kingsley mistook, as Sheffield does, his own deafness to the
historical range of the language for a brilliant rhetorical clinching when
he protested, after quoting from the accounts of miracles in the saints'
Lives which Newman oversaw through the press while still an
Anglican: 'I can quote no more. I really must recollect that my readers
and I are living in the nineteenth century.'[88] Kingsley's rhetoric has no
idea of what it is trying to match because it has no ears for other ways
of speaking English. (Others apart from Newman, and with different
religious allegiances or no religious allegiance at all, felt this deficiency
in Kingsley. Indeed, the great success of the *Apologia* testifies to a
determination across all credal divisions that Newman should be given
the fair hearing that Kingsley had denied him.[89]) Such deafness runs
big risks, including the risk of identifying God's sense of history with
Charles Kingsley's own imagined sensitivity to the *Zeitgeist*. Newman's
joke about Sheffield and 'it's a miracle' faces dense self-assurance
such as Kingsley's and outfaces it. He managed to be witty and adroit
in the satirical polemic he was called to write, but the undertow of
regret and solicitude in his writing should not be missed.

The point of that writing is clear:

I wish to deprive you of your undue confidence in self; I wish to dislodge you
from that centre in which you sit so self-possessed and self-satisfied. Your

---

[88] Kingsley, 'What, then, does Dr. Newman Mean?' (1864), reprinted in Svaglic, 368.
[89] See Svaglic's account of reaction to the Kingsley/Newman controversy, in Svaglic,
pp. xxii–xxv, xliii–xlvii.

fault has been to be satisfied with but a half evidence of your safety; you have been too well contented with remaining where you found yourselves . . . Learn, my dear brethren, a more sober, a more cautious tone of thought. Learn to fear for your souls. It is something, indeed, to be peaceful within, but it is not everything. It may be the stillness of death.[90]

Here, the Newman who had clung to his 'line' among 'educated persons' and flinched from a 'Brummagem' which he thought 'no centre' finds a different angle on that centre. The vicinity of 'self' ('self', 'self-possessed', 'self-satisfied', 'yourselves') to 'centre' starts the possibility that the Established confidence Newman delineates is primarily a form of self-centredness. To be 'self-centred' was in the language at that time not necessarily a bad thing but the word was beginning to gather round itself those pejorative shades which imply the narrow minds and short sight of the merely selfish. The cultural centrality in England of the Anglican Church which seemed one demonstration of its supernatural sanction in the eyes of its ministers and adherents might—such is the drift of Newman's sceptical poise— amount to no more than a gratifying reflection of worldly success with the attendant self-esteem. The 'centre', then, in which the self reposed, on which it relied, would shrink to no more than that self itself; the Anglican Church would be made in the image of its members.

Newman breaks gravely in on the comforts of the Anglican 'Home' but the calm admonitions of this prose do not convey how troublesome it was for Newman to stand apart and in judgement like this, even though the stance was also, he felt, his vocation. The trouble it involved could be put briefly by saying that he was a conservative, as Hopkins was, and that being a rebel for a cause didn't entirely suit him. He was not sufficiently adept at picking up new ways for the taste of other Anglican converts who rose more swiftly to official heights in the Church of Rome than he did, a lack of knack which Stephen Prickett well describes when he notes that 'In contrast to Manning after his conversion, Newman remained emotionally and aesthetically within the tradition which had brought him to Catholicism—the world embraced by the Anglican sensibility of Coleridge and Keble. He remained within a way of feeling that separated him even from many of his fellow converts.'[91] A remark of Cardinal Manning's sufficiently confirms Prickett's judgement; the Cardinal detected something

---

[90] *Anglican Difficulties*, UE i. 95.
[91] Prickett, *Romanticism and Religion*, 170.

dangerous in Newman: 'It is the old Anglican, patristic, literary, Oxford tone transplanted into the Church. It takes the line of deprecating exaggerations, foreign devotions, Ultramontanism, anti-national sympathies. In one word it is worldly Catholicism.'[92] The incautious 'tone of thought' against which Newman warned those he had left behind in the Establishment was held against him as 'the old Anglican . . . Oxford tone' persisting in his own work. Manning's jumble of complaint serves to show that the incomprehension Newman met in the Church of his adoption matched that which he had encountered in the Church he left. The character of that suspicion and misunderstanding also remained constant. Within the Church of Rome too, the messy conglomerates of social habit with strict doctrine resulted in that metaphysical vamping-up of contingent features into essentially significant facts which Newman himself was so expert at diagnosing in the 'Prejudiced Man'. That Manning's catalogue of Newman's supposed failings is a jumble is the most revealing thing about it. 'Old Anglican, patristic, literary, Oxford tone'—the phrase is woven from a tissue of assumptions: 'old Anglican'—Newman used to be an Anglican and still is one at heart; 'Anglican, patristic'—people who study the Fathers need watching because they probably harbour dangerous notions about the authority of the Church; 'patristic, literary'—Newman's attachment to the Church is only that of a scholar and an aesthete, and such intellectuals are notoriously unreliable. 'Oxford', in this context created with all the ingrained artfulness which a prejudiced mind naturally commands, may be regarded as the talismanic synthesis of these (and other) terrors implicit in Manning's phrase. Prickett's judgement on Newman's encounter with such suspicions is sympathetically acute, particularly when he notes a conflict between Newman's cultural inheritance and his religious allegiance, a conflict Hopkins shared.

This conflict produces in both writers an uncertainty about the role and value of habit in human life. The uncertainty may appear at many levels—in Hopkins's creation of idiomatic and remote inflections for the same phrase, or explicitly in Newman's attempt to articulate a philosophy of belief. In the *University Sermons*, Newman's criterion for distinguishing between faith and superstition is that faith 'is kept from abuse, e.g. from falling into superstition, by a right moral state of

---

[92] E. S. Purcell, *Life of Cardinal Manning, Archbishop of Westminster*, 2 vols. (1896, repr., New York, 1973), ii. 323.

mind'.[93] 'A right *moral* state of mind' (my emphasis) must be something more than a state in which one believes only true propositions, because one might hold only true propositions but hold them for bad—in the sense either of wicked or of weak—reasons. It must also be something less than a state in which one believes only true propositions, because if Newman distinguishes faith from superstition merely by saying that faith holds only what is true, then the distinction is nugatory, tells us nothing about faith, except (as Newman cannot have believed) that no individual member of the faithful can ever err in any of his beliefs, a consequence which simply collapses the distinction between faith and superstition. This right moral state of mind consists rather of the conduct of the intelligence than of its contents, and that conduct is formed through habituation to the traditional practice of real communities: 'Texts have their illuminating power, from the atmosphere of habit, opinion, usage, tradition, through which we see them'[94] and 'Most men must and do decide by the principles of thought and conduct which are habitual to them . . .'.[95] John Coulson sums up this aspect of Newman's thought very well when he writes that 'For Newman . . . the way to certitude in matters of belief is not only linguistic (and therefore theological), but social (and therefore ecclesial)'.[96] Newman continues this ethical approach to beliefs in the *Grammar of Assent*, which equally sets out to show 'that a right moral state of mind germinates or even generates good intellectual principles',[97] though the later work tries to make a distinction between real assent to a religious doctrine and what Newman calls 'credence'.

Credences are, roughly, the generally received ideas and patterns of thinking of a historical community in which we happen to find ourselves. They range, for example, for modern English people, from a conviction of the virtue of parliamentary democracy as a political system to the joke that it never stops raining in Manchester. Newman writes eloquently of them:

They give us in great measure our morality, our politics, our social code, our art of life. They supply the elements of public opinion, the watchwords of patriotism, the standards of thought and action; they are our mutual understandings, our channels of sympathy, our means of co-operation, and the

[93] Preface to the 3rd edn. of *University Sermons*, UE, p. xvii.
[94] Sermon on the Feast of the Epiphany, 1839, *University Sermons*, UE 191.
[95] Ibid. 227.
[96] *Newman and the Common Tradition* (Oxford, 1970), 62.
[97] Letter to H. J. Coleridge, 5 Feb. 1871, *Letters and Diaries of J. H. Newman*, xxv. 280.

bond of our civil union. They become our moral language; we learn them as we learn our mother tongue; they distinguish us from foreigners; they are, in each of us, not indeed personal, but national characteristics.[98]

In so far as credences have this supra-personal aspect, they make up the fibre of any possible community, social or ecclesial; they also link each self to its community in the manner of Hopkins's remark that the self consists of 'a centre *and* a surrounding area or circumference'. Credences are, as it were, the radii of a self in a community.

Something is missing from Newman's list of what credences help us understand—creeds. Whatever the cogency of his account of belief in the *University Sermons* and in the *Grammar* (it remains in many essentials unchanged), whatever the appeal of the eloquent piety he invests in that account, it has difficulty explaining or justifying his own conversion, and conversion in any other case. For, of its very nature, the conversion of an Anglican to the Catholic Church entailed a conflict of communities and a friction of credences. There are meta-credences held within particular credential groups about their differences from other groups (the English are more 'empirical' than the French, for example), but this does not help the case where a rational change of credence is in question. If a right moral state of mind is required to distinguish true from false beliefs, and such a state can develop only within actual communities with their patterns of credence, then what can be the rationale of a process in which the convert repudiates so much that he has inherited and declares central elements of what the community holds as truth to be false? More than the outgrowing of a few peripheral *idées reçues* went on in a conversion such as Newman's, for he altered not only major religious beliefs, major historical beliefs, but also produced a critique of how English people arrived (or thought they arrived) at their beliefs, of what the English took 'believing' to be. That is why Newman needed to distinguish between credence and real assent to religious propositions. *Pietas* about credences could no longer be identified with credal piety, if only because amongst the credences of many English people there were several notions about Catholics and Catholicism which were not easy for an English Catholic to assent to, such as the notion that Catholicism is essentially un-English. As R. C. Trench put it, 'It is the anti-national character of the Roman Catholic system which perhaps more than all else revolts Englishmen . . .'[99]

---

[98] *Grammar*, 42.   [99] *On the Study of Words* (1851; 17th, rev. edn., 1878), 318.

Victorian Catholics could appeal against Trench's credence to a past in which England had been Catholic for a thousand years, and this appeal was popular with those whom Manning would have characterized as sharing Newman's 'patristic, literary' tone, with Pugin and Hopkins, for example. Newman made such an appeal himself in his famous sermon at the first Provincial Synod of Westminster in 1852. He is imagining what a Catholic who died before the Synod might have said of the 'Catholic Revival': '. . . I listen, and I hear the sound of voices, grave and musical, renewing the old chant, with which Augustine greeted Ethelbert in the free air upon the Kentish strand.'[100] He would have had to have very good ears. The problem with such appeals is not just that they are doomed to be merely revivalist, that they are neo-Gothic in their sense of history (as shows in Newman's tug on the old 'Saxon liberties' string with 'in the free air' and the linguistic archaising of 'Kentish strand'), but that they depend on the belief that the 'right moral state of mind' necessary to become a Catholic (assuming with Newman after his conversion that the Catholic faith is the true faith) can be acquired by those who have also acquired their credences, which, as Newman said, 'become our moral language', in a country where 'everyone speaks Protestantism'. Such moral health seems to be picked up in a religious atmosphere somehow free of the gravitational drag of an empirical culture in which the individual has come to himself. But a 'moral state of mind' without a fitting 'moral language' is a phantom.

The reason why Newman could not clearly establish the value of habit within his systematic account of belief, crucial though the question of its value was to him, is that the pros and cons of the debate lie at a level other than that at which conceptual logic works. He wrote in the *Grammar* that the essential aim of logic is to treat words irrespective of the variety of their usages and, therefore, irrespective of that semantic multiplicity which can accrue to a word as it passes through many—and sometimes mutually contradictory—applications:

Words, which denote things, have innumerable implications; but in inferential exercises it is the very triumph of that clearness and hardness of head, which is the characteristic talent for the art, to have stripped them of all these connatural senses, to have drained them of that depth and breadth of associations which constitute their poetry, their rhetoric, and their historical life, to have starved each term down till it has become the ghost of itself, and everywhere one and the same ghost . . .[101]

[100] *Sermons preached on Various Occasions* (1870), UE 175.   [101] *Grammar*, 174.

If that is so of logic, then it will be impossible to produce a logic *of* the 'connatural senses' in words, but that is just what is required for a logically consistent answer to the question of the value of habit in human thinking, and especially with reference to the transforming retort made on habit in a conversion. Newman's 'the ghost of itself, and everywhere one and the same ghost' reads like a parody of the canon of Vincent of Lérins which he had, as an Anglican theologian, taken as the criterion of true Christian teaching as distinguished from inauthentic developments: *quod ubique, quod semper, quod ab omnibus creditum est* ['that which everywhere, that which always, that which by everybody has been believed']. The older Newman, having gained 'experience of life', remarked drily: 'If we assume nothing but what has universal reception, the field of our possible discussions will suffer much contraction . . .'.[102]

The recognition that the Vicentine canon is a logical phantasm, dimly though such recognition is made in the *Grammar*, permits us to see the other side of Newman's thought and rhetoric, that side which is urgent for change as a mark of the spirit, fearing that habitual piety may be the 'stillness of death', which in fact led him to conversion, and which celebrates conversion, drawing its energies from a retort upon the moral language of learned credences. The *Essay on Development* adopts a principle which controverts both the Vicentine canon and the identification of our 'moral language' with the 'mother tongue': 'The idea which represents an object or supposed object is commensurate with the sum total of its possible aspects, however they may vary in the separate consciousness of individuals; and in proportion to the variety of aspects under which it presents itself to various minds is its force and depth, and the argument for its reality.'[103] This principle sets the

[102] Ibid. 83. The closeness of an ideal, logical simplicity and consistency of usage as described in *Grammar* to the Vicentine canon is clearer still in the Latin tag Newman attaches to the passage I quote from *Grammar*, 174: 'everywhere one and the same ghost, "omnibus umbra locis" ', where 'omnibus . . . locis' echoes the Vicentine 'ubique', and may recall, through acquired, connatural senses of 'ghost' and 'ubique', the Ghost in *Hamlet* which moves '*Hic & ubique*', *Hamlet*, I. v. 164; Folio l. 853.

[103] UE 34. The first edition of this crucial passage reads: 'An idea ever presents itself under different aspects to different minds, and in proportion to that variety will be the proof of its reality and its distinctness.' The difficulty Newman has in articulating the logic of connatural senses shows clearly in the rough joint here of 'reality', a term of Newman's art, with 'distinctness', a Cartesian inheritance, as also in the slide between the two passages of phrases like 'sum total' and 'in proportion', and the substitution of 'force and depth, and the argument for its reality' in UE for the balder 'proof of its reality' in the first edition. This is one of those occasions when troubled language attests to integrity of thinking.

diversity of the 'connatural senses' of words at the centre of an attempt to distinguish between authentic and inauthentic doctrines (and therefore, between faith and superstition), and it does so at the cost of rendering theology's relation to logic fraught and obscure, for Newman's 'idea' is not the logician's 'concept' and nor is the 'reality' of an idea the truth of a proposition or the validity of an argument, as a logician might understand those terms. Indeed, such adjustment in the understanding of logical inference had to be made from the very nature of his attempt to explain how a man might acquire, let alone change, his religious convictions. If religious beliefs were formed by irrefutable argument from incontestable premisses, then everyone able to follow the reasoning would arrive at the same conclusion, unless inhibited from doing so by self-interest or some other concealed motive (a motive which in a matter of this importance could not but be 'unworthy'). Though Newman believed that he had good reasons to convert to Rome, he could not attribute to mental deficiency the failure of, for example, Keble and Pusey to follow him and his reasons there, and he did not want to ascribe simple baseness of motive to them. The need to account for his own conversion and, at the same time, to do so in such a way as to show that he could have grounds for absolute conviction while others could with integrity fail to share those grounds, is the generative need of much of his work after 1845 and reaches its climax in the *Grammar*.

The kind of thinking Newman's attention to the density of aspect in an 'idea' requires for its application resembles less the stripping of senses Newman attributed to the logician and more the literary activity of understanding and mutually adjusting the various planes on which a poem simultaneously works its meaning—its relation to previous poems by other poets, its rhythmic patternings, its syntactic bearing on the norms of spoken and written syntax in the language, and so on. Theological thinking becomes a kind of wit, as Eliot defined it: 'It involves, probably, a recognition, implicit in the expression of every experience, of other kinds of experience which are possible . . .'.[104] In this approach to the conditions of literary creativity and response, Newman finds a way to place his conversion from his emotional and aesthetic inheritance of credences within that tradition of habits while simultaneously working imaginative retorts on the habits so as to elicit a moral language for his new right moral state of mind. This might

[104] 'Andrew Marvell' (1921), repr. in *Selected Essays*, 303.

strike a logician, following the ideal of logic Newman decribed, as a bizarre activity, one which is perhaps not even strictly feasible, but if we think of the elements of derivation, parody, reversal of values, discovery of new potential in old techniques, as a literary style develops from an author to his successors, we will not find Newman's blend of affinity and critical remoteness strange.

Newman identified a weakness in the strength of the habitually virtuous man who had never ignored the voice of conscience; such a man 'will of all men be least able (as such) to defend his own views, inasmuch as he takes no external survey of himself'.[105] Not that Newman therefore recommended occasional deviations from virtue as a salutary fostering of one's capacity for wit, but the ability to take an 'external survey' of oneself is for him a crucial gain to balance the loss of certainty which comes with the breaking of habits. Newman claims the privilege of judging the issue between Protestants and Catholics better than a Protestant can because he has 'stood on their ground; and would always aim at handling their arguments, not as so many dead words, but as the words of a speaker in a particular state of mind, which must be experienced, or witnessed, or explored, if it is to be understood',[106] and Hopkins makes the same claim for the convert's 'other-man's-point-of-viewishness': 'And surely it is true, though it will sound pride to say it, that the judgement of one who has seen both sides for a week is better than his who has seen only one for a lifetime'.[107] The importance of experience or imagination in under-standing the terms of debate stems from the unpredictable and logically elastic way in which connatural senses cluster around words. The very contingencies of social and historical existence which generate these senses also prevent us from simply inferring what they are like or how they are disposed about a word from an analysis of a word itself. To understand 'what X means by y' is, as I argued in the first chapter, always essential to interpretation because an adequate conception of agency in utterance is necessary for a proper account of that intentionality from which meaning derives. The need for such understanding of words 'as the words of a speaker in a particular state of mind' grows more acute when the speech is passing between antagonists in debate, especially debate as radical and searching as that between Anglicans and converts from Anglicanism. Newman's claim

---

[105] Sermon of 22 Jan. 1832, *University Sermons*, UE 83.
[106] *Present Position*, UE 343.
[107] Letter to Manley Hopkins, 16 Oct. 1866, *Further Letters*, 93.

is, essentially, that he possesses, and strives consistently to exercise, a dramatic imagination in argument; he situates the voice of discussion dramatically as poets such as Wordsworth or Hopkins let the lyric voice be heard dramatically.[108] Such a claim needs an answering style to give it substance, and a principal element of such a style—in Newman's dramatic arguments as in the dramatized lyric—is the printed voice with its ability to evoke the voices of a world of speech which the writing has abandoned without wholly reverting to their accent or succumbing to their pressure. The printed voice grants the opportunity for an 'external survey' of those states of mind in which one is so at home that they scarcely show as anything more than passing but habitual inflections.

When the Society of Jesus sent Hopkins to Ireland, they put on the map for him a condition of exile he already knew. The opening lines of 'To seem the stranger . . .' tell the loss, the rest of the poem goes on to discover the gain in that loss, a gain he described to Patmore as the taking of an 'external survey' of English political self-confidence: 'It is good to be in Ireland to hear how enemies, and those rhetoricians, can treat the things that are unquestioned at home':[109]

> England, whose honour O all my heart woos, wife
> To my creating thought, would neither hear
> Me, were I pleading, plead nor do I: I wéar-
> Y of idle a being but by where wars are rife.
>
> I am in Ireland now; now I am at a thírd
> Remove. Not but in all removes I can
> Kind love both give and get. Only what word
>
> Wisest my heart breeds dark heaven's baffling ban
> Bars or hell's spell thwarts. This to hoard unheard,
> Heard unheeded, leaves me a lonely began.

Given Hopkins's nationalistic ardour, calling England the 'wife' to his thought was more than a colour of rhetoric; a professed celibate, he used the word 'wife' with some emphasis. He wrote on the occasion of Bridges's marriage, 'I have a kind of spooniness and delight over

---

[108] A friend has tartly pointed out to me an objection to the claim that one who has seen both sides of the question is in a better position to judge of the matter: 'So it is equally true that a lapsed Catholic can better judge of these issues than someone who has remained a Catholic?'

[109] Letter of 4 June 1886, *Further Letters*, 367.

married people, especially if they say "my wife", "my husband", or shew the wedding ring"[110]—there is a touch of wit too in his guess at how his enthusiasm might look and sound to others, when he calls his vicarious tenderness 'spooniness'. How do you speak the patriotic devotion which concentrates in 'O' at 'whose honour O all my heart woos'? It may be 'England whose honour O' with a jaunty, carousing tune ('whose honour-O!') or with a sigh on the 'O' as he faces the way England's honour would have sounded to Irish 'enemies, and those rhetoricians' during the Home Rule Crisis (and maybe with an even less utterable despondency if the line faces how England's honour may have *looked* to some of the Irish—as an 'O', a zero). Or the 'O' may be attached to 'all my heart' rather than to 'whose honour', so that it gives a personal verve of dedication in spite of all that is said against England, in spite even of Hopkins's own estranging doubts, his estranging faith.

The vow of celibacy could look like the sour grapes of a rejected suitor; the England he wooed would not hear him. He was bound to sympathize with the Irish as a Catholic people, yet he disliked the nationalist agitation; he was a faithful servant of the Church but gravelled by the involvement of brother-priests with illegal organizations in the cause of Irish independence. The grinding of religious on political loyalties—'thóughts agáinst thoúghts ín groans grínd' and 'We hear our hearts grate on themselves'[111]—works into his verse, into its endured torsions, its composed recoil: 'pleading, plead', 'I: I', 'now; now'. The richness of the euphony that he creates in the internal chimes of his lines looms up against a reader-aloud as a thicket of vocal hurdles, of impacted felicities which baffle themselves and make euphony itself a dilemma for the voice—is one to try to preserve the flow or to bring out these small eddies of the sounds in the verse? There is, for example, almost no possibility of elision between words in the closing lines; the verse compels the voice to make a considerable severance between the words if they are to be distinctly audible. Try saying 'hell's spell': the discrimination asked between voiced and unvoiced 's' gives you pause. The style is remarkably achieved as a supreme dexterity which is the opposite of fluent; it is truly a cutting and straining of the windpipe. You are asked to feel the stretching of

---

[110] 11 Nov. 1884, *GMH/Bridges*, 198.

[111] 'Spelt from Sibyl's Leaves', 1886, Phillips, 175; 'Patience, hard thing . . .', ?1885–6, ibid. 170.

grammatical units across line-ends both as a racking and as a perseverance, to see the lineation as shredding the continuity of phrases, collapsing the verse into wrecked utterances like 'Only what word', cries which search out an eloquence and then, around the line-end, are given hardly foreseeable sustenance.

The writing, though, contains these implicit requests for hearing and vocalization in order to convey the fact that these are things Hopkins says to no human being ('this to hoard unheard') and that when they are heard by God in the silence of His knowledge, they appear to be 'unheeded'; the poem is a supplication for hearing rather than an oratorical performance before an audience of whose attention the poet is sure. The cadence of its final line, that extremely beautiful sound of longing and anticipation which he procures by supplying the English with a new noun, 'a . . . began' (something which was begun in the past, his discipline of self-perfection, but which has lost the novel charm a 'beginning' might have for him), feels the texture of religious waiting, its tremors and its impatiences held within the security of a pattern whose completion is trusted in but which will settle itself not quite as might have been expected, just as the closing rhyme of 'ban'/ 'began' in this sonnet, rhyming a monosyllable and a disyllable, stays slightly out of kilter, off-key.

Both an English patriot and a Catholic priest, Hopkins found himself persistently vexed and troubled. He thought it better not to recognize fully how even his closest literary friend was out of sympathy with him: 'And indeed how many many times must you have misunderstood me not in my sonnets only but in moral, social, personal matters! It must be so, I see now. But it would embitter life if we knew of the misunderstandings put upon us; it would mine at least.'[112] Some of the relevant sense of the intimate buffeting he subjected himself to in his vocation comes out in considering the exchange between Kingsley and Newman which led to the composition of the *Apologia*. Their argument about Catholicism turns on and then into a debate about Englishness. Kingsley accuses Newman, it sometimes seems, not so much of bad reasoning as of treason; both men appeal to the intuitive decency of the English gentleman to confirm their contrary assertions, though in Newman's appeal we can detect the sardonically mild tones of one who has undergone a 'considerable moral and social severance', as when he mock-deprecates his

---

[112] Letter to Bridges, 26 Mar. 1883, *GMH/Bridges*, 177.

membership of a 'most un-English communion'.[113] Kingsley writes
with affability and a show of fair-mindedness to deny the charge that
the English are prejudiced against Catholics: 'If there is (as there is) a
strong distrust of certain Catholics, it is restricted to the proselytizing
priests among them; and especially to those who, like Dr. Newman,
have turned round upon their mother-Church (I had almost said their
mother-country) with contumely and slander.'[114] This is very talented
writing. Kingsley makes a frank admission—'If there is (as there is)'—
and so appears an honest man, but he still retains, in his next
parenthesis, the advantage of slick insinuation—'their mother-Church
(I had almost said their mother-country)'—where the assimilation of
mother-Church and mother-country shows a mind closed to the
questions it pretends, in all honesty, to be raising. Kingsley's prose is
adeptly slack for his own ends rather than careful to do justice to
others. 'Proselytizing priests' with its alliterative thwack is not the
language of impartial consideration, as the *OED* makes clear when we
observe there that 'proselytize' really comes into use in the English
language after the Reformation as a term of abuse for what Catholics
would have thought of as an act of missionary reclamation. As for
'priest', Newman himself had already well described the effect of the
Reformation on that English word (except, as the *OED* notes, in the
North of England):

Into your very vocabulary let Protestantism enter; let priest, and mass, and
mass-priest, and mass-house, have an offensive savour on your palate; let
monk be a word of reproach; let Jesuitism and Jesuitical, in their first intention,
stand for what is dishonourable and vile. What chance has a Catholic against
so multitudinous, so elementary a Tradition.[115]

Newman wrote of Kingsley with an admirable poise of condolence and
satiric bemusement that 'He appears to be so constituted as to have no
notion of what goes on in minds very different from his own. . .'.[116]
Nice and quiet, the hint of relief along with the regret with which
Newman signals that his mind is 'very different' from Kingsley's.

Newman's *Lectures on the Present Position of Catholics in England*
provide an unsolicited testimonial to the profound representativeness
of Kingsley's mind. The man who is not a philosopher, Newman
writes, by which I think he means—as Baudelaire did in his essay on

---

[113] Svaglic, 404.
[114] Kingsley, 'What, then, does Dr. Newman Mean?' repr. in Svaglic, 365.
[115] *Present Position*, UE 73.   [116] Svaglic, 387.

laughter[117]—a man who cannot take an 'external survey' of himself, 'despises other men, and other modes of opinion and action, simply because he does not understand them. He is fixed in his own centre, refers everything to it, and never throws himself, perhaps cannot throw himself, into the minds of strangers, or into a state of things not familiar to him.'[118] Such inability to do anything but see and hear others as you see and hear yourself baffles Catholic speech by producing 'a tradition of nursery stories, school stories, public-house stories, club-house stories, drawing-room stories, pulpit stories; . . . a tradition of selections from the English classics, bits of poetry, passages of history, sermons, chance essays, extracts from books of travel, anonymous anecdotes, lectures on prophecy, statements and arguments of polemical writers, made up into small octavos for class-books, and into pretty miniatures for presents;—a tradition floating in the air; which wc found in being when we first came to years of reason; which has been borne in upon us by all we saw, heard, or read . . .'[119] This list may include an implicit confession from Newman that he had been himself part of this tradition, for 'lectures on prophecy' may refer to his own major anti-Roman work, the *Lectures on the Prophetical Office of the Church* . . . .[120] It is in and athwart the 'vocal life'[121] of such a 'tradition floating in the air' that Hopkins and Newman work for their eloquence.

That eloquence often takes the form of a thrumming accumulation of indignities) suffered (as in Newman's 'nursery stories, school stories . . .' etc.) in imitation of the insistent weight of the anti-Catholic clamour, a weight Manning described: 'monstrous absurdities . . . are

---

[117] 'Ce n'est point l'homme qui tombe qui rit de sa propre chute, à moins qu'il ne soit un philosophe, un homme qui ait acquis, par habitude, la force de se déboubler rapidement et d'assister comme spectateur désintéressé aux phénomènes de son *moi*', in his 'De l'essence du rire . . .' (1845), reprinted in *Curiosités esthétiques* . . . , ed. H. Lemaître (Paris, 1962). [The man who trips over never laughs at his own fall, unless he is of a philosophical turn of mind, a man who has developed, by practice, the ability swiftly to conceive himself as two people at once and so to be present, as a detached spectator, at the experiences of his 'self'.]

[118] *Present Position*, UE 6.

[119] Ibid., UE 88.

[120] Newman would have done well more explicitly to admit that such traditions float in Catholic air too, as he later found to his cost when attempting to establish a university college in Dublin in an atmosphere of ingrained Irish mis-taking such as that described by Hopkins—'how enemies, and those rhetoricians, can treat the things that are unquestioned at home'—or again, in the gruelling quarrels between the Birmingham and London Oratories.

[121] *Present Position*, UE 366.

repeated day by day, as by the monotonous revolution of a mill wheel, which perpetually discharges the same noisy flood. It is of no use to expostulate, to correct, to refute; over and over again, sometimes with a variation of phrase, oftener in the very same words, the same absurdities are poured over us'.[122] Instances of the 'noisy flood' came in *The Times*'s reports of the first Vatican Council, of which Edward Norman remarks that they were 'astonishing for their inaccuracy. The newspaper had sent, as its special correspondent, Thomas Mozley, an Anglican parson (who was also Newman's brother-in-law). He could speak neither Italian nor French.'[123] It cannot have been a great consolation to Newman to receive so graphic a demonstration of the accuracy of his complaints about prejudice in the air from within his own family. Bishop Ullathorne, who was in Rome for the Council, noted, 'I suppose England will believe all this, and it will become part of the Protestant tradition.'[124] Of course Englishmen believed it, they had read it in *The Times*, well-head of credences.

An exemplary retort on such a tradition comes in Newman's jangling catalogue of responses to the re-establishment of the English Catholic Hierarchy in 1850:

Not by an act of volition, but by a sort of mechanical impulse, bishop and dean, archdeacon and canon, rector and curate, one after another, each on his high tower, off they set, swinging and booming, tolling and chiming, with nervous intenseness, and thickening emotion, and deepening volume, the old ding-dong which has scared town and country this weary time; tolling and chiming away, jingling and clamouring and ringing the changes on their poor half-dozen notes, all about 'the Popish aggression', 'insolent and insidious', 'insidious and insolent', 'insolent and atrocious', 'atrocious and insolent', 'atrocious, insolent and ungrateful', 'ungrateful, insolent and atrocious', 'foul and offensive', 'pestilent and horrid', 'subtle and unholy', 'audacious and revolting', 'contemptible and shameless', 'malignant', 'frightful', 'mad', 'meretricious' . . .[125]

The comic imagination harmonizes with outrage, outrage at persecution of the ears by bells, bells of the churches now Anglican, whose reverberations swallow all the air around you so you can scarcely hear yourself speak, tradition massively incumbent on the air. Newman can pierce the din he sets up in his prose to make the joke of 'ringing the changes' play against 'the old ding-dong', to present through the

---

[122] *The Temporal Power of the Pope in its Political Aspect* (1866), in *Anti-Catholicism*, 190.
[123] Norman, 308.
[124] Quoted in Norman, 309.   [125] *Present Position*, UE 76–7.

rhythms he fashions from this opposing uproar 'rector and curate' as themselves the bells 'swinging and booming', the masters of the English clerical atmosphere becoming in his representation of them the tools and victims of their own mechanical clangour. He converts the tune because he does not call it. The passage gives a perfect image of Newman's deprivation in the English tongue, and in being a perfect image it masters deprivation, makes itself eloquent out of 'the shut, the curfew sent', finding a language for his moral state of mind and, what is more, letting people hear themselves through an imaginative retort.

Something more than a counter-argument, a counter-music: 'For an abstraction can be made at will, and may be the work of a moment; but the moral experiences which perpetuate themselves in images, must be sought after in order to be found, and encouraged and cultivated in order to be appropriated.'[126] Newman finds in his passage on the bells images which perpetuate his moral experiences as a convert, which he cultivates and makes his own to retort on the imagination of the Protestant world; Hopkins too sought such images and sounds of converted and retorting imagination. It is testimony to the depth at which Newman influenced Hopkins that the conceptual difficulties of Newman's work so completely translate themselves into the difficult practice of Hopkins's style as he seeks a music in poetry to counter what he heard as the music of Protestant English and finds it in his own style of word pitted against word, a polemic quality in the lyric voice, as the poems accumulate combatant sounds through which the voice will eventually come.

It is here, amidst a tormenting multiplicity of views and voices, the mill-wheel race and the clamour of antagonistic church-bells, that Hopkins's keen attention to the specifically individual qualities of people and things, the celebrated instressing of *haeccitas*, has its place and shows its purpose. For it is, to put it mildly, an achievement of considerable patience to celebrate diversity even when diversity may exact such a cost in hostility and estrangement. When Pope delights in the contrasts of an English scene in *Windsor-Forest*—

> Where Order in Variety we see,
> And where, tho' all things differ, all agree.[127]

—the dappled landscape is a parliamentary idyll in which even

---

[126] *Grammar*, 62.
[127] (1713), repr. in *The Poems of Alexander Pope*, one vol. edn. John Butt (1963), 195.

opposing parties agree to differ; the concept of 'Variety' presents the possibility of a quarrel as only the delectable play of light and shade. Hopkins's pied beauties are more evidently on the verge of being torn apart by their variance, and there is a consequent pressure on Hopkins to wish to resolve all this glitter of distinct selves into clear antitheses of black and white:

> For eárth ǀ her béing has unboúnd; her
> dápple is at énd, as-
> Tray or aswarm, all throughther, in throngs; ǀ self ín self
> steépèd and páshed—qúite
> Disremembering, dismembering ǀ all now. Heart, you round
> me right
> With: Óur évening is óver us; óur night ǀ whélms, whélms,
> ánd will énd us.
> Only the beakleaved boughs dragonish ǀ damask the tool-
> smooth bleak light; black,
> Ever so black on it. Óur tale, O oúr oracle! ǀ Lét life, wáned,
> ah lét life wínd
> Off hér once skéıned stained véined varíety ǀ upon, áll on twó
> spools; párt, pen, páck
> Now her áll in twó flocks, twó folds—bláck, white; ǀ ríght,
> wrong; réckon but, réck but, mínd
> But thése two; wáre of a wórld where bút these ǀ twó tell, eách
> off the óther; of a ráck
> Where, selfwrung, selfstrung, sheathe- and shelterless, ǀ thoúghts
> agáinst thoughts ín groans grínd.[128]

As often elsewhere, Hopkins takes his poem off from the simplest of daily rhythms: the sunset with its iridescent skies is coming to an end, night approaches with its absence of varied colour, an absence against which lights will show up only in sharper contrast. This fading out of shades and sharpening of contrast puts Hopkins in mind of the final separation of mankind, now so variously assorted, into saved and damned, and sets an appreciation of variety to one side as less important than a sense of the severely distinct—'bláck, white; ríght, wrong'. Yet that antithetical clarity eludes the sounds of the poem much as the lines strain to make the antithesis hold. It seems a convenient underscoring of the point that 'white' and 'right' chime so neatly together, but they chime in mid-line. The very clarity of the conceptual and sonic match here obscures the larger structure of the

---

ᵛ   [128] 'Spelt from Sibyl's Leaves', Phillips, 175.

poem, its rhyme-scheme. Hopkins directed that the sonnet be read 'with long rests, long dwells on the rhyme'[129] but this direction is not easy to follow.

Not easy to follow because, in the first place, eight of the fourteen lines seem syntactically to require a run-on: this thwarts the request for 'long dwells'. In the second place, quite what is 'the rhyme' in 'Spelt from Sibyl's Leaves'? There is indeed an abba abba cdcdcd terminal rhyme-scheme but there is also an extreme density of internal rhyme, cross-hatching, for instance, 'white; ríght' here with 'quíte' and 'night' and 'light', shading 'wrong' with 'throngs' and 'selfwrung, selfstrung'. It is as if the moral consciousness of the poem were itself whelmed by the descent of a sonic night which blots out the discriminations that consciousness tries to make. The remarkable, created tussle within the poem, its impeded movement, thought making its way through language as through a muddy and a clogging earth, produces a self-retorting eloquence. So that even when Hopkins most wants to 'get things straight', the lines snap back at themselves:

> párt, pen, páck
> Now her áll in twó flocks, twó folds . . .

This is about sheep and goats, of course, but it does not have the executive confidence we associate with sorting one from the other. The three imperatives do not quite sit together: to 'part' has a pang of separation as well as the pleasure of discernment ('Parting is such sweet sorrow'); to 'pen' is what a farmer does to his animals but also what a poet does to a sonnet, and the pun there wonders about the degree to which such an impulse to make ethical divides might be imaginary and contrived; 'pack' suggests a cramming of the diversity of creation into these two categories, an over-crowding which could lead to a spilling over. 'All in twó' pulls towards 'all in one', a handy compendium, while 'twó folds' has separated itself out from 'twofold'.

This strain between texture and drift in the sonnet sustains the reflection it practises on the status of 'variety' and the desire, at crises where variety threatens to become a chaotic sprawl of differences, to marshal the range of the discrepant into strict categories. Thus 'skéined stained véined varíety' works a complex of connatural senses to match the many aspects of the idea of variety, thereby giving that idea its three-dimensional 'reality' in Newman's sense: 'stained' in relation to 'skeined' sounds innocuously like a reference to the dyeing

---

[129] Letter of 11 Dec. 1886, *GMH/Bridges*, 246.

of wool or flax and has in that sense a pleasant, cottage-industry nuance to it; in relation to 'veined', though, 'stained' begins to acquire the timbre of 'blood-stained'. As the line moves through these words, it both displays an inventive freshness, ever more variations on a theme, and also creates a hubbub, a dinning which wearies of its own inventiveness (*ever more* variations, as who should say 'Oh, not *another* one!'). This double process evokes more fully than Newman did how gruelling the dense plurality of aspect in an idea and its 'reality' could be, though the heaping-up in Hopkins's sonnet—at once a wearisome accumulation and a display of how much he can create and acquire— resembles Newman's catalogue of tolling bells. If it was, for example, a sign of the 'reality' of Catholicism that it could develop so many facets, then the hydra-headedness of Protestantism signalled an equal and opposing 'reality'. The poem checks itself against the desire to rush to judgement in face of such disagreeable variety. For when Hopkins writes 'wáre of a wórld where bút these | twó tell', he means both that the heart should be aware of such a world and also beware of such a world. The compositional task of relating drift and texture (a task which is set again whenever the poem is read aloud) is itself a meditative practice in giving due weight to final Judgement—'be aware that there is a world where right and wrong are seen with absolute clarity to be all that matters'—while also giving due weight to the multiplicity of the interim between the world as it now is and that summation—'beware of acting as if you were the Judge of creation, realize the variety of your world'.

'As kingfishers catch fire . . .' is less grinding than 'Spelt from Sibyl's Leaves', but there too Hopkins does more than revel in diversity as an abstract good while he responds to the plurality of distinct selves in the world:

> As kingfishers catch fire, dragonflies draw flame;
>   As tumbled over rim in roundy wells
>   Stones ring; like each tucked string tells, each hung bell's
> Bow swung finds tongue to fling out broad its name;
> Each mortal thing does one thing and the same:
>   Deals out that being indoors each one dwells;
>   Selves—goes its self; *myself* it speaks and spells,
> Crying *What I do is me: for that I came.*[130]

Certainly, the chimes in these lines are more harmonious than the thickened air of 'Spelt from Sibyl's Leaves' but the chain of rhymes

---

[130] Date of composition not known, probably 1877, Phillips, 129.

from 'kingfishers' through 'ring' and 'string' to 'thing' also stands in its milder way for a variety which is bursting at the seams of an orderly scheme of creation. The tone of 'Each mortal thing does one thing and the same' may be the tone of a man announcing an axiom with some triumph but the line also hangs in an intonational ambiguity which could allow us to hear a lassitude or a bewilderment at this world of selves insistent on themselves (that curious indifference to euphony shown at 'thing does one thing'). The poem attempts to include both variety and the variety of attitudes we might hold towards the diversity of selves. In doing so, it becomes itself something very distinctive. Hopkins claims that everything 'selves'. There is, then, a poignancy in the fact that Hopkins alone used the verb 'to selve' intransitively in nineteenth-century English, as the Supplement to the *OED* concisely notes: '**Selve**. *v. rare* . . . (only G. M. Hopkins)'. That verb is an achievement and an isolation—like selfhood, a 'lonely began'.

Making Yourself Heard

Geoffrey Hill has written that 'the English convert to Rome, however much he might gain, nonetheless suffered an abruption of [a] familiar rhythm . . .'.[131] He was thinking of Hopkins, and referring to rhythm in an imaginatively wide sense, as the rhythm of the Anglican liturgy, of church-going, of the patterned relations between church-going and social life—all the rhythm which made the Church of England 'Home' for Hopkins as for Newman. Some facts are easily stated to support Hill's view: a consequence of the sequestration of Church property and of the penal laws was that most Catholic church buildings in the nineteenth century were new (and many incomplete), if they were hallowed, they were not hallowed by custom or time; in terms of Canon Law, there were no Catholic parishes in England until 1918, all priests were 'missionary priests', and that special continuity across generations worshipping in the same place which is the strength of a parish was missing from the experience of all but a very few English Catholics.[132] Other aspects of interrupted rhythm, though, need more complicated description, as in the case of the pressure and force of 'connatural senses' in exchanges between members of the different communions; any habit makes a 'familiar rhythm' in a life, and there are even rhythms to arguments about whether a habit should be broken.

[131] 'Redeeming the Time' (1973), in *The Lords of Limit*, 100.   [132] Norman, 1.

The lost rhythm in its fond attraction for Newman can be heard
even when he turns against it, heard in his distinctive combination of
parody and nostalgia. He describes the thoughts of a Tractarian pastor
who has decided not to go over to Rome:

> . . . I am doing good in my parish and in my place. The day passes as usual.
> Sunday comes round once a week; the bell rings, the congregation is met, and
> service is performed. There is the same round of parochial duties and
> charities; sick people to be visited, the school to be inspected. The sun shines,
> and the rain falls, the garden smiles, as it used to do . . . I have my daily service
> and my Saints' day sermons, and I can tell my people about the primitive
> Bishops and martyrs, and about the grace of the Sacraments, and the power of
> the Church, how that it is Catholic, and Apostolic, and Holy, and One, as if
> nothing had happened; and I can say my hours, or use my edition of Roman
> Devotions, and observe the days of fasting, and take confessions, if they are
> offered, in spite of all gainsayers.[133]

The passage has, unsurprisingly, a marvellous inwardness with what it
challenges: the move from 'in my parish' to 'in my place' recognizes
how closely bound up together loyalties to a spot of earth, a cure of
souls, and a social situation may be, recognizes it in the rich and
modest pun of 'my place', a pun which is many-angled in that, from
one point of view, it may bespeak complaisance, and, from another,
rootedness; the following sentences lead from liturgical through
charitable to horticultural duties (the first and third of these run
syntactically parallel—'the bell rings . . .', 'the sun shines . . .'), and do
so with a placid sense that all these occupations beautifully consort
together—it is an idyll but also a satirical sketch of a man equally
intent on celebrating the eucharist and pottering about in the
herbaceous border. The self-contentment spills over in a catalogue of
satisfactions ('I have . . . I can . . . I can . . .') but this repletion is tinged
with a self-centredness in which religious observance begins to sound
like an ornament to an individual's own esteem, a hobby in fact ('*my*
daily service', '*my* Saints' day Sermons', '*my* people', '*my* hours', '*my*
edition of Roman Devotions'—my emphases). There is also a canny
and accurate insinuation from Newman that this high Anglican pastor
may not be quite in tune with his congregation: 'take confessions, if
they are offered' suggests that his flock is less eager to enter the
confessional than he would like. Newman acutely sets all this in the
historic present, a tense perfectly judged to give, from one side, the

[133] *Anglican Difficulties*, UE i. 123–4.

pastor's sense of the permanence of such arrangements and, from the other, Newman's own attitude to their historical fragility, an attitude which shows when he drops the present and the dramatic persona, in the clause 'as if nothing had happened'. The something Newman believed had happened was, of course, schism; these familiar Anglican rhythms are in his ears themselves the persisting echo of an abruption.

Hopkins too felt called to hear something else above the beauties of historical rhythm. He wrote to his friend, Urquhart, who was a Puseyite of the sort Newman pictured in the passage just quoted: 'I know that living a moral life, with the ordinances of religion and yourself a minister of them, with work to do and the interest of a catholicwards movement to support you, it is most natural to say *all things continue as they were* and most hard to realise the silence and the severity of God, as Dr. Newman very eloquently and persuasively has said in a passage of the Anglican Difficulties . . .'.[134] Here again, we encounter a type of silence which retorts on the fluency of the world. Hill suggests that Hopkins's poetry shows, and particularly in its rhythms, a practical response to his loss and gain in the abruption of Anglican rhythms. This suggestion is the necessary key for hearing Hopkins's attempt to make his poetry vocal and for realizing that the difficult reserve of his written texts, the occasions when they do not sit easily with spoken English, represent, amongst other things, an attempt to produce in verse something approaching 'the silence and severity of God' with regard to the language in which they are written, to make that silence itself audible. Hill's brilliant insight and in-hearing into the poetry can be supported by noting Hopkins's tendency to write about incomprehension of his poetic work in terms which bring his poetry close to his faith. Thus, to Bridges in 1877: 'My verse is less to be read than heard, as I have told you before; it is oratorical, that is the rhythm is so. I think if you will study what I have here said you will be much more pleased with it and may I say? converted to it. . . . You are my public and I hope to convert you.'[135] Yet the very analogy between his poems and the endeavour of conversion tells as much about why the poems were difficult for Bridges to hear as it does about how Hopkins wanted them to be heard.

Some features of Hopkins's style can be understood as attempts at a repair of and a reparation for the language which he felt had been wounded by the schism from Rome. The tone of solicitous endear-

[134] Letter of 13 June 1868, *Further Letters*, 51.
[135] Letter of 21 Aug. 1877, *GMH/Bridges*, 46.

ment, for example, which comes through the frequent vocal caresses of his lines—the 'ah my dear' of 'The Windhover' or the 'my aspens dear' of 'Binsey Poplars'[136]—brings into the verse a care for the words as maternal as it is pastoral. Hopkins's distinctive way of dwelling on a word, as if he were naming it rather than saying it, has a tenderness in it towards the language itself, both when he repeats the word entire and when he comes closer still to it, takes one of its phonemic elements and broods over that; it seems he feels the language needs to be cossetted back into harmony with itself, as a child's rage or grief may need to be placated by repeated endearings. These verbal or phonemic repetitions, which sometimes halt the verse or gasp in it, as at the end of 'To seem the stranger . . .' or 'Spelt from Sibyl's Leaves', may also have the effect of calmative stroking: 'the sweetest, sweetest spells' of 'The Caged Skylark', or, in the same poem, the tone repetition produces of parental surprise and pride in a child's enterprise at 'Why, hear him, hear him babble and drop down to his nest'.[137]

His most riskily cherishing poem is 'The Bugler's First Communion' where he represents the poetic voice joining in the young excitement and iteration of the boy, as an adult joins in the cries and emphases of a child:

> A bugler boy from barrack (it is over the hill
> There)—boy bugler, born, he tells me, of Irish
>   Mother to an English sire (he
> Shares their best gifts surely, fall how things will),
>
> This very very day came down to us after a boon he on
> My late being there begged of me, overflowing
>   Boon in my bestowing,
> Came, I say, this day to it—to a First Communion.[138]

It is an extremely idiosyncratic piece of work but idiosyncrasy is its subject as well as its failing. The incident Hopkins describes is fraught with oddities for much of the poem's imaginable audience: the practice of the reservation of the Blessed Sacrament to which Hopkins refers at 'Forth Christ from cupboard fetched' is not widespread in the Anglican Church; the extreme importance Hopkins attaches to the First Communion depends on an understanding of the sacrament which itself might have seemed absurd or even sacrilegious to many

[136] 30 May 1877, Phillips, 132; 13 Mar. 1879, Phillips, 142.
[137] Aug. 1877, Phillips, 133.
[138] ? July–Aug. 1879, Phillips, 146–7.

readers, had the poem had many readers. The poem, like the bugler who figures in it, is a cultural hybrid. He is Anglo-Irish, an awkward position for a member of the English army in the late 1870s, and the poem has a nervy agility which equally stems from having a foot in more than one camp, both in its blend of sprung rhythm with an abba quatrain, and in its sensed division of linguistic and liturgical allegiances. The syntax too has a double quality as both a hectic rush of conversational enthusiasm and as a deliberated oddity of poetic style. The ellipses and interpolations are indeed those of speech, but, written down here, they become observable curios (consider the impossibly fast tempo at which the poem would need to be delivered to make the syntax sound authentically spoken—'impossibly' because the piece would be incomprehensible to the ear at such a rate). This double aspect of the syntax points to the estranged homeliness which characterizes the mature Hopkins, as does the double standing of the rhymes—'Irish'/'sire (he/sh'—which appear evidently contrived to the eye and sound like casual felicities to the ear. The poem works like the phenomenon of beams of shadow meeting in the east at sunset, which Hopkins observed in 1883, and of which he wrote, 'It is merely an effect of perspective, but a strange and beautiful one'[139]; the page arranges this perspective as the dramatic scene of a self-audition, it does not wholly cede to these jerks and gushes of the enthusing voice nor does it stand off from them with cold sobriety. Print in the divergent demands it makes on vocalization conducts an 'external survey' at the same time that it strives for intimate expression. Hopkins here, though extravagant, resembles Newman in Newman's imagination of an Anglo-Catholic pastor; their writing is both inward and distant, as if both passages were in *style indirect libre* and indeed a consciousness of the way in which print ambiguously retains spoken utterances, of the lack of coincidence between the spoken and written forms of a language and what can be made out of that lack, is an essential element of *style indirect libre*.

Hopkins's spiritual earnestness may topple into the ridiculous in rhymes like 'boon he on' and 'Communion', rhymes which are tolerable only in the self-bantering style of a Byron or a Browning. Yet the Hopkins who recognized 'spooniness' in himself was also capable of a solemn self-bantering, the sort of thing which is evident in 'The shepherd's brow, fronting forked lightning . . .', as, for instance, in the

---

[139] Letter to *Nature*, 12 Nov. 1883, *GMH/Dixon*, 162.

Prufrockian bringing-together of a self-inciting sense of spiritual crisis with the clutter of a tea-table:

> And I that die these deaths, that feed this flame,
> That . . . in smooth spoons spy life's masque mirrored: tame
> My tempests there, my fire and fever fussy.[140]

He could fetch Christ from cupboard and by the same token knew about storms in tea-cups. In 'The Bugler's First Communion' preciosity of rhyme has a further point. If you pronounce the words so that 'Communion' produces a full rhyme with 'boon he on', you have to dwell with a fulsome sense of rarity on the 'Communion'; if you opt for a less bizarre pronunciation, in which 'Communion' only half-rhymes with 'boon he on', you make the word very familiar, slightly off-hand. The combination of rarity and familiarity is what Hopkins sought, for it is a *first* communion to the bugler and a habitual act for the priest, habitual but twice repristinate in this context, both because of the boy's zeal and because Hopkins strives to match that zeal in the verse with his far-fetched rhyme. The Catholic belief in the Real Presence makes the daily species of bread become the substance of divinity; the peculiarities of Hopkins's style here, in its fantastic veering from the supernatural to the mundane, answer to the peculiarities of that doctrine, as Hopkins faces us with it in the workings of the language. It would be natural to say that 'The Bugler's First Communion' is not a successful poem, that it determines to flout the dictional sensibilities of a reader with its calculatedly wild clashes of different levels of the language, its extreme setting of eye and ear against each other in the rhymes, and that it is not really a case of solemn self-bantering but something more unbalanced than that: a ceremonious self-mockery. That seems to me fair enough, provided we are clear about the grounds for Hopkins's failure in his attempts. We have to realize that our judgements of dictional consistency, of permissible ingenuity in rhyme, form part of a literary sense of proportion which is itself related to wider senses of the balanced and the fitting. A religious doctrine such as the Catholic doctrine of the Real Presence may itself seem to have little to do with literary judgement, but, as that doctrine associates itself over time with cultic practices and attendant types of religious art, it will eventually inform even apparently remote areas of a believer's or a non-believer's judgement. The doctrine of the Real Presence does not consort well

---

[140] 3 Apr. 1889, Phillips, 183; the ellipsis is in the original.

with certain senses of literary decorum, but that does not cast doubt only on the doctrine. Clearly, though, a poet works in the actuality of a language with whatever a culture has made of that language; he does not write in the dialect of a theological orthodoxy. Hopkins may be guilty here of refusing the conditions of his own art, of being deaf to the way he must be heard, if at all, by an English audience, even a Catholic one.

In a letter to Bridges in 1882, he recognized the necessity of 'finding the ear of an audience' in a work of literature. He called this direction towards an audience 'bidding': 'a nameless quality which is of the first importance . . . I sometimes call it *bidding*. I mean the art or virtue of saying everything right *to* or *at* the hearer, interesting him, holding him in the attitude of correspondent or addressed or at least concerned, making it everywhere an act of intercourse . . .'.[141] Even as he states his dictum, Hopkins's very emphasis sounds shaky; saying something '*to*' somebody implies a different relation with the interlocutor than saying something '*at*' him; similarly, to be 'correspondent' is less than to be 'addressed', though both entail more attention than being 'at least concerned', for you may be concerned in what is said without listening to it at all. There are circumstances in which such straight talk may be brow-beating or button-holing rather than the true 'act of intercourse' he describes. Browning had felt the troubles you could get into in such directness of address.[142] And the troubles would, in Hopkins's case, be just those Browning said attended on telling people that they had got things all wrong, and that their truth was a lie, as Newman felt in the dilemma of how to reason persuasively with old friends like Keble and how to articulate the activity of reason in the process of conversion without weakening his own case or wronging those who had a different estimate of its strength.

The question of address can be put more sharply if we consider the relation between Hopkins's desire for 'bidding' in literary composition and his commitment in his priestly vocation to the different demands of spiritual 'calling'. He is explaining to his congregation what the word 'Paraclete' means:

. . . a Paraclete is one who calls us on to good. One sight is before my mind, it is homely but it comes home: you have seen at cricket how when one of the batsmen at the wicket has made a hit and wants to score a run, the other doubts, hangs back . . . how eagerly the first will cry / Come on, come on!—a

---

[141] Letter of 4 Nov. 1882, *GMH/Bridges*, 160.   [142] See above, p. 206.

Paraclete is just that, something that cheers the spirit of man, with signals and with cries, all zealous that he should do something and full of assurance that if he will he can, calling him on, springing to meet him half way, crying to his ears or to his heart: This way to do God's will, this way to save your soul, come on, come on![143]

The 'calling on' of the Paraclete and the 'bidding' of the artist may have a common urgency and even a shared direction to their urging, but calling on and bidding in some circumstances fall apart. The Paraclete, the Holy Spirit, spoke through the Apostles to each man in his own language without loss of its own integrity of voice; bidding in artistic actuality may not always find itself able to do this, because men's ears and their hearts may not be at one for the artist as they are for Hopkins's Paraclete 'crying to his ears or to his heart', where the 'or' indicates that it does not matter whether the Spirit speaks to the ears *or* to the heart because the Paraclete's words cannot, as human words can, be deflected on their way from the ears to the heart. The 'or' in the passage stands for divine certainty; the 'or' in Hopkins' '*to* or *at* the hearer' and in his 'correspondent or addressed or at least concerned' marks rather the troubles of human utterance. The true conduct of 'calling on' in England demanded more reserve and tact in the 'bidding', the sort of reserve and tact Newman showed in his dealings with Hopkins when they were preparing for Hopkins's reception into the Church: 'Dr. Newman was most kind, I mean in the very best sense, for his manner is not that of solicitous kindness but genial and almost, so to speak, unserious. And if I may say so, he was so sensible. He asked questions which made it clear for me how to act . . . but in no way did he urge me on, rather the other way.'[144] A wonderful tribute to Newman's intelligence, epitomizing what is best in his polemic writings—a manner not of solicitous kindness but of almost unserious geniality, an asking of questions which enable the hearers to hear themselves rather than his voice, and so help them to become clear on how to act—and the more compelling as a tribute coming from Hopkins who so much less maintained a calm in his calling and persuasions. The poet's difficulty in his public art was keener than Newman's in private conversation with a would-be convert (a public art in that the essence of the poet's dealing with the language is to work in it as the language of a community of which he is only a part, even when his writing is, as Hopkins's was, secluded).

[143] Sermon of 25 Apr. 1880, *GMH Sermons*, 70.
[144] Letter to Bridges, 24 Sept. 1866, *GMH/Bridges*, 5.

Hitting the right note for the listener might demand a forsaking of the true key for the poet, and where the true key might be within the 'connatural senses' of words associated with a religious doctrine such forsaking would be unthinkable for a poet such as Hopkins. He underwent difficult choices set by the fact that a poet needs to be biddable, pliant to his culture, in order to bid it or please it with his bidding. 'The Bugler's First Communion', for instance, seems a poem torn between a dictional intransigence required by fidelity to doctrine and a supple rhythm of persuasion. As a result of such strife, Hopkins was sometimes subject to attacks of self-righteousness during which he transferred dogmatic certainty into the realm of artistic practice, writing to Bridges of his work: '. . . if you do not like it it is because there is something you have not seen and I see. That at least is my mind, and if the whole world agreed to condemn it or see nothing in it I should only tell them to take a generation and come to me again.'[145] (He was writing about his music here, but this is only the most emphatic of such remarks about his artistic work including the poems.)

Hopkins's achievement is not unchallengeable, as he sometimes slipped into believing it to be, with the unchallengeability of the Holy Spirit: 'When . . . it is said that the Paraclete *will convince the world* of three things it is meant that he will convince the world of its being wrong about these things, will convince it of himself being right about them, will take it to task about them, reprove it, and so bring the force and truth of his reproof home to it as to leave it no answer to make'.[146] G. M. Hopkins telling 'the whole world' to think again sounds dismayingly like the Paraclete convincing the world that it is wrong and he is right. That many readers did after a generation or so come back to his poems and declare him right as against Bridges is scarcely the point (his confidence in his abilities as a musician has not been endorsed by posterity). Rather, the point is the complicated blurring and overlay of poetic and religious conviction—at times straining against each other, at times in harmony, at times lending each other possibly delusive support. When Newman spoke of the innumerable concurrent evidences which might lead to faith, or of the 'reality' of an idea as derived from its capacity to absorb and reflect many aspects of development, he was responding to the need to grant the actual informality of reasoning in concrete processes of coming to believe. The many angles from which assents converge to form a real assent are

[145] Letter on April Fool's Day, 1885, ibid. 214.
[146] Sermon of 25 Apr. 1880, *GMH Sermons*, 72.

then better understood on the model of artistic composition than of consecutive inference. For Hopkins, the most particular instance of such analogy between his artistic and his religious work was the eliciting of voice from the scant evidences of reason or of print—'voice' being for both Hopkins and Newman a key religious concept. He spoke of Bridges' need to hear the poems' voice before he was converted to them; equally, he described the mysterious crystallization into conversion of reasons which did not sway others as the hearing of a voice:

> . . . all converts agree in feeling that they are led by God's particular will. They are bound to go, it will be sin to say, God calls them, bids them etc: 'I hear a voice you cannot hear' etc. We who are converts have all heard that voice which others cannot or say they cannot hear.[147]

It was because he heard in his writing a voice others could not or said they could not hear that Hopkins felt he had to try typographical means to indicate the accents of speech: 'I do myself think, I may say, that it would be an immense advance in notation (so to call it) in writing as the record of speech, to distinguish the subject, verb, object, and in general to express the construction to the eye'.[148] The placing of '(so to call it)' immediately after 'an immense advance' puzzles the prospects he opens up: what else does he think it should be called? Perhaps the parenthesis refers to what follows, meaning 'in writing as the, as it were, record of speech'. That gives the game away for only some writing tries to be the record of speech, and Hopkins's poetry is not that kind of writing, however much it supplicates the ear and voice of a reader/hearer. A 'record' is made of what is already past but the life in Hopkins's poetry, as in all vital poems, goes on and is to come; writing is the material in which to imagine a future on the voice and in Hopkins's poetry it is not a 'record' but a 'began'.

The poems would gain little by such visible constructions, even if they could be unambiguously built—we do not want the construction of 'England whose honour O all my heart' rendered visible for the eye, not, anyway, if what Hopkins has in mind is a machinery for selecting a single voicing from the several which the line may carry, because the point of the line is its suspension between alternative voices, and the simultaneity in which it then permits us imaginatively to hear all the voicings at once as the rich and pained texture of his attitude.

[147] Sermon of 21 Sept. 1879, ibid. 25.
[148] Letter to Bridges, 6 Nov. 1887, *GMH/Bridges*, 265.

Or take the construction of the lines which conclude, 'Not, I'll not, carrion comfort . . .'. Hopkins has answered his own question as to why God has treated him so fiercely by saying that the ferocity was a form of winnowing his chaff from his grain, and that, in fact, the process was not without its pleasures, for he has been heartened to applause:

Nay in all that toil, that coil, since (seems) I kissed the rod,
Hand rather, my heart lo! lapped strength, stole joy, would laugh, cheer.

Cheer whóm though? The héro whose héaven-handling flúng me, fóot tród
Me? or mé that fóught him? O whích one? is it eách one? That níght, that year
Of now done darkness I wretch lay wrestling with (my God!) my God.[149]

These closing lines pivot in part on the ambiguity of 'cheer' between 'hearten' and 'applaud'; that ambiguity asks the following questions: whom did I really support in the conflict, Christ or myself? and did my heart derive its cheer from, as well as give its cheers to, Christ or my resistance to him? does 'cheer' mean the same in both its occurrences, or is there an antanaclasis on the word? if there is, what my heart cheered (applauded) may not be what its applause pleased—that is, I may have applauded myself but, in His mercy, Christ will have taken this self-glorification as just a stage in my conforming to him; alternatively, what I applauded may have been Christ but the effect of that support of him may have been that he heartened me. It would not be an advantage to these lines to have their construction made any clearer to the eye, for it would be only a delusive clarity, the questions here being part of a permanent discipline of self-examination, and not in need of solution—a life's integrity consists in maintaining them (though at times it is natural to ask 'O whích one?'). Similarly, the ambiguity of 'since (seems) I kissed the rod' holds a very self in question; 'seems' might qualify 'since' which would make the line mean 'it seems that ever since I submitted to the Jesuit discipline there has been nothing but toil and coil', but the next line suggests that it qualifies rather 'rod', with the sense 'since I did what seemed to be kissing the rod but was in fact kissing my hand—for what appears to be a harsh discipline is not so but an act of delighted chivalry'. There is the further possibility that '(seems)' discriminates between 'rod' and 'hand' with the sense 'what seems like submission to punishment is really a loving contact with God'. Hopkins had already considered this

[149] 1885–?1887, Phillips, 168.

relation of 'rod' and 'hand' in the opening stanzas of 'The Wreck . . .' which move from feeling God's 'finger' through the 'lashed rod' of lightning in an angered sky which shows God's wrath on to a courtly gesture towards a more placable nature which also figures the Divine:

> I kiss my hand
> To the stars, lovely-asunder
> Starlight, wafting him out of it . . .

No visible construction of such textures of avowal and reticence, of a life contained in lines, would be possible; such a notation, even were it possible, would deprive the life and the line of their eventual-ness, the time in which each comes to be understood. We need to pass through the various interpretations of 'seems', so that the cost is counted before the prize is claimed.

As soon as Hopkins had made his declaration in favour of the delineation of syntax and meaning by visible construction, he was forced back along a series of qualifications:

And I daresay it will come. But it would, I think, not do for me: it seems a confession of unintelligibility. And yet I don't know. At all events there is a difference. My meaning surely *ought* to appear of itself; but in a language like English, and in an age of it like the present, written words are really matter open and indifferent to the receiving of different and alternative verse-forms, some of which the reader cannot possibly be sure are meant unless they are marked for him. Besides metrical marks are for the performer and such marks are proper in every art. Though indeed one might say syntactical marks are for the performer too. But however that reminds me that one thing I am now resolved on, it is to prefix short prose *arguments* to some of my pieces. These too will expose me to carping, but I do not mind. Epic and drama and ballad and many, most, things should be at once intelligible; but everything need not and cannot be.[150]

'But', 'And yet', 'At all events', 'but', 'Besides', 'Though indeed', 'But however', 'but', 'but': the syntax suggests a man groping for a light-switch in the dark. He *feels sure* in the fifth sentence quoted that his meaning ought to be self-evident; he thinks it impossible in the tenth sentence quoted that all meanings, and, by implication, at least some of his, should be self-evident. The central realization in the passage is of a fact which makes it impossible to regard writing as simply 'the record of speech': 'in a language like English, and in an age of it like the present, written words are really matter open and indifferent to the

[150] Letter to Bridges, 6 Nov. 1887, *GMH/Bridges*, 265.

receiving of different and alternative verse-forms'. That is so because
the same set of written words may be variously uttered; the inflection
of a phrase cannot be absolutely governed because stress goes where
emphasis lies, and emphasis lies on what people think important. What
counts in the community governs the poet's numbers, and not vice
versa, though a poet may create 'new compositions of feeling'[151] and in
so doing elicit new communities.

∠ It baffles poets when the page yields only with reluctance their
intended voices back to readers; this is a specific form of the general
surprise people feel when they are mis-taken. Given the frequency of
misunderstanding, the surprising thing is the surprise. Perhaps people
live in hope that misunderstanding, like death, is something that
happens to *other* people. It strikes a poet particularly, though, because
the pleasure of poetry dwells in an exactness of words, blurs show up
more clearly in the reading of poetry, especially in the reading of it
aloud, like stains which are the more evident the whiter the table-linen
is. Tennyson's intermittent worries about this matter have already
been mentioned. Other poets, such as Pound, were more persistently
bothered: 'Every one has been annoyed by the difficulty of indicating
the *exact* tone and rhythm with which one's verse is to be read.' Pound,
like Hopkins, resorted to typographic methods to make the sound of
his meaning clear—'ALL typographic disposition, placings of words *on*
the page, is intended to facilitate the reader's intonation, whether he be
reading silently to self or aloud to friends.'[152] But Hopkins was able to
discover in just this apparent plight of the voice on the page the figure
and texture of his best intentions, intentions which went beyond what
he individually planned (in ways which, for example, Pound with his
eighteenth-century individualism could not plumb in his 'intended . . .
facilitate . . . intonation', where the controlling intent is thought to be
simply helpful to the reader, with no sense of the clash of selves—and
of things beyond the self—which might be involved in the questionable task of voicing lines).

The same technical intricacies represent different issues for
different poets, Hopkins and Pound in this instance. In part, this is
because the connatural senses of the word 'voice' are differently
disposed for the two writers. In Hopkins's work, the drama of voicing

---

[151] Wordsworth, letter to John Wilson of 7 June 1802, in *Wordsworth Letters: Early
Years*, 355.
[152] 'The Island of Paris', *The Dial*, Dec. 1920; letter of Feb. 1939 to Hubert
Creekmore, in *Letters of Ezra Pound 1907–1941*, ed. D. D. Paige (1951), 418.

exemplifies the activity of faith in search of understanding. Here again, his practice as a writer actualizes in implicit detail matters which were the subject of explicit argument for Newman. As Newman urged, the special claim of the Catholic Church was to be the living bearer of the meaning of the Scriptures, writings which were not a text for the Catholic, but a voice:

Now, in the first place, what is faith? it is assenting to a doctrine as true, which we do not see, which we cannot prove, because God says it is true, who cannot lie. And further than this, since God says it is true, not with His own voice, but by the voice of His messengers, it is assenting to what man says, not simply viewed as a man, but to what he is commissioned to declare, as a messenger, prophet, or ambassador from God.[153]

Men were 'to be saved by hearing'; there was 'an essential difference between the act of submitting to a living oracle, and to his written words'.[154] The question Charles Reding puts in reply to Catholic claims in *Loss and Gain* as against the Anglican Church is '*which* was the voice of Christ?'[155] The search for true developments, for a distinction between faith and superstition, which was so powerful a motor in driving Newman from the Anglican Church, and in taking people such as Hopkins along with him, found its rest, Newman thought, in a voice, the voice of the Church as incarnating 'a living, present authority'[156] on matters of faith. Newman's fertility in historically substantiated scepticisms about Christian doctrines was to lead him to the fullest acceptance of the apostolic succession as the only surety of faith, even though he had some reservations about the manner in which the development of apostolic succession into the dogma of Papal infallibility was declared. His questions about religious authority can be put like this: the text of Scripture underdetermines the meaning of Scripture, as any written text underdetermines an unequivocal vo.cing; it does so for various reasons, such as that doctrines (for example, the Trinity) which have been held for centuries to be essentials of the faith, cannot be found explicitly and unambiguously presented in the texts; the text must then be read within a tradition of interpretation, that tradition requires a community to carry it, and such a community must have authority for its claim to rule on interpretation. This traditional authoritative community of interpretation is the Church, and Newman eventually took this

---

[153] *Discourses addressed to Mixed Congregations* (1849), UE 194–5.
[154] Ibid. 199, 200.   [155] UE 112.   [156] *Grammar*, 229.

Catholic doctrine to its fullness by accepting it in its Roman Catholic form. The Church perpetuates the Incarnation, which is one reason for calling it the Body of Christ; its particular importance for a man of Newman's cast of mind was that, as the Body of Christ, it was also his voice. The Church found for the biblical texts just that voice whose loss rendered them fragmentary and ambiguous; it was in fact the 'living voice' of Christ. These arguments of Newman's had tremendous force for Hopkins; they are the dimension of historical faith in his poetic activity. It is not that Hopkins's practice as a writer follows from acceptance of Catholic doctrines. If there had been a strict relation of implication between doctrine and practice, there would have been many more writers who adopted something like Hopkins's manner, and there were not. The doctrine no more entails the practice than the actually connatural senses of a word from some speakers are implicit in the 'meaning' of that word and discoverable by logical analysis of that 'meaning'. Yet such senses guide a person's understanding of his own activity quite as firmly as do those senses which are more strictly articulable, giving body in Hopkins's case to his apprehension of what it was for him to have a 'voice'.

Not that Hopkins confused his own voice in poetry with that of the Church, but that the tribulation of making a Catholic voice heard in nineteenth-century English helped make exemplary and purposeful his own efforts towards audible writing, as the Incarnation had helped him to understand the place of inconsequence in his life:

I think that the trivialness of life is, and personally to each one, ought to be seen to be, done away with by the Incarnation—or, I shd. say the difficulty wh. the trivialness of life presents ought to be. It is one adorable point of the incredible condescension of the Incarnation . . . that our Lord submitted not only to the pains of life, the fasting, scourging, crucifixion etc. or the insults, as the mocking, blindfolding, spitting etc, but also to the mean and trivial accidents of humanity. . . . It seems therefore that if the Incarnation cd. *versari inter* trivial men and trivial things it is not surprising that our reception or non-reception of its benefits shd. be also amidst trivialities.[157]

Reception and non-reception of his writing equally take place amidst trivialities, across insults, through submission, though these terms must be understood as applying to Hopkins's career in the language as well as to his life. Consider, for example, Hopkins's 'one adorable point of the incredible condescension of the Incarnation'. Both

[157] Letter to E. H. Coleridge, 22 Jan. 1866, *Further Letters*, 19–20.

'adorable' and 'incredible' have a strict and a colloquially exaggerating sense: 'adorable' as 'worthy of adoration' which would be applied by Hopkins only to a religious mystery or to God and as 'extremely attractive, charming'; 'incredible' as 'that which cannot be believed' and as 'extremely great'. His phrase requires us to take 'adorable' in the strict sense and 'incredible' in the colloquially exaggerating sense (he does not mean that it is part of the charm of the Incarnation that nobody could believe in it). A critic might fix on this as slack writing, but this is a letter after all and we should allow for some relaxation in the rigour of the writing. The vagary of usage from strict to loose shows two things: that the language has in it a tendency to take words of religious or metaphysical import and use them for more worldly purposes and secondly, that Hopkins speaks both dialects of the language, having learned the worldly idiom first and imposed upon that the theological definitions of the Catholic convert. Here again, the socially conversant and the spiritually elevated melt down into each other, linguistic usage holding together in an uneasily suspensed solution connatural senses whose diverging implications might make them seem easy to distinguish. Newman celebrated this practical density as the 'poetry' of words, but it is not clear that such potential slithers of meaning as appear in Hopkins's letter can always be dextrously 'managed' by a poet. When Hopkins writes '(my God!) my God' or 'Enough! the Resurrection . . .',[158] we read in the grind of colloquial against religious idiom both a mark of compositional prowess and also the toil of that prowess, its reluctant implication in a sociality of language where the usage of the same word may range from devout to nonchalant. Such moments of the arranged clash of idiom are like Hopkins's marks for accent—they show an effort to overcome a difficulty but they remain signs of that effort as of its overcoming.

He wrote of 'The Loss of the Eurydice', 'Stress is the life of it',[159] and the remark is true of all his best writing. 'Stress' occurs in Hopkins not just as a rhythmic pulse but as a perpetually created and creative pain; the verse has stress because it is under stress. He returns to the word in the widest variety of contexts in his preaching: 'a commonwealth is the meeting of many for their common good, for which good all are solemnly agreed to strive and being so agreed are then in duty bound to strive, the ruler by planning, the ruled by performing, the sovereign

---

[158] Phillips, 168; 'That Nature is a Heraclitean Fire and of the comfort of the Resurrection', 26 July 1888, Phillips, 181.
[159] Letter to Bridges, 21 May 1878, *GMH/Bridges*, 52.

by the weight of his authority, the subject by the stress of his obedience'[160] or again, on the comfort to be derived from belief in Providence: 'If we feel the comfort little, there, my brethren, is our fault and want of faith; we must put a stress on ourselves and make ourselves find comfort where we know the comfort is to be found.'[161] He was similarly drawn to the word 'strain' as punning between 'effort' and 'tune'. He wrote in his preface to his own poems that his lines were to be scanned continuously, passing on from one to the next, so that 'all the stanza is one long strain'.[162] John Hollander has written well of 'the remarkable relation between Gerard Manley Hopkins's technical vocabulary for describing and naming some of his prosodic concepts, and the character of the imagery in his own poems. Surely the field of the latter yields up the former, and surely the expressive force of that vocabulary far exceeds its strictly conceptual utility',[163] though I think he is wrong to be sure that Hopkins must have encountered 'stress' and 'strain' first in his imagery, and the moral experiences which perpetuate themselves in imagery, and then turned them into principles of his verse and its prosody.

There seems no evidence for this order of priority, but Hollander's remark none the less remains acutely illuminating of Hopkins's ability to produce a lyricism in which a certain laboriousness holds pride of place, a lyricism of bodies which have been moulded by their work, tissue instinct with profession, as when he writes of a dead sailor in 'The Loss of the Eurydice':

> They say who saw one sea-corpse cold
> He was all of lovely manly mould,
>   Every inch a tar,
> Of the best we boast our sailors are.
>
> Look, foot to forelock, how all things suit! he
> Is strung by duty, is strained to beauty,
>   And brown-as-dawning-skinned
> With brine and shine and whirling wind.
>
> O his nimble finger, his gnarled grip!
> Leagues, leagues of seamanship
>   Slumber in these forsaken
> Bones, this sinew, and will not waken.[164]

---

[160] Sermon of 11 Jan. 1880, *GMH Sermons*, 56.
[161] Sermon of 14 Dec. 1879, ibid. 47–8.
[162] About 1883, repr. in Gardner and MacKenzie, *The Poems of Gerard Manley Hopkins*, 48.   [163] *Vision and Resonance*, 5.   [164] Apr. 1878, Phillips, 137

The lines take up from the colloquial fluency of 'Every inch a tar', a fluency which Hopkins attributes to an imagined community with its idiomatic solidarities and quick admirations ('They say', 'we boast' with its chiming and grinding 'our best'), but the lines move on to sayings which come less easily to the common tongue. Consider how the bluffness of 'Every inch a tar' is converted when set against 'Leagues, leagues of seamanship'. The smallness of that every inch of the sailor's body put against the leagues his skill covered, leagues of sea from which his seamanship could not save him, renders the tiny space the man's body now occupies in contrast to the extent of the ocean and the length of his career, and from that contrast there springs the tone of elegiac wonder in the grieving repetition, 'Leagues, leagues . . .'. 'Every inch' moves from the weightless commendation of the land-lubbing community to the particular literalness of this dead body, somehow more measurable than the living man, as corpses generally seem, because there is nothing to them but what can be observed and tabulated. Hopkins works from the plangency of 'these forsaken | Bones' to the clipped exactitude, the singling out, of 'this sinew', as he has in imagination eyed the corpse from 'foot' to, not 'head', but 'forelock', that loving meticulousness which so characterizes his verse in its exceptional combination of the lush and the exacted. These verses seem to me to achieve what 'The Bugler's First Communion' fails to secure, a created sense of perspective in the language. And their quality—which is not maintained throughout the poem—comes specially from the genius with which Hopkins finds a rhythmic counterpart to the dictional workings of the verse. The rhymes are pondered in the first two stanzas quoted, the lines rest on them, but the last three lines of the third stanza throw this formal poise about, casting grammatical units across the line-end, making the verse metre to the eye only, so that stress falls on the first word of the following line rather than on the patterned close of the rhyme, convulsing the verse into choked phrases in a rebuff and uptake of attentive grief.

[Stress is a crux for Hopkins because it is the mark of selfhood, and his faith requires him to see the self as both an obstacle to God's will and what that will requires as an offering] On the one hand, 'Stress . . . is the making a thing more, or making it markedly, what it already is; it is the bringing out its nature';[165] in so far as God loves the creation, He is in the absolute particularity of His love pleased with the marked

---

[165] Letter to Patmore, 7 Nov. 1883, *Further Letters*, 327.

individuality of his creatures. Yet, on the other hand, Hopkins could imagine Hell as a condition of emphatic selfhood: 'the understanding open wide like an eye, towards truth in God, towards light, is confronted by . . . that act of its own, which blotted out God and so put blackness in the place of light . . . Against these acts of its own the lost spirit dashes itself like a caged bear and is in prison, violently instresses them and burns, stares into them and is the deeper darkened.'[166] Milton was Hopkins's master in passages like this with their shocking aptitude to make a syntax for the self-absorption and self-bafflement of the damned.

Hopkins found a practical means in his poetry to contain these conceptually divergent values of the self by the extremely individual manner in which he created his surrender to the condition of the language as he learned it:

. . . when I consider my selfbeing, my consciousness and feeling of myself, that taste of myself, of *I* and *me* above and in all things, which is more distinctive than the taste of ale or alum, more distinctive than the smell of walnutleaf or camphor, and is incommunicable by any means to another man (as when I was a child I used to ask myself: What must it be to be someone else?). Nothing else in nature comes near this unspeakable stress of pitch, distinctiveness, and selving, this selfbeing of my own.[167]

The passage itself shows this selfhood it puzzles over in such characteristic turns as 'the taste of ale or alum' which puts together along an assonantal chain the 'ale' of hearty Englishness and the 'alum' of industrial exploitation—the 'alum' which was used to adulterate bread ('And chalk and alum and plaster are sold to the poor for bread'[168]). The phrase 'unspeakable stress of pitch' serves well to describe those moments in the poems where Hopkins puts metrical stress on words the voice would not naturally pick out:

> Not, I'll not, carrion comfort, Despair, not feast on thee;
> Not untwist—slack they may be—these last strands of man
> In me ór, most weary, cry *I can no more*. I can . . .[169]

---

[166] Retreat notes, 1881, *GMH Sermons*, 138.
[167] Retreat notes, 1880, ibid. 123.   [168] *Maud*, I. i, R ii. 522.
[169] Phillips, 168. The words 'I can no more' come to Hopkins both from *Antony and Cleopatra*—'Now my Spirit is going, | I can no more' (IV. xv. 58–9; Folio, ll. 3069–70)—and from *The Dream of Gerontius*—'I can no more; for now it comes again, | That sense of ruin, which is worse than pain' (UE 319). The two sources compose together a relevant field of question about suicide and patience at death; contrast the double literary inheritance from Keats and the Bible which I discussed above with reference to 'Easter Communion'.

The stress on 'or' here is not itself unspeakable, but its pitch of selfhood is unspeakable in the usual spirit of English without turning selfhood into idiosyncrasy. That stress has a point, as Hopkins has. The lines resolve not to commit suicide ('Not untwist . . . these last strands of man | In me') nor even to cry out in despair ('cry *I can no more*'), and the weight put on 'or' stresses the gravity of the disjunction between the two alternatives, not because crying out in despair is so much less sinful than committing suicide, as Hopkins's contemporaries would mostly have thought, but because they are to him both acts of serious disobedience to the will he is supposed to have given himself up to. It seems odd of Hopkins to make no distinction between final self-destruction and a natural moan such as 'I can no more' but perhaps the oddity would seem less if we considered that one sense 'I can no more' might have had for Hopkins could have been the laying down of his vocation and leaving the Society of Jesus. He could have given up on his duties as on himself. Taking this possibility, the violent emphasis on 'or' still directs a reader to a conflict of ethical attitudes, for it might seem much less a thing to fail in a religious vocation than to kill oneself, but that is not how things seemed to Hopkins, who wrote to his mother in a happier style after taking his vows 'I have bound myself to our Lord for ever to be poor, chaste, and obedient like Him and it delights me to think of it.'[170] Being so bound, suicide was an untwisting of the man in him, but a defeat in his vocation would have untwisted the Christ in him.

The oddity of the stress on 'or' arises from the fact that this word arranges a meeting between Hopkins's sense of his own conduct of his self and the more general sense in the community of how people ought to behave. This is a 'Miltonic' combat with the language and not a 'Shakespearian' collaboration with it, and it may fail to convince those who feel that English ought to be kept up. The unspeakable stresses Hopkins sometimes calls for elicit a music against the propensities of the language because the poetry is in part at odds with the attitudes of the language-community; these stresses do not ignore the weight of what they pull against as they try to create a literary grace that answers to a conviction of religious grace, grace which Hopkins called 'divine stress . . . the counter stress which God alone can feel'.[171]

Among the philological notes in Hopkins's early journals, there is one on the word 'skill':

[170] Letter to his mother, 10 Sept. 1870, *Further Letters*, 113.
[171] Retreat notes, 1881, *GMH Sermons*, 154, 158.

Skill etc.

Primary meaning, to divide, cut apart. *Skill*, discernment. To *keel*, to skim. *Keel*, that part of a ship which cuts a way through the water.[172]

It is appropriate to the complex of severance and reconciliation which is the ground of Hopkins's style that as early as 1864 he should have sensed skill as a form of division—a process of making distinctions which confers distinctness, a sifting which results in the skilled person's himself being sifted. His skill in verse consists of an arranged contact and parting with the norms of speech in Victorian English, so that, as Geoffrey Hill has crucially observed, 'The achievement of sprung rhythm is its being "out of stride" if judged by the standards of common (or running) rhythm while remaining "in stride" if considered as procession, as pointed liturgical chant or as shanty.'[173]

The literary judgement of skill itself became for Catholic writers a matter bound in with their faith, because it was so frequent a reproach against English Catholicism that it was, as compared with the Anglican Church, culturally impoverished or insignificant. Charles Reding's Oxford tutor complains to him about Catholic style:

'But look at their books of devotion,' insisted Carlton; 'they can't write English.'

Reding smiled at Carlton, and slowly shook his head to and fro, while he said, 'They write English, I suppose, as classically as St. John writes Greek.'

Here again the conversation halted, and nothing was heard for a while but the simmering of the kettle.[174]

A smart reply, as it translates Carlton's complaint into a fundamental question about the Western literary tradition in which classical models and standards co-exist with Christian attitudes and intents which they do not always perfectly meet, but the smartness belongs to the partisanship of the convert in the making, not to the Catholic familiar with the deficiencies of English Catholics in the way of literacy about a culture which was both their own and not their own. W. G. Ward remarked to Benjamin Jowett, 'When a Catholic meets a Protestant in controversy, it is like a barbarian meeting a civilized man',[175] and this is one of the leading Catholic intellectuals of the century in correspondence with one of the leading Protestant intellectuals. The situation of the

---

[172] Journal entry, 1864, *GMH Journals*, 31.
[173] 'Redeeming the Time', *The Lords of Limit*, 98.
[174] *Loss and Gain*, UE 371.
[175] The remark was made in 1858, and is quoted in Norman; 212.

educated convert from Anglicanism was even more fraught, for in him
Ward's 'barbarian' and 'civilized man' met in one person. These
difficulties arose most signally for Newman during his years as Rector
of what eventually became University College, Dublin, an institution
which uneasily attempted to meet the demands both of an Oxonian
humanistic culture and of the rigorist Irish hierarchy. It is apt that
Hopkins finally went to work in the College which had imaged for
Newman some of the strains in a convert's allegiances. These
differences of culture bring some theological problems once again into
the sphere of literature. Stephen Prickett has observed the paradoxical
nature of Newman's definition of real assent: 'It is of the nature of real
assent that it is *both* deeply personal to the point of incommunicability,
and simultaneously arises from a linguistic and metaphorical community.'[176] The screw is turned even tighter when real assent springs
from more than one community, expresses itself in a hybrid and
conflictual language; in such circumstances, the 'deeply personal' may
seem, in its self-divisions, at times scarcely 'personal' at all. In this, real
assent resembles Hopkins's 'unspeakable stress of pitch' and may
require at times such a style as Hopkins's for its expression.

Newman's terms in the *Grammar* help to explain why this is so, and
particularly his uncertainty with the phrase 'common measure'. His
arguments about the social texture of human reasoning compel him to
admit that 'where there is no common measure of minds, there is no
common measure of arguments'.[177] Because this is so, 'it becomes a
necessity if it be possible, to analyze the process of reasoning, and to
invent a method which may act as a common measure between mind
and mind, as a means of joint investigation, and as a recognized
intellectual standard,—a standard such as to secure us against
hopeless mistakes, and to emancipate us from the capricious *ipse dixit*
of authority . . .'[178] and Newman believes that such a standard might be
found in logic: 'One function indeed there is of Logic . . . which the
Illative Sense does not and cannot perform. It supplies no common
measure between mind and mind . . .'[179] Newman appears to imply
here that, if a common measure of minds is possible at all, then logic
can provide such a measure. But this is actually more than logic can do
for us. Newman himself had already given the reasons for its
insufficiency in matters of concrete reasoning when he spoke of its
clearing words of their 'connatural senses' for the sake of definitional

[176] Prickett, op. cit. 199–200.
[177] *Grammar*, 266.   [178] Ibid. 170–1.   [179] Ibid. 233.

clarity. Logic does not supply a common measure between mind and mind but between certain classes of propositions. As the mind also operates with a great many things other than explicit propositions, and works in areas where the 'connatural senses' of words are themselves at the heart of debate—areas such as the meeting of Catholic and Protestant Englishmen in controversy—logic will not save the day, or solve the issues of the day. As the *Grammar* often argues, the whole man reasons in such cases as arriving at religious beliefs, and logic may not be even the skeleton of such a whole man. When Newman stressed the importance of treating propositions in argument on these matters 'not as so many dead words, but as the words of a speaker in a particular state of mind', he set a task which logic alone could not perform, a task more suited to that dramatically situated lyricism which I have been describing.

Newman gives an excellent counter-instance to his own claim for logic when discussing what he calls 'natural inference': 'A parallel gift is the intuitive perception of character possessed by certain men, while others are as destitute of it, as others again are of an ear for music. What common measure is there between the judgements of those who have this intuition, and those who have not?'[180] Much the same might be said of those who have an ear for the voice of a religious calling and those who have not. Eventually, it is not logic but Providence on which Newman relies to reconcile what was for him the supreme divergence of English minds:

Here, I say again, it does not prove that there is no objective truth, because not all men are in possession of it; or that we are not responsible for the associations which we attach, and the relations which we assign, to the objects of the intellect. But this it does suggest to us, that there is something deeper in our differences than the accident of external circumstances; and that we need the interposition of a Power, greater than human teaching and human argument, to make our beliefs true and our minds one.[181]

Hopkins's rhythmic practice as Geoffrey Hill describes it, at once in and out of stride depending on the standard of judgement, could be seen and heard as an attempt to meet the sort of quandaries Newman feels in his writing about the existence, somewhere, of a 'common measure' not actually present for his writing. A measure is both a 'recognized intellectual standard' and a tune, and it is therefore possible for a poetic measure to turn the clash and fret of competing

[180] Ibid. 215.   [181] Ibid. 242.

standards into a pleasure like that of musical counterpoint. Hopkins's lines can sound at once deviant and true, true in the sense in which we say (bringing sound and intellectual rectitude together) that a remark 'rings true'; they achieve tonally a composed doubleness of voice which acknowledges the ways of speaking and the ways of life which the verse alters and seeks to convert.

That doubleness of 'voice' is possible only on the page, for any physical voicing will at least move, though it may not settle, in one direction or the other. So, for example, the points in the mature verse where the habitual order of the words of spoken English is wilfully, but not self-wilfully, inverted witness to the bearing of this Catholic poet towards the world, and the character of the reaction of that bearing upon him:

> When will you ever, Peace, wild wooddove, shy wings shut,
> Your round me roaming end, and under be my boughs?
> When, when, Peace, will you, Peace?—I'll not play hypocrite
>
> To own my heart: I yield you do come sometimes; but
> That piecemeal peace is poor peace. What pure peace allows
> Alarms of wars, the daunting, wars, the death of it?[182]

'To own my heart', as Hopkins himself wrote,[183] means 'to my own heart': a paraphrase should read 'I won't be deceitful about what goes on in my heart, I do sometimes experience peace'. Bending 'to my own heart' into 'to own my heart' produces the characteristic Hopkins strain of simultaneous effort and fluency. Grammatically, we feel the words under pressure; rhythmically, they achieve a smoother melody than they could without the inversion. The voicing of the lines must decide between equally and divergently demanding duties: to stress the grammatical strain, to follow the melodic arch. What differentiates an inversion like this from 'Then fail'd the tongue' in 'The Escorial' is that the inversion introduces here an apt searching of the word 'own' in its relation to 'heart'. If we take, as the inverted syntax permits us, 'own' to be a verb meaning 'to confess' or 'to possess', we can paraphrase as 'I won't falsify my experience so as to seem the more simply and completely in possession of it' or as 'I won't falsify the intermittences of my heart so as to be able to say something more immediately striking about it'. There are other ambiguities in these lines; what matters is not the ambiguity but its source—the

---

[182] 2 Oct. 1879, Phillips, 149.
[183] Letter to Bridges, 21 Aug. 1884, *GMH/Bridges*, 196.

simultaneously common and uncommon measure of the words which negotiates between communities within the language. Similarly, 'I yield' is archaic, and so deviates from the speech of Hopkins's day, when it is used here to mean 'I acknowledge', but the deviance permits the uttering of a religious truth: when the self submits to the Divine Will, it experiences spiritual peace: 'I' yields and therefore 'You' comes. This is also true of Hopkins's individual creation for and against the spirit of the English language.

'The Wreck of the Deutschland'

In his valuable account of the genesis of 'The Wreck of the Deutschland', Norman Weyand wrote that on the occasion of the wreck, 'Hopkins, together with all England, was evidently moved'.[184] It would also be true to say that Hopkins felt moved by the evidence in that wreck of how apart he was from 'all England'. Everybody read the story, of course. The Victorians were fond of narratives about shipwrecks and read avidly such works as W. H. D. Adams's *Great Shipwrecks: a Record of Perils and Disasters at Sea* (1877), *Constable's Miscellany of Original and Selected Publications*, Vol. 89: *Shipwrecks and Disasters at Sea* (1883), *Great Shipwrecks of Queen Victoria's Reign* (Liverpool, 1887), and Uncle Hardy's *Notable Shipwrecks* (1883). Shipwrecks were matters of concern as well as entertaining in the way that disasters are; Parliament had select committees to investigate the causes of shipwrecks and to study seaworthiness in 1826, 1839, 1843, and 1873.[185] The literature of the period bears the impress of such concern, in Poe's *The Narrative of Arthur Gordon Pym* and Clarke's *For the Term of His Natural Life*, in *David Copperfield* and *Armadale*, in Tennyson's *Enoch Arden* and 'The Wreck'. Hopkins shared the habit of reading about disasters, as his letter to his mother announcing the conception of his first mature masterpiece makes clear:

I am writing something on this wreck, which may perhaps appear but it depends on how I am speeded. It made a deep impression on me, more than any other wreck or accident I ever read of.[186]

---

[184] 'The Historical Basis of *The Wreck of the Deutschland* and *The Loss of the Eurydice*', in *Immortal Diamond*, ed. Norman Weyand (1949), 353.

[185] I am indebted here to A. W. Brian Simpson's *Cannibalism and the Common Law* (Chicago, 1984).

[186] Letter of 24 Dec. 1875, *Further Letters*, 135.

The habit of reading about disasters—'any other wreck or accident I ever read of'—underlines part of the poem he eventually wrote, that part of it which is concerned with knack, with professional dextrousness born of long familiarity, and particularly underlies Hopkins's remarkable calibration against each other of the knack of sailors and the knack of a poet (for, sorrowful though 'The Wreck of the Deutschland' is, it also stands as a poem of announced mastery, like *The Rape of the Lock*, in which we sense a poet's knowledge that he is coming into his own). The closeness of poetic and maritime skill in the work is already hinted at when Hopkins uses an idiom of voyaging in this letter to his mother—'it depends on how I am speeded'.

We need to recall the frequency of these newspaper reports and the bizarrerie of the details they sometimes contained to locate the range of tone from the elevated to the roustabout in Hopkins's poem. Coventry Patmore described one wreck which surprisingly failed to catch the journalistic eye:

The storms have made immense chaos here. Two or three days ago a French schooner laden with Hollands and musical instruments was wrecked at Fairlight, and the shore was strewn with corpses and pianos. All the crew was drowned except one, who was found wandering up the cliffs with his wits gone. Three hundred wreckers (the Hastings fishermen chiefly) tapped the Hollands and some died of the drink; and the ruin of the parade and several houses is a sight to see. No London newspaper seems to have heard of the wreck.[187]

'Strewn with corpses and pianos' parcels up lost lives and lost commodities in a way that shocks us now, and was shocking then, though it is to Patmore's credit that he did not conceal the facts of this and similar cases. Wrecking was a frequent crime in the period, as was the pillaging of wrecks. The wreck of the *Deutschland* was pillaged too, and the reports in *The Times* which Hopkins read (on 8, 9, 10, 11, and 13 December 1875), from which his poem began, concentrated on such heartless scavenging and on the absence of proper rescue services. They were concerned with man's inhumanity to man, as Hopkins was in lines such as 'Nor rescue, only rocket and lightship, shone'. But Hopkins's poem turns on more than this, it turns on the question of God's inhumanity to man, for it conducts a theodicy as it describes an accident.

He preached a sermon in 1880 which brings together the

[187] Champneys, ii. 273.

theological searching-out of God's providence from brutal happenings and his own laborious creation of a double aspect in the hearing of his mature poems as at once deviant from and deeply true to the nature of the English language:

God knows infinite things, all things, and heeds them all in particular. We cannot 'do two things at once', that is cannot give our full heed and attention to two things at once. God heeds all things at once. He takes more interest in a merchant's business than the merchant, in a vessel's steering than the pilot, in a lover's sweetheart than the . . . lover, in a sick man's pain than the sufferer, in our salvation than we ourselves.[188]

His superiors disapproved of the word 'sweetheart' in a sermon, and orthodoxy may indeed require a less jumpy style than Hopkins worked in, but the strength of 'The Wreck of the Deutschland' as a poem responsive to shores strewn with 'corpses and pianos' lies just in this ability to leap between levels of English as between levels of fact. Doing two things at once is actually the structural principle of the poem as it tosses words back and forth between conflicting, connatural senses. The most obvious thing about it is that it is in two parts, and between those two parts Hopkins creates an interim in which poetically there can be heard the workings of that 'eventual-ness' which I have previously described at work in his mature style.

'The Wreck of the Deutschland', like that other masterpiece of theological inquiry at sea, 'The Rime of the Ancyent Marinere', is a gigantic echo-chamber.[189] Words recur from part to part with altered connatural senses and, as one drops a stone down a well, waiting for the splash to determine the depth, so in the poem it is the distance between occurrences which most needs to be fathomed. Thus, in the third stanza, Hopkins writes of his heart flashing 'from the flame to the flame then' and the context makes plain that the line means 'escape from the flames of Hell by appeal to Christ's redemptive sacrifice', setting the flames of supernatural punishment against the ardour of atoning love. In the eleventh stanza, Death has the first words of the second part:

> 'Some find me a sword; some
> The flange and the rail; flame,
> Fang, or flood' goes Death on drum . . .

[188] Sermon of 25 Oct. 1880, *GMH Sermons*, 89.
[189] The phrase is from William Empson's introduction to *Coleridge's Verse: A Selection*—'the whole poem is an echoing-chamber, taking to a wild extreme the ballad technique of repetition', 65.

Death, aptly, is literal-minded and this-worldly, his 'flame' just part of a catalogue of fatalities, blankly enumerated, and quite without the dimensions of a divine plan which 'flame' had had earlier in the poem. A similar afflictive stripping of theological orientation from terms occurs when

> I steady as a water in a well, to a poise, to a pane,
> But roped with, always, all the way down from the tall
>   Fells or flanks of the voel, a vein
> Of the gospel proffer, a pressure, a principle, Christ's gift.

becomes

> One stirred from the rigging to save
> The wild woman-kind below,
> With a rope's end round the man, handy and brave—
> He was pitched to his death at a blow . . .

Hopkins's security in his vocation, as he dedicates himself to become worthy of the promises of Christ and is maintained by divine grace, topologically and topographically contrasts with the sailor's hardihood which has only man-made ropes to rely on. The landscape in which Hopkins writes, with its streams trickling down hillsides to replenish wells, is re-imagined as a ship, the hills masts and the streams rigging. The juxtaposition does not carry only the pious moral that men should trust only in God; it also reflects on the comfortable circumstances in which moralizing like that usually goes on, and how frail such moralizing seems when confronted with terrible occasions of need in which people must, however hopelessly, try to do something for each other and themselves. Hopkins and the mariner in their diverse ways are both perfected in their occupations; Hopkins grows towards a 'poise' and the sailor is 'pitched'—'pitch' being for Hopkins a musical term with a metaphorical sense of formed will as well as a term for violent physical throwing. The sailor is 'pitched to his death' but also 'pitched' in his death: he comes to an integrity as he dies, because he dies for others. Reading across the poem in this way,[190] we are not asked to regard the later occurrences of words as superseding what has come before, or revealing the truth of what was previously implicit. Nor do the usages offer 'alternative viewpoints' on the same issue, viewpoints between which we should choose. The occurrences are the

---

[190] Compare also 'wafting' (l. 35) into 'waft' (112), 'flood' (55) into 'flood' (128), 'at bay' (56) into 'bay' (95), and so on.

constituents of a span of timed thought, and it is the span we are asked to realize.

The poem does not aim at the exposition of settled doctrine but at the creation of a stressed time, a time in which we may, as Newman wrote, 'habitually search out and lovingly hang over the traces of God's justice'.[191] Newman's phrase 'habitually search' perfectly aligns with Hopkins's aim in the poem, for the phrase brings together the values of familiarity ('habitually') with a converting retort on the familiar ('search') in a manner stylistically and religiously true to this writing. 'The Wreck of the Deutschland' lovingly hangs over the traces of God's justice in its composed dwelling on certain words as also in its vital sense that what can be discerned in this wreck is only 'traces', fugitive signs of a permanent scheme. Just the fact that they are only 'traces' requires the person who would read them to hang over them lovingly, for such committed attention half-creates the patterns it perceives. Nothing in the text, or in the world, unambiguously indicates God's justice, any more than a text can unambiguously indicate a living voice. There is a 'want of witness' in creation[192]—a lack of evidence and a need for testimony.

This might seem a modernist view of theodicy and one which would not have recommended itself to Hopkins. It is not. It is the classic account of Saint Augustine when he describes how through years of significant perplexity he came to hear providence as 'a kind of eloquence in events';[193] it is also the account Hopkins gave of his own searchings of pattern as he sought for pattern:

. . . search the whole world and you will find it a million-million fold contrivance of providence planned for our use and patterned for our admiration.

But yet this providence is imperfect, plainly imperfect. The sun shines too long and withers the harvest, the rain is too heavy and rots it or in floods spreading washes it away; the air and water carry in their currents the poison of disease; there are poison plants, venomous snakes and scorpions; the beasts our subjects rebel . . . ; at night the moon sometimes has no light to give, at others the clouds darken her; she measures time most strangely and gives us reckonings most difficult to make and never exact enough; the coalpits and oilwells are full of explosions, fires, and outbreaks of sudden death, the sea of

---

[191] Sermon of 8 Apr. 1832, UE 107.   [192] See above, p. 281.

[193] The whole passage from *Concerning the City of God against the Pagans*, Book XI, ch. 18, is relevant, particularly Augustine's comparison of the contrasts in world history to 'the kind of antithesis which gives beauty to a poem'. I quote from the translation of Henry Bettenson (Harmondsworth, 1972), 449.

storms and wrecks, the snow has avalanches, the earth landslips; we contend
with cold, want, weakness, hunger, disease, death, and often we fight a losing
battle, never a triumphant one; everything is full of fault, flaw, imperfection,
shortcoming; as many marks as there are of God's wisdom in providing for us
so many marks there may be set against them of more being needed still, of
something having made of this very providence a shattered frame and a broken
web.[194]

To hold together the 'million-million fold contrivance of providence'
and the fact that such contrivance is 'plainly imperfect' requires doing
two things at once. Hopkins attempts such a benign double-dealing in
the poem by creating it as a 'shattered frame'; it seems at once an act of
ordering and an irruption of disorderly forces onto that act (as I
described previously with respect to 'Spelt from Sibyl's Leaves'), in
stride and out of stride. The poem offers an aesthetic activity of doing
two things at once which shall afford both a model of theological
understanding and a form of that 'common measure' between unlike
minds, man's and God's as well as various men's, which Newman
desired. It makes much, for example, of the likeness of words to each
other, of mercy to mastery or Calvary to chivalry, but it does not
superstitiously rest on those resemblances as fixed and encoded signs
of God's purpose, rather it employs the resemblances as objects for a
meditational patience, a time spent scanning the words and what lies
between them. This patience is a conduct of theodicy, not, that is, a
philosophical argument to prove God's justice, but a way of spending
time with the question of his justice, a way of living with questions
which may, practically, answer them.

What was different about the wreck of the *Deutschland* such that it
impressed Hopkins more deeply than any other wreck or accident he
had read of? When he lists catastrophes in his sermon on providence,
he speaks of 'storms and wrecks' as part of a general imperfection in
divine providence as men usually understand it, but human law as well
as divine law had a hand in this disaster. The five nuns were at sea
because they were victims of Bismarck's Falk laws against Catholics in
Germany, exiles for their faith as Hopkins was to feel himself exiled in
'To seem the stranger . . .'. Two years before Hopkins wrote the poem,
the Catholic paper, *The Freeman's Journal*, had compared England to
Bismarck's Germany when it commented on the fact that in England
an Elizabethan statute forbidding the introduction of Papal documents
into the country had been successfully invoked in a recent law-suit:

[194] Sermon of 25 Oct. 1880, *GMH Sermons*, 90.

The all-important fact, however, remains—that the first common law tribunal in the land has decided that every rescript, every document, every letter sent by His Holiness the Pope to his spiritual subjects in this country with reference to their spiritual affairs, is an illegal document. . . . In the many cruel laws with which Prince Bismarck is afflicting Catholic Germany there is none so stern and sweeping as this Act of Elizabeth which has been suddenly disentombed from the bloodstained code of the second Tudor Queen. . . . Such a state of the law . . . turns Emancipation into a mockery, destroys religious equality at a blow, calls the penal code again into life. . . .[195]

This is excitable stuff, but probably the Hopkins who wrote to his mother about the persecution of Spanish Jesuits 'To be persecuted in a tolerant age is a high distinction'[196] responded to such fervour.

It matters that there was this human element of ill will involved in the fate of the nuns for the destructive forces unleashed and undergone were not something Hopkins saw as simply opposed to human civilization but rather something which formed part of civilization's self-imperilling essence. He had written to Bridges that civilization 'as it at present stands in England . . . is itself in a great measure founded on wrecking';[197] he was referring to the treatment of the working classes and, more generally, to that toll of exploitation on which prosperity is built whose best image in the nineteenth century was seafaring and what could happen to those who went to sea. Perhaps Hopkins also remembered a passage in Newman where, for different reasons, Newman identified the English Establishment itself as a wreck when explaining why he continued to try to persuade Anglicans to come over to Rome:

It is this keen feeling that my life is wearing away, which overcomes the lassitude which possesses me, which scatters the excuses which I might plausibly urge to myself for not meddling with what I have left for ever, which subdues the recollection of past times, and which makes me do my best, with whatever success, to bring you to land from off the wreck, who have thrown yourselves from it upon the waves, or are clinging to its rigging, or are sitting in heaviness and despair upon its side. For this is the truth: the Establishment, whatever it be in the eyes of men, whatever its temporal greatness and its secular prospects, in the eyes of faith is a mere wreck.[198]

Newman's prose is full of storms and harbours; the Anglican

---

[195] (1873), repr. in *Anti-Catholicism*, 88–9.
[196] Letter of 7 Feb. 1869, *Further Letters*, 106.
[197] Letter of 2 Aug. 1871, *GMH/Bridges*, 28.
[198] *Anglican Difficulties*, UE i. 4.

Establishment was a 'mere wreck' and, by the same token, coming into
the Church of Rome 'was like coming into port after a rough sea'.[199]
Beautiful though Newman's rhetoric is at such a moment in the
Apologia, it seems narrowed and placid in its sense of conversion when
we compare it with Hopkins's recognition that Catholics too could
have experience of wrecks.

This separation from England needs to be borne in ear if we are
properly to hear what happens in the poem when Christ is revealed:

> But how shall I . . . make me room there:
> Reach me a . . . Fancy, come faster—
> Strike you the sight of it? look at it loom there,
> Thing that she . . . There then! the Master,
> *Ipse*, the only one, Christ, King, Head . . .

'*Ipse*' is Latin for 'his very self' but Hopkins did not put in English that
supreme stress on the supreme self of Christ. The most intimate word
appears strange and possibly estranged in his poem, as 'selve' is odd in
his vocabulary. '*Ipse*' is set apart both by being Latin and by being in
italics. As Hopkins noted, 'Italics do look very bad in verse. But people
will *not* understand where the right emphasis is',[200] the remark proves
its own truth by needing italics on '*not*'. Italics serve two principal
functions in English writing; they mark a word as foreign or they ask
for it to be stressed. Hopkins's *Ipse* does both: it marks its stress as
alien. Hopkins believed that Christ was 'the right emphasis' for
English people, and the Christ of the Catholic Church with its Latin
liturgy, but how was he to put that belief to the English, especially
when they believed they were already in the right?

This unvoiceable density of '*Ipse*' on the page makes an opportunity
for the English to hear their own voice: the word cannot be
pronounced as if it were an English word (to rhyme, say, with 'tipsy')
for then it would set off a jarring of rhymes in the stanza with 'she',
'the', 'only', 'He', 'extremity', and that would diminish the uniqueness
of its occurrence as 'the only one'. If, on the other hand, it is
pronounced as in Church Latin, it sounds neither in the language of
the English nor of the Anglican Church; it has no ceremonial place.
The dilemma of speaking that word is rich with the accidents of cultic
history, as we can see if we contrast this moment in Hopkins with the
close of a poem Newman wrote while he was still an Anglican,
'Absolution':

---

[199] Svaglic, 214.   [200] Letter to Bridges, 26 Jan. 1881, *GMH/Bridges*, 120.

'Look not to me—no grace is mine;
But I can lift the Mercy-sign.
This wouldst thou? Let it be!
Kneel down, and take the word divine,
    ABSOLVO TE.'[201]

It is no wonder that Newman was so acute about the High Church
pastor who wanted to hear confessions which his flock did not want to
make; he had been there himself. The Latin formula of absolution
does not rhyme with the English 'Let it be!', not even capitals can
make it do so. If 'TE' is to rhyme with 'be', then it must be pronounced
as if it were 'tea', and that, though a very English word, is scarcely a
'word divine'. The Anglican Newman attempts to make the Roman
liturgy and the English literary language consort but produces only a
false concord; Hopkins's *Ipse* makes no such attempt, it is a retort on
English, a retort which looks forward to the poem's hope from out the
wreck, that (eventually) this '*Ipse* . . . King' should be 'Our King back,
Oh, upon English souls!' The prepositions there 'back . . . upon' are
deliberately awkward, and made more so by the gauche interruption of
'Oh'. Had he written 'back over' or 'back in', the words would sit more
easily, but his is an honest awkwardness: a conversion is a tricky
business, and it resembles a retort in that it turns a life—whether an
individual's or a nation's—'back upon' itself. *Ipse* may be 'the Master'
but it is something other than what is usually appreciated as poetic
'mastery' which happens when Hopkins writes the word.

'The Wreck of the Deutschland' celebrates at its heart an act which
is also a passivity, a masterful surrender patterned on Christ's
passional suffering unto death—the nun's calling:

She to the black-about air, to the breaker, the thickly
Falling flakes, to the throng that catches and quails
    Was calling 'O Christ, Christ, come quickly':
The cross to her she calls Christ to her, christens her wild-worst Best.

'To' in these lines, that word of direct address, of Hopkins's 'bidding',
veers in direction and sense, as if the nun turned now here, now there.
The suspensive construction of the sentence holds her utterance to
one clear purpose, one addressee, but her purpose goes through a
thick, refractive atmosphere. She was calling 'to the black-about air, to
the breaker'—not speaking her message to them but as if appealing to
Christ to come out of this material null and save her from it. She was

---

[201] Dated 14 Dec. 1832 by Newman, UE 84.

calling 'to the throng that catches and quails', to the people on deck, and she calls to them in a manner different from her call 'to the black-about air'. She called to them, presumably, in encouragement or admonition, and they heard her with mixed feelings; the throng 'catches' her drift and its breath, it 'quails' from the danger of shipwreck, and from her message amidst that danger, and from the danger in her message which converts the habit of self-preservation into the requirement of self-sacrifice. She calls the cross to her (she summons it) and she calls that which is a cross to her (her sufferings in this wreck) her Christ (her saviour) because she patterns her sufferings on the crucifixion and accepts them as signs that she is joined to her 'Best' in a pained and redemptive love.

The Times, as Hopkins read it, gave a different version of her cry: 'the chief sister, a gaunt woman 6ft. high, calling out loudly and often "O Christ, come quickly!" till the end came.'[202] Hopkins has added his own voice to her call by adding his own characteristic, repetitive caress on a word turning 'O Christ' into 'O Christ, Christ'. He also removed the unfortunate clash in the newspaper's report of 'come' against 'came'—'calling out loudly and often "O Christ, come quickly!" till the end came'—which makes it sound as if what she wanted was Christ but what she got was her 'end'. Hopkins would have said that she died in Christ. Actually, the nun, being German, probably cried neither 'O Christ, Christ, come quickly' nor 'Oh Christ, come quickly'; her speech may have been foreign to either English report. She may not even have been speaking German. Hopkins would have known that the cry to the Divine to come quickly is a liturgical cry in the Office of the Church, especially in evening prayers and in the season of Advent when the *Deutschland* went down; 'Dominum in adiutorium meum intende, domine ad adiuvandam me festina' ('Lord, hear my cry, O Lord, make haste to aid me'). She was perhaps not crying at all, she was saying her office. It is not likely that the *Times* reporter would have heard that in the nun's words but Hopkins, sharing a faith and a language of faith with her, translates her cry into his English lines through a Latin which he shares with her, foreign though his English and her German were to each other when they spoke. The greatness of the art in this poem permits us to believe that he heard both the nun's cry and the reporter's deafness to its sense; he was generous enough to attend at once to *The Times* and to the *Graduale Romanum*. The exceptional verse elicits a common measure.

x

[202] Report of 11 Dec. 1875, in Weyand, op. cit. 368.

These are accidents of how language has happened to people; the story of Hopkins, as the story he tells in this masterpiece, is that of 'any other wreck or accident I ever read of'. A masterpiece is also an accident. People *might* always have understood each other but then they would lack a prime material for imaginative creation—the possibility of being mistaken. In 'The Wreck of the Deutschland', the cultural multiplicity of its sources corresponds to a fantastic theological and tonal amplitude (theology and tone are at one in the poem because it finds the 'reality' of an idea). Take

> One stirred from the rigging to save
> The wild woman-kind below,
> With a rope's end round the man, handy and brave—
> He was pitched to his death at a blow,
> For all his dreadnought breast and braids of thew:
> They could tell him for hours, dandled the to and fro
> Through the cobbled foam-fleece. What could he do
> With the burl of the fountains of air, buck and the flood of the wave?
>
> They fought with God's cold—
> And they could not and fell to the deck
> (Crushed them) or water (and drowned them) or rolled
> With the sea-romp over the wreck.

The parentheses are meticulously faithful to brutal facts as Hopkins had always been from 'The Escorial' on. Why is such pain put in parenthesis? '(Crushed them)' and '(and drowned them)'—are these side-issues or so terrible as they confront the imagination that they can be spoken, or quasi-spoken in the printed voice, only under the breath? Voicing these lines, a reader is thrown between a reporter's indifferent noting and a truly participative intake of dismay. What Hopkins shockingly calls 'God's cold' falls on the verse itself. In his sermon on providence he noted that 'we contend with cold', perhaps remembering 'They fought with God's cold', but what remains a perplexity in the preacher's voice as he tries to fit evil and pain into a loving scheme was made an achieved absence of voice in the lines of the poem. The victims of the wreck 'could not', as Hopkins puts it with fierce abruptness. 'Could not' what? Could not kindle that cold, could not triumph, could not get free, perhaps. The abruptness of the modal verb is an absolute of failure; it slams human ability against something deadly and unremitting like the sea. It is meant as a picture of God. Yet Hopkins created this incapacity, and his capacity thus to create turns the God of this wreck from a blank demon into an incarnate lord, one flesh with what suffers. It is as if God were the reader of the poem, poised between vocal expertise and having nothing to say. The reader 'can not' as the victims 'could not': readers and victims become together.

The stanzas have many near-sarcasms about human enterprise, on the sea or in verse. Consider the fracturing of 'handyman' into 'the man, handy', or the pivot of religious and practical concern on the word 'save' in 'to save | The wild woman-kind below'. This sailor took on more than he could cope with when he tried to 'save' the women, but it was not wrong of him to try. He had a 'dreadnought breast'. The *OED* tells us that for nineteenth-century England a 'dreadnought' was 'a thick coat or outer garment worn in very inclement weather'. Hopkins's phrase makes the man seem to take a confidence in his own chest such as a manufacturer of dreadnought overcoats might have advertised as on offer to a purchaser (a slogan is imaginable: 'You can face anything in a Dreadnought'). This is a confidence in human skill, luxuriant in its own prowess as the decorative 'braids of thew' shows. But it comes up hard in the lines against something other: 'for *all* his dread*nought* breast' (my emphasis). It is an all or nothing matter, as death always is, but the sailor was not prepared to see it that way in his 'dreadnought'. None the less, this male skill, at the ready to save women, was treated as merely childish by the power of the elements; he with all his confidence was 'dandled' for hours, like a baby in the hands of superior force. 'Dandled' is ferociously accurate, cruel even, but it is the cruelty only of coming to know the human place in this world; the placing of that word is at the same time a being-placed for the skill which found it.

These lines undergo, as the whole poem does, the paradox of trying to imagine God's cold power as evidence of his love. Where the voice is troubled or made to lack by the text, there a 'want of witness' occurs. The appeal for a voice is most acute where no voice can sound what is being said, but the poem shows that what cannot be spoken need not be passed over in silence; the unspeakable can be written down and dwelt upon. Nobody could voice what Hopkins meant by an invention like 'sea-romp': it is a frisky titanism that bounds about and elbows the human aside as a trivial bit in its big games, but also, because 'romp' imaginatively derives from the Latin 'rompere', 'to shatter', it counts the human cost in these elemental shows of strength, measures breakage as the word itself breaks up.

Words are created to turn in the storms of human purpose. Only by being buffeted do they carry between people. 'The Wreck of the Deutschland' finds reconciliation just when it recoils:

> Sister, a sister calling
> A master, her master and mine!—
> And the inboard seas run swirling and hawling;
> The rash smart sloggering brine
> Blinds her; but shé that weather sees óne thing, one;
> Has óne fetch ín her: she rears herself to divine
> Ears, and the call of the tall nun
> To the men in the tops and the tackle rode over the storm's brawling.

'She rears herself to divine' sounds at first as if 'divine' were a verb, as if she were about to prophesy, but the lineation cuts the words down to size, humbles the verb to an adjective—'she rears herself to divine | Ears'. She does not speak out, she asks to be heard. That happens to be the case of Hopkins himself as a poet, and it is exemplary. 'The Wreck of the Deutschland' is also what it itself hearkens to, 'an ark | For the listener'.

[Work like this is always about to turn into wreck.] What is called 'mastery' in the enthusiasm of literary description was made, by Hopkins as by many other poets, in a material and constant maladjustment of two states of natural language with regard to each other, made in that no-man's-land which is the region between speech and writing. It is a terrain of rare impartiality but also a land of scars, an image of perfection actually disfigured. Hopkins wrote to his friend, Canon Dixon, in 1886:

See how the great conquerors were cut short, Alexander, Caesar just seen. Above all Christ our Lord: his career was cut short and, whereas he would have wished to succeed by success—for it is insane to lay yourself out for failure, prudence is the first of the cardinal virtues, and he was the most prudent of men—nevertheless he was doomed to succeed by failure; his plans were baffled, his hopes dashed, and his work was done by being broken off undone. However much he understood all this he found it an intolerable grief to submit to it. He left the example: it is very strengthening, but except in that sense it is not consoling.[203]

It is apt that these words also describe Hopkins's work, a poet's work, in a language which matters to him because it does not belong only to him. The language is a witness, and poetic work bears witness in and to a language and the world of those who use it. There are in fact

[203] Letter of [?13] July 1886, *GMH/Dixon*, 137–8.

differences about what words make evident; those differences are the matter of poetry in its 'million-million fold contrivance' which remains 'plainly imperfect'.

Hopkins wasn't only a poet. A fragment of one of his sermons survives:

This is the house of God, this is the gate of heaven. None but you . . . can say this. See the old churches, temples of God turned to dens of thieves. Pious according to their lights but are children of those who stoned the prophets, i.e. of those who persecuted the people of God. This is the house of God not only because God recognises this only but because God comes here, gate of heaven not only because sacraments of baptism and penance are administered here but because there is a key here which unlocks heaven and brings Christ from heaven out. This key is in my lips. I have the power etc[204]

Poetry is not written in this key, and, as a poet, Hopkins could not have had such confidence in his lips.

[204] Sermon of 14 Sept. 1879, *GMH Sermons*, 235.

BIBLIOGRAPHY

AARSLEFF, H., *The Study of Language in England, 1780–1860* (Princeton, 1967).

ALLINGHAM, W., *A Diary*, ed. H. Allingham and D. Radford (1907). [Cited as Allingham in notes.]

ALTIERI, C., *Act and Quality* (Brighton, 1981).

ANSCOMBE, G. E. M., *Intention* (Oxford, 1957, 2nd edn., 1963).

ARCHER, W., *Real Conversations* (1904).

ARMSTRONG, I. (ed.), *The Major Victorian Poets: Reconsiderations* (1969).

—— (ed.), *Writers and their Background: Robert Browning* (1974).

ARNOLD, M., *Complete Poems*, ed. K. Allott (1965).

—— *The Complete Prose Works of Matthew Arnold*, ed. R. H. Super (11 vols.; Ann Arbor, 1960–77).

ST AUGUSTINE, *Concerning the City of God against the Pagans*, trans. H. Bettenson (Harmondsworth, 1972).

AUSTIN, J. L. *Philosophical Papers*, ed. J. O. Urmson and G. J. Warnock (Oxford, 1961, 2nd edn., 1970).

—— *How to Do Things with Words*, ed. J. O. Urmson (Oxford, 1962).

BAGEHOT, W., *The Collected Works of Walter Bagehot*, ed. N. St-John Stevas (15 vols.; 1965–86).

BARON, M., 'Speaking and Writing: Wordsworth's "Fit Audience"', *English*, 32 (Autumn, 1983).

BARTHES, R., *Le grain de la voix* (Paris, 1981).

BAUDELAIRE, C., 'De l'essence du rire . . .', repr. in *Curiosités esthétiques . . .*, ed. H. Lemaître (Paris, 1962).

BAYLEY, J., *An Essay on Hardy* (Cambridge, 1978).

BECKETT, S., *Molloy* (1950), trans. S. Beckett and P. Bowles (1955, repr. 1959).

—— *Collected Shorter Prose 1945–1980* (1986).

BERRY, F., *Poetry and the Physical Voice* (1962).

BLOUNT, B. G. and SANCHES, M. (eds.), *Sociocultural Dimensions of Language Change* (New York, 1977).

BOLINGER, D. (ed.), *Intonation* (Harmondsworth, 1972). [Cited as Bolinger in notes.]

BOSWELL, J., *Life of Johnson*, ed. G. B. Hill, rev. and enlarged, L. F. Powell (6 vols.; 1934, 1964–71).

BOULGER, J. D., *Coleridge as Religious Thinker* (New Haven, 1961).

BRADLEY, F. H., *Ethical Studies* (1876, 2nd edn., 1927).

BRIAN SIMPSON, A. W., *Cannibalism and the Common Law* (Chicago, 1984).

BROWNING, E. B., *Poetical Works*, ed. F. G. Kenyon (1897). [Cited as Kenyon in notes.]

BROWNING, R. and BARRETT, E. B., *The Letters of Robert Browning and Elizabeth Barrett Barrett, 1845–1846*, ed. E. Kintner (2 vols.; Cambridge, Mass., 1969).

BROWNING, R.. *The Poems*, ed. J. Pettigrew, supplemented and completed by T. J. Collins (2 vols.; Harmondsworth, 1981). [Cited as P in notes].

—— *The Ring and the Book*, ed. R. D. Altick (Harmondsworth, 1971). [Cited as Altick in notes.]

—— *Letters of Robert Browning: Collected by T. J. Wise*, ed. T. L. Hood (Yale, 1933, repr., Dallas, 1973).

—— and BLAGDEN, I., *Dearest Isa: Robert Browning's Letters to Isabella Blagden*, ed. E. C. McAleer (Austin, 1951).

—— and WEDGWOOD, J., *Robert Browning and Julia Wedgwood: A Broken Friendship as Revealed in Their Letters*, ed. R. Curle (1937). [Cited as Curle in notes.]

BRUCE, F. F., *The Epistle to the Hebrews* (1964).

CARLYLE, J. W., *Jane Welsh Carlyle: A New Selection of her Letters*, ed. T. Bliss (1949).

CARLYLE, T., *Reminiscences* (2 vols.; 1881), ed. C. E. Norton (1887), repr., 1 vol., ed. I. Campbell (1972).

CARROLL, L., *The Penguin Complete Lewis Carroll* (Harmondsworth, 1982).

CASEY, J. P., *The Language of Criticism* (1966).

—— (ed.), *Morality and Moral Reasoning* (1971).

CERCIGNANI, F., *Shakespeare's Works and Elizabethan Pronunciation* (Oxford, 1981).

CHAMPNEYS, B., *Memoirs and Correspondence of Coventry Patmore* (2 vols.; 1900). [Cited as Champneys in Notes.]

COLERIDGE, S. T., *Biographia Literaria* (1817), ed. J. Engell and W. J. Bate (2 vols.; Cambridge, Mass., 1983).

—— *Specimens of the Table Talk of Samuel Taylor Coleridge*, ed. H. N. Coleridge (1837, 3rd edn., 1851).

—— *Poetical Works*, ed. E. H. Coleridge (1912).

COLLEY, A. C., *Tennyson and Madness* (Athens, Georgia, 1983).

COLLINS, P., *Reading Aloud* (Lincoln, 1972).

COULSON, J., *Newman and the Common Tradition* (Oxford, 1970).

COWPER, W., *The Poems of William Cowper*, ed. J. D. Baird and C. Ryskamp (2 vols.; Oxford, 1980).

CROCE, B., *Estetica* (Bari, 1902), trans. Douglas Ainslie as *Aesthetic as Science of Expression and General Linguistic* (1909, rev. edn., 1922).

CROW, C., *Paul Valéry and the Poetry of Voice* (Cambridge, 1982).

CRYSTAL, D., *Prosodic Systems and Intonation in English* (Cambridge, 1969).

—— *The English Tone of Voice: Essays in Intonation, Prosody, and Paralanguage* (1975).

DAVIE, D., *The Poet in the Imaginary Museum*, ed. B. Alpert (Manchester, 1977).

DAVIS, P., *Memory and Writing from Wordsworth to Lawrence* (Liverpool, 1983).

DERRIDA, J., *Of Grammatology* (Paris, 1967), trans. G. C. Spivak (Baltimore, 1976).

—— *Writing and Difference* (Paris, 1967), trans. A. Bass (1978).

—— 'Signature Event Context', in *Marges de la Philosophie* (Paris, 1972), trans. S. Weber and J. Mehlman, *Glyph*, 1 (Baltimore, 1977), 192. [Cited as SEC in notes.]

—— 'LIMITED INC a b c . . .', trans. S. Weber, *Glyph*, 2 (Baltimore, 1977). [Cited as INC in notes.]

DE SAUSSURE, F., *Course in General Linguistics*, trans. Wade Baskin (1959; rev. edn., 1974).

DICKENS, C., *Dr. Marigold* (1865), repr. in *Christmas Stories* (Oxford, 1956).

DIXON, P., ' "Talking Upon Paper": Pope and Eighteenth Century Conversation', *English Studies*, 46/1 (Feb. 1965).

DUMMETT, M. E., *Frege: Philosophy of Language* (1973; 2nd edn., 1981).

ELIOT, T. S., *Selected Essays* (1932, 3rd enlarged edn., 1951, repr. 1976).

—— *The Idea of a Christian Society* (1939).

—— *Knowledge and Experience in the Philosophy of F. H. Bradley* (1964).

—— *The Complete Poems and Plays* (1969).

EMPSON, Sir W., *Seven Types of Ambiguity* (1930, 2nd edn., 1947, repr., Harmondsworth, 1961).

—— *Some Versions of Pastoral* (1935).

—— *The Structure of Complex Words* (1951, 3rd edn., 1977).

—— *Using Biography* (1984).

—— and PIRIE, D. (eds.), *Coleridge's Verse: A Selection* (1972).

EVERETT, B., *Poets in their Time* (1986).

FITZGERALD, E., *The Letters of Edward FitzGerald*, ed. A. M. and A. B. Terhune (4 vols.; Princeton, 1980). [Cited as Terhune in notes.]

—— *Letters and Literary Remains*, ed. W. A. Wright (3 vols.; 1889).

FLAUBERT, G., *Correspondance*, ed. J. Bruneau (2 vols.; Paris, 1973).

FREUD, S., *Civilization and its Discontents* (Vienna, 1930), trans. Joan Riviere, rev. and ed. James Strachey (1963).

GAY, P., *The Bourgeois Experience*, vol. i, *Education of the Senses* (New York, 1984); vol. ii, *The Tender Passion* (New York, 1986).

GELB, I. J. *A Study of Writing* (Chicago, 1952, rev. edn., 1963).

GRICE, H. P., 'Utterer's Meaning and Intention', *Philosophical Review* (1969).

—— 'Logic and Conversation', in P. Cole and J. Morgan (eds.), *Syntax and Semantics*, iii: *Speech Acts* (New York, 1965).

—— 'Presupposition and Conversational Implicature', in P. Cole (ed.), *Radical Pragmatics* (New York, 1981).

GURR, A., *The Shakespearean Stage 1574–1642* (Cambridge, 1980).

HACKER, P. M. S., *Insight and Illusion: Wittgenstein on Philosophy and the Metaphysics of Experience* (Oxford, 1972).

HALLAM, A. H., *The Writings of A. H. Hallam*, ed. T. H. Vail Motter (New York, 1943). [Cited as Motter in notes.]

—— *The Letters of Arthur Henry Hallam*, ed. J. Kolb (Columbus, Ohio, 1981). [Cited as Kolb in notes.]

HARDY, T., *The Complete Poems*, ed. S. Hynes (3 vols.; Oxford, 1982–5). [Cited as Hynes in notes.]

—— *The Personal Writings of Thomas Hardy*, ed. H. Orel (1967). [Cited as Orel in notes.]

—— *The Personal Notebooks of Thomas Hardy*, ed. R. H. Taylor (1978).

—— *The Life and Work of Thomas Hardy*, ed. M. Millgate (1984) from *The Early Life of Thomas Hardy* (1928) and *The Later Years of Thomas Hardy* (1930). [Cited as Hardy, *Life* in notes.]

HAVELOCK, E. A., *Preface to Plato* (Cambridge, Mass., 1963).

HAZLITT, W., *The Complete Works of William Hazlitt*, ed. P. P. Howe (21 vols.; 1930–4).

HEGEL, G. W. F., *Hegel's Philosophy of Right* (1821), trans. T. M. Knox (Oxford, 1952).

HENDERSON, E. J. A. (ed.), *The Indispensable Foundation: A Selection from the Writings of Henry Sweet* (1971).

HERRNSTEIN SMITH, B., *On the Margins of Discourse: The Relation of Literature to Language* (Chicago, 1978).

HILL, G. W., *The Lords of Limit: Essays on Literature and Ideas* (1984).

HOLLANDER, J., *Vision and Resonance* (New York, 1975).

HOPKINS, G. M., *The Poems of Gerard Manley Hopkins* (Oxford, 1918, 4th edn., rev. and enlarged, ed. W. H. Gardner and N. H. MacKenzie, Oxford, 1967, corrected, 1970).

—— *The Letters of Gerard Manley Hopkins to Robert Bridges*, ed. C. C. Abbott (Oxford, 1935, rev. edn., 1955). [Cited as *GMH/Bridges* in notes.]

—— *The Correspondence of Gerard Manley Hopkins and Richard Watson Dixon*, ed. C. C. Abbott (Oxford, 1935, 2nd, rev. imp., 1955). [Cited as *GMH/Dixon* in notes.]

—— *Further Letters of Gerard Manley Hopkins . . .* , ed. C. C. Abbott (Oxford, 1938, 2nd, rev. and enlarged edn., 1956). [Cited as *Further Letters* in notes.]

—— *The Sermons and Devotional Writings of Gerard Manley Hopkins*, ed. C. Devlin (Oxford, 1959). [Cited as *GMH Sermons* in notes.]

—— *The Journals and Papers of Gerard Manley Hopkins*, ed. H. House and G. Storey (Oxford, 1959). [Cited as *GMH Journals* in notes.]

—— *Gerard Manley Hopkins*, ed. C. Phillips (Oxford, 1986). [Cited as Phillips in notes.]

HUTTON, R. H., *Literary Essays* (1871, 3rd enlarged edn., 1888). [Cited as Hutton, *Essays* in notes.]

—— *Aspects of Religious and Scientific Thought*, ed. E. M. Roscoe (1899).

IRVINE, W., and HONAN, P., *The Book, the Ring and the Poet* (1975).

JAMES, H., *The Middle Years* (1917), repr., *Autobiography*, ed. F. W. Dupee (1956).

JOHNSON, S., *The Rambler* (1750–2), ed. W. J. Bate and A. B. Strauss (New Haven, 1969).

—— *The History of Rasselas, Prince of Abissinia* (2 vols.; 1759) ed. G. Tillotson and B. Jenkins (1971).

—— 'Preface' to *Shakespeare* (1765), repr., *Johnson on Shakespeare*, ed. A. Sherbo (2 vols.; New Haven, 1968).

—— *Lives of the English Poets* (1779–81, repr., 2 vols, 1906).

JOHNSON, W. S., *Sex and Marriage in Victorian Poetry* (Ithaca, 1975).

JONES, J., *Practical Phonography* (1701).

JOYCE, J. A., *Ulysses* (Paris, 1922), ed. H. W. Gabler *et al.* (3 vols.; New York and London, 1984).

JUMP, J. D. (ed.), *Tennyson: The Critical Heritage* (1967). [Cited as Jump in notes.]

KEATS, J., *The Poems of John Keats*, ed. M. Allott (1970, 3rd, corrected, impression, 1975).

—— *The Letters of John Keats 1814–1821*, ed. H. E. Rollins (2 vols.; Cambridge, Mass., 1958).

KEBLE, J., *The Christian Year: Thoughts in Verse for the Sundays and Holidays throughout the Year* (2 vols.; Oxford, 1827).

KENNY, A., *Action, Emotion and Will* (1963).

LASLETT, P., OOSTERVEEN, K., and SMITH, R. M. (eds.), *Bastardy and its Comparative History* (1980).

LEAVIS, F. R., *New Bearings in English Poetry: A Study of the Contemporary Situation* (1932, repr., Harmondsworth, 1963).

—— *Revaluation* (1936, repr., Harmondswroth, 1964).

—— 'Poet as Executant', *Scrutiny*, 15/1 (1947).

—— *The Common Pursuit* (1952, repr., Harmondsworth, 1964).

—— *Reading out Poetry* (Belfast, 1979).

LITZINGER, B., and SMALLEY, D. (eds.), *Browning: The Critical Heritage* (1970).

LONGFORD, E., *Victoria R.I.* (1964).

LORD, A. B., *The Singer of Tales* (Cambridge, Mass., 1960).

MACCABE, C., *James Joyce and the Revolution of the Word* (1978).

MCGREGOR, O. R., *Divorce in England: A Centenary Study* (1957).

MACKENZIE, N. H., *A Reader's Guide to Gerard Manley Hopkins* (1981).

MACKINNON, D. M., *Borderlands of Theology and Other Essays*, ed. G. W. Roberts and D. E. Smucker (1968).

MCLUHAN, M., *et al.*, *Gerard Manley Hopkins: A Critical Symposium by the Kenyon Critics* (1945, enlarged edn., 1975).

MALLARMÉ, S., *Œuvres complètes*, ed. H. Mondor and G. Jean-Aubry (Paris, 1945).

MAURICE, F. D., *Theological Essays* (1853, 4th edn., 1881).

MEREDITH, G., *Poems*, ed. P. B. Bartlett (2 vols.; New Haven, 1978). [Cited as Bartlett in notes.]

MILL, J. S., *Dissertations and Discussions Political, Philosophical and Historical* (4 vols.; 1859).

—— *Autobiography* (1873, repr. from MS, 1971), ed. J. Stillinger.

MILLER, J. R., *Home Making of the Ideal Family Life* (n.d.).

MILLGATE, M., *Thomas Hardy: A Biography* (1982).

MILROY, J., *The Language of Gerard Manley Hopkins* (1967).

MILTON, J., *Paradise Lost*, ed. A. Fowler (1968, corrected edn., 1980).

MOORE, G., *Ave* (1911).

MOORE, K., *Victorian Wives* (1974).

MOORES, D. F., *Educating the Deaf: Psychology, Principles, and Practices* (Boston, 1978).

NEWMAN, J. H., *The Tamworth Reading-Room . . .* (1841).

—— *Fifteen Sermons preached before the University of Oxford* (1843). [Cited as *University Sermons* in notes.]

—— *An Essay on the Development of Christian Doctrine* (1845).

—— *Loss and Gain* (1848).

—— *Discourses addressed to Mixed Congregations* (1849).

—— *Lectures on Certain Difficulties felt by Anglicans in Submitting to the Catholic Church* (1850; repr. and rev., 2 vols., as *Certain Difficulties felt by Anglicans in Catholic Teaching Considered*, 1879, 1876).

——*Lectures on the Present Position of Catholics in England* (1851).

——*Apologia pro vita sua* (1864–5, variously revised until about 1886), ed. M. J. Svaglic (Oxford, 1967). [Cited as Svaglic in notes.]

——*Verses on Various Occasions* (1865).

——*An Essay in Aid of a Grammar of Assent* (1870), ed. I. Ker (Oxford, 1985). [Cited as *Grammar* in notes.]

——*Sermons preached on Various Occasions* (1870).

——Uniform Edition of the Works (36 vols.; 1868–81). [Cited as UE in notes.]

——*The Letters and Diaries of John Henry Newman*, ed. C. S. Dessain *et al.* (31 vols.; 1961, in progress).

NORMAN, E. R., *The English Catholic Church in the Nineteenth Century* (Oxford, 1984). [Cited as Norman in notes.]

——(ed.), *Anti-Catholicism in Victorian England* (1968). [Cited as *Anti-Catholicism* in notes.]

NYSTRAND, M. (ed.), *What Writers Know: The Language, Process, and Structure of Written Discourse* (New York, 1982).

OHMANN, R., 'Speech, Literature and the Space Between', *New Literary History*, 4 (1974).

OLDHAM, J. B., 'On the Difficulties and Obscurities Encountered in a Study of Browning's Poems', *Browning Society Papers*, ii (1899–90).

ONG, W. J., *The Presence of the Word* (New Haven, 1967).

——*Orality and Literacy: The Technologizing of the Word* (1982).

PAGE, N. (ed.), *Tennyson: Interviews and Recollections* (1983). [Cited as Page in notes.]

PAGET, Sir R., *Human Speech* (1930).

——*Babel, or The Past, Present, and Future of Human Speech* (1930).

PATMORE, C., *The Poems of Coventry Patmore*, ed. F. Page (Oxford, 1949). [Cited as Patmore, *Poems* in notes.]

——'Essay on English Metrical Law' (1857), repr. from the rev. text of 1894, *Coventry Patmore's 'Essay on English Metrical Law': A Critical Edition with a Commentary*, ed. Sister M. A. Roth (Washington, 1961).

——*Principle in Art, Religio Poetae and other Essays* (1889; one vol. edn., 1913). [Cited as *Principle . . .* in notes.]

POOLE, A., *Tragedy: Shakespeare and the Greek Example* (Oxford, 1987).

POPE, A., *The Poems of Alexander Pope*, ed. J. Butt (1963).

POUND, E. L., 'The Island of Paris', *The Dial*, Dec. 1920.

——*Letters of Ezra Pound 1907–1941*, ed. D. D. Paige (1951).

PRATT, M. L., *Towards a Speech Act Theory of Literary Discourse* (Bloomington, 1977).

PRICKETT, S., *Romanticism and Religion: The Tradition of Coleridge and Wordsworth in the Victorian Church* (Cambridge, 1976).

PROUST, M., *Du coté de chez Swann* (1913); *A l'ombre des jeunes filles en fleurs* (1918); *Le côté de Guermantes* (1920); *La prisonnière* (1923); *Le temps retrouvé* (1926); all repr. in *A la recherche du temps perdu*, ed. P. Clarac and A. Ferré (3 vols.; Paris, 1954).

——*Pastiches et mélanges* (1919), repr., *Contre Saint-Beuve*, ed. P. Clarac and Y. Sandre (Paris, 1971).

PURCELL, E. S., *Life of Cardinal Manning, Archbishop of Westminster* (2 vols.; 1896, repr., New York, 1973).

RADER, R. W., *Tennyson's 'Maud': The Biographical Genesis* (Berkeley and Los Angeles, 1963).

RICKS, C. B., *Tennyson* (1972).

—— *The Force of Poetry* (Oxford, 1984).

RICŒUR, P., 'The Model of the Text', *New Literary History*, 5 (Autumn 1973).

ROSSETTI, C., *Poems*, ed. R. W. Crump (2 vols.; Baton Rouge, 1979).

ROSSI, M., *L'intonation: de l'acoustique à la sémantique* (Paris, 1981).

ROWELL, G., *Hell and the Victorians* (Oxford, 1974).

RYLE, G., *The Concept of Mind* (1949, repr., Harmondsworth, 1963).

SEARLE, J. R., *Speech Acts* (Cambridge, 1969).

—— 'Reiterating the Differences: A Reply to Derrida', *Glyph*, 1 (Baltimore, 1977).

—— KIEFER, F., and BIERWISCH, M. (eds.), *Speech Act Theory and Pragmatics* (Dordrecht, 1980).

SHAKESPEARE, W., *The First Folio . . .* (1623), repr. prepared by C. Hinman (New York, 1968).

—— *The Works of William Shakespeare*, ed. W. G. Clark and W. A. Wright (1873).

—— *Hamlet*, First Quarto (1603, repr., Menston, 1969).

—— *Hamlet*, Second Quarto (1604–5, repr., 1940).

—— *Hamlet*, ed. H. Jenkins (1982).

SHELLEY, P. B., *Poetical Works*, ed. T. Hutchinson (1905, repr., 1943).

SIDNEY, Sir P., *The Defence of Poesie* (1595, repr., Menston, 1968).

SKINNER, Q. R. D., 'Hermeneutics and the Role of History', *New Literary History*, 7/1 (Autumn 1975).

SONNINO, L. A., *A Handbook to Sixteenth-Century Rhetoric* (1968).

SPERBER, D. and WILSON, D., *Relevance* (Oxford, 1986).

SPINOZA, B., *Ethica in ordine geometrico demonstrata* (1675), trans. as *Ethics* by W. H. White and A. H. Stirling, ed. J. Gutman (New York, 1949).

STEELE, J., *An Essay towards establishing the Melody and Measure of Speech* (1775, repr., Menston, 1969).

TANNEN, D. (ed.), *Spoken and Written Language: Exploring Orality and Literacy* (Norwood, NJ, 1982).

TENNYSON, ALFRED Lord, *The Poems of Tennyson*, ed. C. B. Ricks (1969, 2nd edn., 3 vols.; 1987). [Cited as R in notes.]

—— *The Letters of Alfred Lord Tennyson*, ed. C. Y. Lang and E. F. Shannon, Jr. (3 vols.; in progress, 1982). [Cited as Lang and Shannon in notes.]

—— *Tennyson: Poems and Plays* (Oxford, 1953, repr., 1965).

TENNYSON, Sir CHARLES, *Alfred Tennyson* (1949).

TENNYSON, HALLAM Lord, *Alfred Lord Tennyson: A Memoir by his Son* (2 vols.; 1897). [Cited as *Memoir* in notes.]

TENNYSON, HALLAM (ed.), *Studies in Tennyson* (1981).

TRENCH, R. C., *On the Study of Words* (1851, 17th, rev. edn., 1878).

—— *English Past and Present* (1855, 4th, rev. edn., 1859).

TUCKER, A., *Vocal Sounds* (1773, repr., Menston, 1969).

——VACHEK, J., *Written Language: General Problems and Problems of English* (The Hague, 1973).

VALÉRY, P., *Cahiers* (29 vols.; Paris, 1957–61).

WESLING, D., 'Difficulties of the Bardic: Literature and the Human Voice', *Critical Inquiry*, 8/1, (Autumn 1981).

WEYAND, N. (ed.), *Immortal Diamond* (1949).

WILSON, E. (ed.) *Shaw on Shakespeare* (1961).

WINTERS, Y., 'The Audible Reading of Poetry', *Hudson Review*, 4/3 (Autumn 1951).

WISEMAN, CARDINAL N., *The Social and Intellectual State of England* (1850).

WITTGENSTEIN, L., *Philosophical Investigations* (1953; 2nd edn., 1958), trans. G. E. M. Anscombe.

—— *On Certainty*, ed. G. E. M. Anscombe and G. H. von Wright, trans. Denis Paul and G. E. M. Anscombe (Oxford, 1974).

WORDSWORTH, W., *Poems*, ed. J. O. Hayden (2 vols.; Harmondsworth, 1977). [Cited as Hayden in notes.]

—— *The Prelude* (1805–50), ed. M. H. Abrams, S. Gill, and J. Wordsworth (1979).

WORDSWORTH, W., and COLERIDGE, S. T., *Lyricals Ballads* (1798, 1800), ed. R. L. Brett and A. R. Jones (1963).

WORDSWORTH, W., and COLERIDGE, S. T., *Lyrical Ballads* (1798, 1800), ed. R. L. Brett and A. R. Jones (1963).

edn., rev. C. L. Shaver (2 vols.; 1967). [Cited as *Wordsworth Letters: Early Years* in notes.]

YEATS, W. B., *Reveries over Childhood and Youth* (1915), repr., *Autobiographies* (1955).

—— *The Letters of W. B. Yeats*, ed. A. Wade (1954).

—— *Essays and Introductions* (1961).

—— *Explorations*, selected by Mrs W. B. Yeats (1962).

—— *W. B. Yeats: Uncollected Prose*, vol. i, ed. J. P. Frayne (1970).

—— YORK, R., *The Poem as Utterance* (1986).

INDEX

Inherent diff. of Str. Linguists 34

Theorists Morality, p. 35

ing. 33

Lit. & Speech, 36

* Good on Keats, p. 36 ! —

       Then on 37 he theorizes &
         looks at objections : p. 37 *

Illocutionary re. description … 38.

Lit. crit. — turns readers into an audience — p. 38

Searle 39

* 42: Speech act : relevance to written documents? Good & Austin.

* 43 Tone of voice : ambiguity : functional. *
            "not unambiguously Represented by writing."

p. 48 passes
p. 48 ff. Deride / idiom …

p. 49 "Breaking free" *

Vs. Herrnstein Smith p. 50

p. 51 writing detaches from context, speech doesn't

density of argument makes for slow reading

Excellent particular reading:

1. p. 170 End of ALT section!
2. p. 318-319 on "spelt from Sybil's leaves"
3. p. — on "the Voice"
4. p. 324 on Boufe Bay

5. 355 on parens.

"The pause allows"...
          "pause shows"   p 157.
  "The pause" p. 7   234
                                  259
                                  275
                                  352

                                           ┌──────────────
                                           │ oddities
                                           │
                                           │ footling?! p.15
                                           └──────────────

        p. 311 the printed ̈all"
        p. 314 on Kingsley
        p. 325 print

                            " dramatize": supplication to
                              an imagined voice...

* print, p. 13 v. good   * dram. monolog. p.24
* writing is an inherently ambiguous
  notation, p. 17   │ diff. of a Notation
  vs. Herrnstein Smith p.19 │ p. 28 ff.

  Tone of voice 23
  vs. Crystal, p. 27: tough! Metre/dram. monolog.
  Steele – musical Notation   pp. 74-5
  Grice, "Implicature," p. 30

  IAGO pp. 30 - 33
  * p. 33 'Speaker / context